Soil Fertility and Plant Nutrition for Sustainable Cropping Systems

Soil Fertility and Plant Nutrition for Sustainable Cropping Systems

Guest Editors

Daniel Menezes-Blackburn
Bingjie Jin
Dong-Xing Guan

MDPI

Basel • Beijing • Wuhan • Barcelona • Belgrade • Novi Sad • Cluj • Manchester

Guest Editors

Daniel Menezes-Blackburn
Department of Soils, Water and Agricultural Engineering
Sultan Qaboos University
Muscat
Oman

Bingjie Jin
Institute of Urban Environment
Chinese Academy of Sciences
Xiamen
China

Dong-Xing Guan
College of Environmental and Resource Sciences
Zhejiang University
Hangzhou
China

Editorial Office
MDPI AG
Grosspeteranlage 5
4052 Basel, Switzerland

This is a reprint of the Special Issue, published open access by the journal *Sustainability* (ISSN 2071-1050), freely accessible at: https://www.mdpi.com/journal/sustainability/special_issues/PP79QA0FX7.

For citation purposes, cite each article independently as indicated on the article page online and as indicated below:

Lastname, A.A.; Lastname, B.B. Article Title. *Journal Name* **Year**, *Volume Number*, Page Range.

ISBN 978-3-7258-7785-0 (Hbk)
ISBN 978-3-7258-7786-7 (PDF)
https://doi.org/10.3390/books978-3-7258-7786-7

Contents

About the Editors

Daniel Menezes-Blackburn

Daniel Menezes-Blackburn is an Associate Professor at Sultan Qaboos University in Oman. His research expertise spans applied microbiology, plant nutrition, and enzymology, focusing on bacterial phytases, soil organic phosphorus, soil salinity, biochar, DGT methods, and feed enzymes. Dr. Blackburn's work on sustainable agricultural practices has provided many innovative solutions for soil management and nutrient cycling in different soil types and climatic regions.

Bingjie Jin

Bingjie Jin is a Postdoctoral Fellow in Environmental and Soil Biogeochemistry. He previously worked as an Assistant Researcher at the Institute of Urban Environment, Chinese Academy of Sciences, in 2023, before moving to Zhejiang University as a Postdoctoral Fellow in 2025. Dr. Jin's research interests focus on nutrient cycling processes driven by different microbiomes in soil–plant systems. His specific interests and projects focus on the following: 1) microbially driven nitrogen and phosphorus cycling in soil–plant systems and composting systems; 2) the microbiological mechanisms driving organic sulfur hydrolysis in paddy soils. He has authored more than 10 peer-reviewed publications in well-reputed journals such as *Soil Biology and Biochemistry* and *Biology and Fertility of Soils*, and has been granted two patents. His research has been supported by the Young Scientists Fund (Type C) of the National Natural Science Foundation of China and the General Project of the China Postdoctoral Science Foundation.

Dong-Xing Guan

Dong-Xing Guan is an Associate Professor in Environmental and Soil Biogeochemistry. Dr. Guan obtained his Ph.D. in Environmental Science from Nanjing University, China, in 2016. Prior to joining Zhejiang University, he held positions in Environmental Geochemistry, first as a postdoc with Nanjing University and then as an Associate Professor with Tianjin University. Dr. Guan's research interests are oriented around how contaminants and nutrients cycle in the environment. His specific interests and projects focus on the following: 1) biogeochemical and interfacial processes of trace elements in the water–soil–biota continuum; 2) bioavailability and ecological remediation of chemical contaminants in soils and sediments. He has authored over 100 peer-reviewed publications in well-reputed journals such as *Environmental Science and Technology* and *Soil Biology and Biochemistry*, as well as contributed to one book chapter in addition to 12 licensed patents and software copyrights.

Preface

The accelerating pressures of climate change, soil degradation, and resource competition have reshaped the global discourse on soil fertility management. No longer confined to yield maximization, the concept of soil fertility has evolved into a holistic framework that integrates biological, chemical, and physical processes to sustain both productivity and environmental stewardship. This Special Issue, "*Sustainable Soil Fertility Management: Recent Advances and Future Directions*", brings together diverse contributions that highlight the critical intersections of agronomy, microbiology, soil chemistry, and farmer-centered decision-making.

The collected works underscore several transformative themes. First, soil biological indicators are advancing from experimental tools to decision-relevant variables, linking nutrient cycling with climate mitigation. Second, sustainability in agriculture requires balanced nutrient strategies that extend beyond nitrogen optimization to include potassium, magnesium, sulfur, and phosphorus, while recognizing the role of microbial processes in nutrient mobilization. Third, ecological land management practices—such as cover cropping, set-aside strategies, and organic amendments—demonstrate measurable benefits for biodiversity, resilience, and long-term soil health. Finally, the pathway from innovation to impact depends on farmer engagement, institutional frameworks, and supportive policies that enable the adoption of conservation practices.

Together, these contributions point toward holistic soil fertility management dynamically adjusted both for yield and also for sustainability outcomes, including greenhouse gas emission mitigation, biodiversity restoration, and contaminant risk reduction. The studies presented here provide both mechanistic insights and practical pathways, reinforcing the need for interdisciplinary collaboration and long-term standardized research.

This Special Issue is suitable for researchers, practitioners, and policymakers alike, who look forward to embracing soil as the underpinning of food security, climate resilience, and ecological integrity. The advances documented here represent recent scientific progress on harmonizing soil fertility and soil health, biological monitoring with precision agriculture, and sustainable soil management.

Daniel Menezes-Blackburn, Bingjie Jin, and Dong-Xing Guan
Guest Editors

sustainability

MDPI

Editorial

Sustainable Soil Fertility Management: Recent Advances and Future Directions

Daniel Menezes-Blackburn [1,*], Bingjie Jin [2,3] and Dong-Xing Guan [3]

[1] Department of Soils, Water and Agricultural Engineering, Sultan Qaboos University (SQU), Al-Khoud, Muscat 123, Oman
[2] Institute of Urban Environment, Chinese Academy of Sciences, Xiamen 361021, China; dxguan@zju.edu.cn (D.-X.G.)
[3] College of Environmental and Resource Sciences, Zhejiang University (ZJU), Hangzhou 310058, China
* Correspondence: danielblac@squ.edu.om

Sustainable crop production is increasingly constrained by climate change, soil degradation, and competition for water and nutrients [1–6]. These pressures have accelerated the transition from yield-focused soil fertility concepts toward integrated soil health frameworks that incorporate biological, chemical, and physical processes [7,8]. At the same time, conventional input-intensive systems have exposed limitations in nutrient-use efficiency and increased risks of nutrient losses and environmental degradation [4,5,9]. Addressing these challenges requires integrated nutrient management strategies that combine agronomic practices, environmental regulation, and biological processes [10–14]. The contributions to this Special Issue address these themes from complementary angles, with an emphasis on the soil biological regulation of nutrient cycling, farmer decision-making, and interface between fertilization and contaminant mobility in the soil. Additionally, this collection further examined the role of biological amendments in sustaining crop productivity under monoculture systems and the importance of land management practices, such as set-aside strategies, in restoring soil biodiversity and biological fertility. The articles included in this Special Issue are listed in List of Contributions:

Latini et al. (2026) [Contribution 1] systematically review whether soil microorganisms drive the trade-off between soil carbon (C) sequestration and non-CO_2 greenhouse gas emissions (N_2O and CH_4) in European agricultural soils [1,8,12]. Importantly, this review suggests that plant richness and diversity can align climate and fertility objectives by accelerating C storage while constraining N_2O emissions, although the outcomes are context-dependent. Consequently, this review reframes soil biological monitoring not as a descriptive add-on but as an essential infrastructure for credible assessments of soil-based climate mitigation in European cropping systems.

Harasim and Kwiatkowski (2025) [Contribution 2] evaluated the interaction between farming systems (organic, integrated, and conventional) and irrigation scheduling in shaping soil physicochemical properties, microbial indicators, and enzymatic activity in a spring wheat system. Their findings support the argument that soil quality outcomes cannot be predicted based on input intensity alone. The integrated system, characterized by reduced fertilizer and crop protection inputs, produced relatively balanced soil parameters. The organic system supported beneficial microbial community structures and favorable physical properties. This study illustrates the need for water–nutrient stewardship frameworks that jointly optimize fertility and soil physical resilience under climate variability. Biological indicators are not only descriptive tools but are also functional drivers that, when inte-

https://doi.org/10.3390/su18073359

grated with balanced chemical fertilization and precise water management, may create a synergistic effect on crop resilience and carbon sequestration [10,14].

Hoidal et al. (2025) [Contribution 3] addressed the farm-scale and social dimension by examining soil health practices and decision drivers across diversified vegetable farms [10,14]. The survey revealed a high interest in soil health and widespread cover crop use relative to regional baselines; however, it also identified a recurring management vulnerability: large-volume compost applications are often made without a clear understanding of nutrient concentrations or soil test integration, creating the potential for nutrient accumulation and off-farm losses. The study also identified structural barriers to cover cropping and reduced tillage linked to intensive production schedules and labor constraints. Organic certification was associated with the higher adoption of several soil health practices, suggesting that institutional frameworks and market incentives can shape management. Overall, this study highlights that sustainable nutrient management depends on information access, interpretability of diagnostics, and supportive policies that compensate farmers for the time and risks associated with conservation practices. Furthermore, as global policy shifts toward regenerative mandates, future frameworks must move beyond yield stability to value soil as a multifunctional asset for biodiversity and global cooling.

Janke et al. (2024) [Contribution 4] provide a soil microbiome-centered evaluation of organic management and intercropping in arid-zone perennial orchard systems. Over multiple years, organic management increased soil organic matter, Olsen P, and water-holding capacity, whereas microbial enumeration and respiration measurements indicated higher activity relative to conventional management. High-throughput sequencing further revealed increased microbial richness and diversity. A key contribution of this study is the demonstration that perennial horticultural systems can exhibit strong microbial and structural responses to management shifts, despite arid constraints. This study strengthens the case for diversification and organic inputs as mechanisms to advance soil biological functioning without necessarily compromising orchard productivity [8,13,14].

Andrzejewska et al. (2025) [Contribution 5] address a persistent gap in agronomic nutrient management: the tendency to optimize N fertilization while underestimating the role of potassium (K), magnesium (Mg), and sulfur (S) in regulating N capture and utilization. Through a multi-year field experiment on winter wheat, the authors demonstrated that K source and Mg/S supplementation strongly influenced yield stability and nitrogen use efficiency (NUE) indices, with Korn–Kali-based fertilization producing comparatively stable grain yields across years. Beyond yield responses, this study advances a sustainability narrative by linking balanced fertilization to reduced residual inorganic N at harvest, thereby lowering the risk of post-season nitrate losses. This work is a practical reminder that closing the "N gap" tends to be a multi-nutrient problem, where secondary macronutrients and K nutrition determine the efficiency frontier of N use [6,7,9].

Andrade-Sifuentes et al. (2024) [Contribution 6] addressed phosphorus scarcity and low P bioavailability in organic amendments by integrating phosphate-solubilizing rhizobacteria, phosphate rock, and Eisenia fetida during the vermicomposting of horse manure. Their experiments demonstrated synergistic increases in assimilable phosphate forms, indicating that biological processing can convert geogenic P sources into plant-available pools within organic substrates. This study links nutrient recycling with targeted functional biology, moving beyond generic composting toward engineered organic amendments with defined nutrient outcomes. Together, the studies of Andrzejewska et al. (2025) and Andrade-Sifuentes et al. (2024) demonstrated that both mineral and biological nutrient strategies converge toward improving nutrient-use efficiency through complementary mechanisms [11,13].

https://doi.org/10.3390/su18073359

Han et al. (2024) [Contribution 7] extended the understanding of the impact of sustainable nutrient management on environmental risk by investigating how different nitrogen (N) fertilizer forms influence arsenic (As) mobility in paddy soils. The study showed that N inputs altered soil properties and generally reduced soluble As. These results highlight that fertilizer selection affects not only crop nutrition but also soil redox processes, enzymatic activity, and contaminant behavior, which is particularly relevant in flooded rice systems [9,14]. This study strengthens the broader sustainability proposition that nutrient recommendations should be evaluated using integrated soil chemistry–microbiology frameworks rather than nutrient response curves alone. While certain fertilizers may reduce arsenic mobility, the pathway from innovation to impact is often hindered by the complexity and heterogeneity of living soil systems and their knowledge gaps.

Fabiani et al. (2026) [Contribution 8] evaluate how different set-aside management systems influence soil biological fertility and biodiversity in agricultural soils. Their study showed that set-aside practices could substantially increase soil biological quality by increasing bacterial diversity and supporting the recovery of soil microarthropod communities. These results highlight the ecological value of temporary land-resting strategies in restoring soil food web complexity and ecosystem functions, particularly in intensively managed agricultural landscapes [3,8,14]. By demonstrating measurable increases in biological indicators of soil health, this study reinforces the role of land management practices as a complementary approach to fertilization strategies in sustainable soil fertility management.

Rymuza et al. (2026) [Contribution 9] investigated maize productivity under long-term monoculture conditions and demonstrated that biological amendments can mitigate yield declines associated with climatic variability and simplified cropping systems [10,14]. Field experiments showed that microbial and biostimulant-based products increased grain yield and thousand-grain weight compared with untreated controls, while also increasing economic returns. These findings highlight the potential of biological soil inputs to support crop resilience and productivity in intensive cereal-based systems.

Three overarching conclusions emerged from these contributions. First, soil biological indicators are advancing from experimental measurements to decision-relevant variables, especially when climate mitigation and nutrient cycling are treated as coupled outcomes [8,12]. Second, agronomic sustainability depends on nutrient balance and management integration across water, chemistry, and biology, rather than on the isolated optimization of a single input [7,10,11]. Third, the pathway from innovation to impact requires farmer-centered evidence, including practical diagnostics and targeted incentives [10,14]. Fourth, several studies in this Special Issue demonstrate that biological amendments and ecological land management strategies can stabilize crop yields and restore soil biodiversity under intensive agricultural systems, including monoculture cropping and degraded soils [3,8].

Future research should prioritize long-term standardized studies that simultaneously measure SOC, non-CO_2 emissions, microbial function, and nutrient loss [1,12]. Multi-nutrient frameworks for NUE should be expanded to include mechanistic diagnostics and evaluations under variable rainfall regimes [7,9]. The development of engineered organic amendments and biofertilizers requires field trials to quantify their yield, nutrient budgets, and environmental trade-offs [11,13]. Finally, nutrient management must incorporate contaminant mobility and food safety risks, particularly for flooded systems and regions with toxic metal contamination, to ensure that sustainability transitions do not substitute one environmental harm for another [9,14]. The future of sustainable soil management lies in the harmonization of biological monitoring with precision agriculture. The work assembled in this Special Issue points to the next generation of fertility management

https://doi.org/10.3390/su18073359

strategies that will need to be dynamically adjusted for not only yield but also sustainability, considering greenhouse gas emissions and toxic metal bioavailability.

Author Contributions: Conceptualization, D.M.-B., B.J. and D.-X.G.; writing—original draft preparation, D.M.-B.; writing—review and editing, D.M.-B., B.J. and D.-X.G. All authors have read and agreed to the published version of the manuscript.

Conflicts of Interest: The authors declare no conflicts of interest.

List of Contributions

1. Latini, A.; Di Gregorio, L.; Valkama, E.; Costanzo, M.; Maenhout, P.; Suhadolc, M.; Vitali, F.; Mocali, S.; Lagomarsino, A.; Bevivino, A. Do Soil Microbes Drive the Trade-Off Between C Sequestration and Non-CO_2 GHG Emissions in EU Agricultural Soils? A Systematic Review. *Sustainability* **2026**, *18*, 319. https://doi.org/10.3390/su18010319.
2. Harasim, E.; Kwiatkowski, C.A. Effect of Farming System and Irrigation on Physicochemical and Biological Properties of Soil Under Spring Wheat Crops. *Sustainability* **2025**, *17*, 6473. https://doi.org/10.3390/su17146473.
3. Hoidal, N.; Bugeja, S.M.; Lindenfelser, E.; Pagliari, P.H. Soil Health Practices and Decision Drivers on Diversified Vegetable Farms in Minnesota. *Sustainability* **2025**, *17*, 1192. https://doi.org/10.3390/su17031192.
4. Janke, R.R.; Menezes-Blackburn, D.; Al Hamdi, A.; Rehman, A. Organic Management and Intercropping of Fruit Perennials Increase Soil Microbial Diversity and Activity in Arid Zone Orchard Cropping Systems. *Sustainability* **2024**, *16*, 9391. https://doi.org/10.3390/su16219391.
5. Andrzejewska, A.; Przygocka-Cyna, K.; Grzebisz, W. Balanced Fertilization of Winter Wheat with Potassium and Magnesium—An Effective Way to Manage Fertilizer Nitrogen Sustainably. *Sustainability* **2025**, *17*, 6705. https://doi.org/10.3390/su17156705.
6. Andrade-Sifuentes, A.; Peña-Uribe, G.d.J.; Sáenz-Mata, J.; Quezada-Rivera, J.J.; Palacio-Rodríguez, R.; Muro-Pérez, G. The Integration of Phosphorus-Solubilizing Rhizobacteria, Eisenia fetida and Phosphorus Rock Improves the Availability of Assimilable Phosphorus in the Vermicompost. *Sustainability* **2024**, *16*, 7576. https://doi.org/10.3390/su16177576.
7. Han, B.; Chen, W.-Q.; Jiao, Y.-Q.; Yang, R.; Niu, L.-L.; Chen, X.-R.; Ji, C.-Y.; Yin, D.-X. Effects of Nitrogen Fertilizer Application on Soil Properties and Arsenic Mobilization in Paddy Soil. *Sustainability* **2024**, *16*, 5565. https://doi.org/10.3390/su16135565.
8. Fabiani, A.; Mocali, S.; Binazzi, F.; Del Duca, S.; Vitali, F.; Landi, S. Effect of Different Set-Aside Management Systems on Soil Biological Fertility and Biodiversity of Bacterial and Microarthropod Communities. *Sustainability* **2026**, *18*, 2575. https://doi.org/10.3390/su18052575.
9. Rymuza, K.; Radzka, E.; Kapela, K.; Gugała, M. Yield Formation and Stability of Maize Under Monoculture in Response to Biological Amendments, Weather Variability and Cultivar Maturity. *Sustainability* **2026**, *18*, 2542. https://doi.org/10.3390/su18052542.

References

1. Hui, D.; Ray, A.; Kasrija, L.; Christian, J. Impacts of Climate Change and Agricultural Practices on Nitrogen Processes, Genes, and Soil Nitrous Oxide Emissions: A Quantitative Review of Meta-Analyses. *Agriculture* **2024**, *14*, 240. [CrossRef]
2. Sugumar, T.; Shen, G.; Smith, J.; Zhang, H. Creating Climate-Resilient Crops by Increasing Drought, Heat, and Salt Tolerance. *Plants* **2024**, *13*, 1238. [CrossRef] [PubMed]
3. Szymańska, M.; Gubiec, W.; Smreczak, B.; Ukalska-Jaruga, A.; Sosulski, T. How Does Specialization in Agricultural Production Affect Soil Health? *Agriculture* **2024**, *14*, 424. [CrossRef]
4. Cooper, R.Z.; Ergas, S.J.; Nachabe, M. Multi-Decadal Nutrient Management and Trends in Two Catchments of Lake Okeechobee. *Resources* **2024**, *13*, 28. [CrossRef]
5. Zhang, Y.; Lei, J.; Peng, Y.; Chen, X.; Li, B.; Chen, Y.; Xu, Y.; Farooq, T.H.; Wu, X.; Wang, J.; et al. Impact of Intercropping on Nitrogen and Phosphorus Nutrient Loss in *Camellia oleifera* Forests on Entisol Soil. *Forests* **2024**, *15*, 461. [CrossRef]
6. Guan, D.-X.; Menezes-Blackburn, D.; Li, G. The Importance of Mineral Elements for Sustainable Crop Production. *Agronomy* **2024**, *14*, 209. [CrossRef]
7. Xing, Y.; Xie, Y.; Wang, X. Enhancing Soil Health through Balanced Fertilization: A Pathway to Sustainable Agriculture and Food Security. *Front. Microbiol.* **2025**, *16*, 1536524. [CrossRef] [PubMed]

https://doi.org/10.3390/su18073359

8. Chen, Q.; Song, Y.; An, Y.; Lu, Y.; Zhong, G. Soil Microorganisms: Their Role in Enhancing Crop Nutrition and Health. *Diversity* **2024**, *16*, 734. [CrossRef]
9. Ayamba, B.E.; Abaidoo, R.; Opoku, A.; Ewusi-Mensah, N. Mechanisms for Nutrient Interactions from Organic Amendments and Mineral Fertilizer Inputs under Cropping Systems: A Review. *PeerJ* **2023**, *11*, e15135. [CrossRef] [PubMed]
10. Chandel, N.; Kumar, A.; Kumar, R. Towards Sustainable Agriculture: Integrating Agronomic Practices, Environmental Physiology and Plant Nutrition. *Int. J. Plant Soil Sci.* **2024**, *36*, 492–503. [CrossRef]
11. Sande, T.J.; Tindwa, H.J.; Alovisi, A.M.T.; Shitindi, M.; Semoka, J. Enhancing Sustainable Crop Production through Integrated Nutrient Management: A Focus on Vermicompost, Bio-Enriched Rock Phosphate, and Inorganic Fertilisers—A Systematic Review. *Front. Agron.* **2024**, *6*, 1422876. [CrossRef]
12. Al-Ismaily, S.; Lal, R.; Kuzyakov, Y.; Chorover, J.; Ba Abood, F.; Al Maghatasi, B.; Blackburn, D. Biochar Raises Soil Health and Reduces Greenhouse Gas Emissions in Arid Lands. *Front. Soil Sci.* **2026**, *6*, 1740397. [CrossRef]
13. Alzate Zuluaga, M.Y.; Fattorini, R.; Cesco, S.; Pii, Y. Plant–Microbe Interactions in the Rhizosphere for Smarter and More Sustainable Crop Fertilization: The Case of PGPR-Based Biofertilizers. *Front. Microbiol.* **2024**, *15*, 1440978. [CrossRef] [PubMed]
14. Al-Shammary, A.A.G.; Al-Shihmani, L.S.S.; Fernández-Gálvez, J.; Caballero-Calvo, A. Optimizing sustainable agriculture: A comprehensive review of agronomic practices and their impacts on soil attributes. *J. Environ. Manag.* **2024**, *364*, 121487. [CrossRef] [PubMed]

https://doi.org/10.3390/su18073359

sustainability

MDPI

Article

Effect of Different Set-Aside Management Systems on Soil Biological Fertility and Biodiversity of Bacterial and Microarthropod Communities

Arturo Fabiani [1], Stefano Mocali [1], Francesco Binazzi [2], Sara Del Duca [3], Francesco Vitali [1] and Silvia Landi [2,*]

[1] Research Centre for Agriculture and Environment, Council for Agricultural Research and Economics, Via di Lanciola 12A, 51025 Firenze, Italy; arturo.fabiani@crea.gov.it (A.F.); stefano.mocali@crea.gov.it (S.M.); francesco.vitali@crea.gov.it (F.V.)

[2] Research Centre for Plant Protection and Certification, Council for Agricultural Research and Economics, Via di Lanciola 12A, 51025 Firenze, Italy; francesco.binazzi@crea.gov.it

[3] Research Centre for Genomics and Bioinformatics, Council for Agricultural Research and Economics, Via San Protaso 302, 29017 Fiorenzuola D'Arda, Italy; sara.delduca@crea.gov.it

* Correspondence: silvia.landi@crea.gov.it

Abstract

Soil health and agricultural sustainability are primarily threatened by organic matter depletion and a decline in biodiversity. To mitigate these processes, managing set-aside systems remains a simple practice to preserve biodiversity and functions essential to soil fertility. The objective of this study was to evaluate the effectiveness of medium-term set-aside management in preventing soil degradation using changes in microbial and microarthropod communities, both of which are involved in organic matter degradation. Three different set-aside managements, with mowing (May or July) and without mowing, were compared to conventional rotation in three sites located in North, Central, and South Italy. The microbial community was analyzed through both biochemical methods, such as the assessment of soil respiration and microbial biomass, and molecular techniques, such as Denaturing Gradient Gel Electrophoresis. Soil fertility was assessed by the Biological Fertility Index, while soil biodiversity was assessed with microbial and microarthropod indicators. All set-aside management showed a statistically significant separation from conventional crop rotation for both microbial and microarthropod communities. Indicators highlighted a lower efficiency of set-aside management in organic matter use compared to conventional crop rotation, while microbial and microarthropod biodiversity increased in all set-aside managements. The QBS-ar index showed good soil quality standards in each set-aside management, benefiting the euedaphic microarthropods, whereas soil microbial respiration highlighted a higher microbial activity in the same experimental context.

Keywords: soil biodiversity; diversity indices; microbial community structure; microbial community activity; microarthropods community structure

1. Introduction

In recent decades, the expansion of intensive agriculture has exerted increasing pressure on soil ecosystems, altering belowground biological communities [1,2]. The intensive use of mineral fertilizers and pesticides, deep mechanical tillage, and simplified crop rotations have been associated with either a reduction in the diversity and abundance of soil microarthropods or with effects on the structure as well as functional capacity of microbial communities [3,4]. Soil biodiversity represents one of the fundamental pillars of

https://doi.org/10.3390/su18052575

agroecosystem functioning and biological soil fertility, including a wide range of organisms ranging from microorganisms to soil-dwelling invertebrates. Soil microbial biodiversity plays a key role in biogeochemical cycles, soil fertility, and the sustainability of agricultural production [5,6]. Microbial communities regulate essential processes such as organic matter decomposition, nutrient mineralization, nitrogen fixation, and soil structure formation [7,8]. However, these processes do not occur in isolation but are closely interconnected with the activity of soil fauna. Soil microarthropods represent a fundamental component of the belowground biodiversity and play a key role in controlling microbial communities and regulating ecosystem processes [9,10]. Through litter fragmentation, soil mixing, and selective predation on bacteria and fungi, microarthropods influence nutrient availability and organic matter dynamics, thereby enhancing microbial activity and promoting the stabilization of soil aggregates [11,12]. The loss of soil biodiversity, encompassing both microorganisms and arthropods, can compromise agroecosystem resilience, reducing soil ability to provide essential ecosystem services and increasing reliance on external inputs to sustain agricultural productivity [13,14]. In this context, agricultural policies play a critical role in shaping land management practices and their impacts on soil biodiversity. The Common Agricultural Policy (CAP) 2023–2027 of the European Union represents a key framework aimed at promoting sustainable agriculture, explicitly recognizing soil protection and biodiversity conservation as strategic objectives. Nine Good Agricultural and Environmental Condition (GAEC) standards have been defined within the CAP. Specifically, GAEC Standard 8 is designed to protect soil biological communities and enhance biodiversity through the maintenance, at farm level, of a proportion of arable land to non-productive areas and features. There are several agronomic techniques suitable to comply with this standard, including ecological set-aside, crop rotations, pastures or cover cropping. The set-aside has been widely applied in Europe from 1988 to 2008, but its actual effectiveness in maintaining or enhancing soil biodiversity is still a matter of controversy. Recent studies highlight how set-aside practices may contribute to mitigate the simplification of soil food webs induced by intensive farming, thus enhancing resistance to environmental stressors such as climate change, soil degradation, and long-term fertility decline [15,16]. Several authors studied the effect of set-aside on the soil microarthropod community, which evidenced an increase in terms of abundance and diversity of the populations and a better soil biological quality [17–19]. In a Mediterranean agroecosystem, set-aside practices have been shown to affect microbial communities and modify the composition of microarthropod community, highlighting the importance of land-use and disturbance regimes in shaping soil microbial and microarthropod populations [20]. Mocali et al. [21] investigated the effects of set-aside management on soil microbial communities and ground-dwelling arthropods: results indicated an increase in the abundance and diversity of microarthropod populations, whereas no significant differences in microbial biodiversity were highlighted between set-aside and conventional land.

The aim of this work was to evaluate the potential effectiveness of different medium-term set-aside management systems on the biological fertility of soil across different geographical areas of Italy. The experimental sites were located in North, Central and South Italy with different soils and climate conditions. Three set-aside management systems, with mowing in May or July and without mowing, were compared to conventional rotation. The effect of the set-aside management was evaluated by (i) the soil microbial activity and biodiversity indicators, (ii) the soil microarthropod structure and biodiversity indicators, and (iii) the relationship between environmental variables and biological indicators. The hypothesis of this study was that set-aside management may improve the soil biological fertility as well as bacterial and microarthropods biodiversity.

https://doi.org/10.3390/su18052575

2. Materials and Methods

2.1. Site and Treatment Description

The selected experimental fields used in this study were located in Caorle (Veneto, North Italy), Fagna (Tuscany, Central Italy), and Metaponto (Basilicata, South Italy). They had been managed as set-aside since 2008, after being previously used for cereal crops with various conventional rotations. Specifically, the Caorle experimental site was selected in the farm "Valle Vecchia" of Veneto Agricoltura (45°37′51.21″ N–12°58′10.29″ E) on a lagoon plain at 1 m a.s.l. According to the Köpper climate types, this area was classified as Cfb (temperate oceanic climate), and the characteristic crop rotation was maize/sorghum (Table S1) [22]. The Fagna experimental site was located in the "Centro Sperimentale di Fagna" farm of CREA in Scarperia, Florence (43°58′53.28″ N–11°21′01.15″ E), on a hill at 253 m a.s.l. The climate was classified as Cfa (Humid Subtropical Climate), and the wheat or barley (1 year)/alfalfa (4 years) rotation was applied. The Metaponto experimental site was located in the farm "Pantanello" of CREA in Metaponto, Matera (40°23′12.13″ N–16°47′40.08″ E), on a plain at 4 m a.s.l. The climate is classified as Csa (Hot-Summer Mediterranean Climate), and the durum wheat (3 years)/fallow (1 year) rotation was applied. Soil chemical properties for each site were reported in Table 1.

Table 1. Soil chemical properties of the topsoil (0–20 cm) in Caorle, Fagna, and Metaponto sites after five years of set-aside application. CR, crop rotation; MJ, set-aside with mowing in July; MM, set-aside with mowing in May; NM, set-aside without mowing.

Soil Fertility	Caorle				Fagna				Metaponto			
	CR	MJ	MM	NM	CR	MJ	MM	NM	CR	MJ	MM	NM
EC (μS cm^{-1})	0.5 ± 0.05	0.4 ± 0.05	0.6 ± 0.14	0.4 ± 0.04	0.3 ± 0.02	0.3 ± 0.02	0.3 ± 0.02	0.3 ± 0.03	0.4 ± 0.05	0.4 ± 0.04	0.5 ± 0.07	0.5 ± 0.06
Soil pH	8.3 ± 0.02	8.4 ± 0.04	8.2 ± 0.04	8.5 ± 0.05	8.3 ± 0.02	8.2 ± 0.03	8.4 ± 0.04	8.3 ± 0.03	8.6 ± 0.09	8.4 ± 0.04	8.4 ± 0.04	8.3 ± 0.06
Olsen P (mg kg^{-1})	28.2 ± 2.3	8.4 ± 0.7	9.8 ± 2.3	7.9 ± 0.3	6.4 ± 1.0	8.2 ± 0.8	8.7 ± 1.1	10.4 ± 1.3	10.7 ± 0.8	16.4 ± 1.2	20.0 ± 1.4	18.8 ± 1.3
TOC (g kg^{-1})	1.1 ± 0.04	1.1 ± 0.04	1.3 ± 0.04	0.8 ± 0.02	0.9 ± 0.03	1.2 ± 0.04	1.1 ± 0.03	1.1 ± 0.03	1.2 ± 0.02	1.4 ± 0.03	1.5 ± 0.04	1.4 ± 0.02

The same experimental design was applied in each site to compare three different set-aside management systems with the conventional crop rotation: (i) set-aside with mowing in May (MM) without removing the vegetation cover (agronomically more suitable period, but not allowed by the Directive 147/2009 CE to encourage bird breeding); (ii) set-aside with mowing in July (MJ), avoiding the removal of the vegetation cover (an agronomically less suitable period, but permitted by the Directive 147/2009 CE to encourage bird breeding); (iii) set-aside without mowing (NM), with "wild" plots (without mowing) and thus similar to the abandonment; (iv) conventional crop rotation (CR), specific to each site and characterized by different degrees of soil disturbance (as reported above). The total experimental area in each site was 2 ha (corresponding to 0.5 ha per treatment) with an ecotonal belt on at least one side.

2.2. Soil Sampling Design

In each study area, in 2012 and 2013, corresponding to the final years of the experiment (years 4 and 5), soil samples were collected in spring (before mowing) and autumn (after mowing) at three different points in each plot, spaced at approximately 20 m, 40 m, and 60 m from the ecotonal belt. For each examined parameter, three samples were collected for every four treatments and four times, corresponding to a total of 48 samples.

https://doi.org/10.3390/su18052575

To characterize microbial biomass, respiration, and eubacterial biodiversity, soil samples were taken at 0–20 cm depth. To characterize the soil microarthropod structure, an additional set of soil samples was collected close to the previous ones. The sampling was performed using a special 10 cm^3 corer for mesofauna sampling, as reported by Parisi et al. [23]. Soil samples were placed in a plastic bag and stored at 4 °C for microbial samples and at room temperature for microarthropod samples. A sub-sample of about 50 g was prepared from each soil sample and stored at −20 °C for the molecular analysis.

2.3. Soil Microbial Community Analysis

The direct extraction of DNA from the soil was carried out using the FastDNA® SPIN Kit for Soil (MP Biomedicals, Irvine, CA, USA) following the manufacturer's instructions. The extracted DNA was amplified using bacterial 16S rDNA (GC986f–UNI1401r) [24]-specific primers. Amplification reactions were performed according to Baiano et al. [25]. DGGE analysis was performed with the INGENYphorU-2 System (Ingeny International BV, Goes, The Netherlands) loading 500 ng of amplified DNA on a 6% polyacrylamide gel (acrylamide/bis ratio 37.5:1) under denaturation conditions (urea, 7 mol L^{-1}; 40% formamide with a denaturing gradient ranging from 42 to 58%). The gels were run in 1× TAE buffer at 80 V for 17 h at a temperature of 60 °C and stained with SYBR® Green (Molecular Probes®, Eugene, OR, USA; dilution 1:10.000 in 1× Tris-acetate-EDTA—TAE—buffer) for 30 min in the dark. Digital pictures were performed using a ChemiDoc™ Imaging System (BioRad, Hercules, CA, USA). The richness of the obtained bacterial DGGE bands (S_bact), together with biodiversity indices (Shannon diversity, H_bact; evenness, J_bact; Simpson dominance, D_bact), were calculated.

Regarding microbial activity indices, microbial respiration was assessed according to Isermeyer [26]; soil respiration was expressed as mg C-CO_2 kg^{-1} soil d^{-1} and measured in a closed environment after 1, 3, 7, 10, 15, 21, and 28 days for each soil using DL50-Rondolino titrator (Mettler Toledo, Columbus, OH, USA). Cumulative respiration (CumR) represents the amount of total C-CO_2 production after 28 days of incubation. Basal respiration (BasR) represents the value of mineralized C in a definite period when a steady-state condition has been reached (this condition is established when the CO_2 production rate does not change over time, under constant conditions, generally after 28 days). Microbial biomass carbon MBC (mg C kg^{-1} soil) was measured by the fumigation extraction method on air-dried soils which were conditioned at −33 kPa water tension and preincubated for 10 days in open glass jars at 30 °C. The extraction solution was K_2SO_4 0.5 M, and the calculation formula was MBC (g C/g soil) = EC × 2.64, where EC is the difference between the amount of C of fumigated and non-fumigated soils [27]. Several derived parameters were then calculated in accordance with Mocali et al. [28]: (i) qCO_2 (metabolic quotient), calculated by the ratio BasR/MBC [(mg C-CO_2 basal mg MBC^{-1}) h^{-1}]; (ii) qM (mineralization quotient), expressed as mg C-CO_2 kg^{-1} soil h^{-1} and calculated from cumulative respiration after 28 days by the relationship: CumR/TOC; (iii) the Biological Fertility Index (BFI) was calculated as proposed by Renzi et al. [29], and here briefly reported: BFI = BasR + CumR + MBC + SOM + qCO_2 + qM, where SOM (%) is soil organic matter, obtained from TOC multiplied by a conversion factor of 1.724. The biological fertility index is scored in increasing classes of soil fertility. Class I, BFI < 9, is stressed soils with very low fertility; class II, 9 < BFI < 12, is pre-stressed soils; class III, 13 < BFI < 18, is intermediate fertility soils; class IV, 19 < BFI < 24, is good fertility soils; and class V, BFI > 24, is very high fertility soils.

2.4. Soil Microarthropod Community Analysis

Microarthropods were extracted using modified Berlese–Tullgren funnels following the standard methodology [23], observed at the stereomicroscope, and identified at

https://doi.org/10.3390/su18052575

the order level. The edaphic microarthropod communities were characterized using the (i) individual abundance (I_meso), (ii) richness (S_meso) determined by counting the number of taxa, (iii) biodiversity indices (Shannon diversity, H_meso; evenness, J_meso; Simpson dominance, D_meso), and (iv) soil biological quality (QBS-ar index) according to Parisi et al. [23] in order to evaluate the soil biological quality by arthropod adaptation to the edaphic habitat.

2.5. Multi-Level Biotic Response Statistical Analysis

For each site, the overall effect of treatment on the microbial and microarthropod community and activity indices were tested using the Kruskal–Wallis test. For statistically significant results (p-value < 0.05), pairwise differences between treatments were tested using Dunn's post hoc test with Benjamini–Hochberg (BH) multiple comparison correction.

Multivariate methods were performed to assess variation in soil biota indices (i.e., microbial activity, bacterial and microarthropods diversity) and soil physicochemical properties in different experimental contexts with different set-aside managements. Non-metric multidimensional scaling ordinations (NMDS) on the Euclidean distance matrix on the obtained indices and soil properties were performed using the vegan (version 2.6-6.1) [30] R package [31]; to remove scale effect between different indices, each value was divided by the maximum value of that variable, using decostand() command in the vegan package (method = max; margin = 2). The effect of different experimental factors on the multivariate set of indices was tested using PERMANOVA with 999 permutations. The overall analysis, integrating soil biota indices, environmental parameters (soil physicochemical properties) and experimental factors (year, season, management, site), was obtained using Multi Factor Analysis (MFA) obtained with the FactoMineR package (version 2.11) [32]. In addition, within sites, Canonical Correspondence Analysis (CCA) was performed using all soil biota indices with respect to experimental/environmental constraints (sand, clay, EC, pH, TOC, Olsen P, and management) using the vegan R package. Management was converted to 4 numerical pseudo-variables for the analysis. The model and terms significance was assessed using ANOVA, while the contribution of constrains to the canonical axes was assessed using canonical regression coefficients obtained with the scores function in vegan. During model fitting, one of the constrains (NM management) was automatically excluded by the cca function because it was aliased with the constrains of other management practices due to high collinearity.

3. Results

3.1. Bacterial Community and Microbial Activity Indices

The results of bacterial diversity and microbial activity are reported in Table 2. Overall, samples from the Caorle, Fagna, and Metaponto sites showed a high degree of similarity in terms of bacterial diversity. All biodiversity indices (S_bact, J_bact, H_bact, and D_bact) displayed very similar values across all management practices and sites.

At the Caorle site, no significant differences were detected among treatments for microbial biomass carbon (MBC), basal respiration (BasR), cumulative respiration (CumR), qCO_2, or the biological fertility index (BFI). In contrast, qM values were significantly higher in MJ and NM soils than in CR and MM soils (Table 1; $p < 0.05$). Specifically, MJ and NM soils showed qM values ranging from 4.405 to 4.823, with no significant difference between them, whereas lower and comparable values were observed in CR (2.892) and MM (2.924).

In Fagna, microbial respiration parameters varied among management systems. Soils under conventional crop rotation (CR) exhibited the lowest CumR value (311.009 mg C–CO_2 mg^{-1} soil), which was significantly lower than those observed in the other treatments (Table 1; $p < 0.001$). In contrast, MJ, MM, and NM soils showed higher

CcumR values than CR soils (+42.7%, +62.5%, and +50.3%, respectively), and BFI followed a similar pattern. No significant differences were observed in basal respiration across treatments. The lowest qM value was recorded in CR soils (3.343), although it did not differ significantly from MJ (3.914) and NM (4.163). Notably, qM values in NM soils were comparable to the highest values observed under MM management (4.702). Consistently, BFI reached its highest value under MM management (+17.5% compared to CR), while MJ and NM soils also showed higher BFI values (+14.0% and +14.6%, respectively).

Table 2. Bacterial community and microbial activity indices in soils managed with different practices (CR, crop rotation; MJ, mowing in July; MM, mowing in May, NM, no mowing) in the three sites. Table reports mean (and standard error in brackets) of each variable–management combination in the three different sites. The effect of the different management practices on the indices was assessed using the Kruskal–Wallis test and *p*-values were reported in table (column KW *p*-value). For significant results (evidenced in bold) a post hoc comparison between different management practices was performed using the Dunn test with Benjamini–Hochberg correction for multiple testing. Results of the Dunn test are reported as letter notations.

	CR	MJ	MM	NM	KW *p*-Value
Caorle					
S_bact	18.25 (0.836)	18.25 (1.41)	18.333 (1.394)	18.182 (1.086)	0.9690
J_bact	0.944 (0.007)	0.930 (0.01)	0.919 (0.015)	0.921 (0.018)	0.6880
H_bact	2.727 (0.031)	2.662 (0.046)	2.635 (0.061)	2.649 (0.048)	0.3990
D_bact	0.074 (0.002)	0.079 (0.003)	0.082 (0.004)	0.082 (0.004)	0.2950
MBC	82.292 (12.128)	80.729 (10.724)	90.388 (13.987)	79.569 (13.853)	0.9880
BasR	6.047 (0.702)	10.337 (1.668)	9.079 (1.51)	9.123 (1.751)	0.2590
CumR	302.741 (28.432)	468.79 (64.282)	380.024 (50.779)	398.079 (58.99)	0.2610
qCO_2	3.973 (0.753)	8.511 (2.94)	6.256 (1.944)	12.334 (5.767)	0.7960
qM	**2.892 (0.256) b**	**4.405 (0.518) ab**	**2.924 (0.351) b**	**4.823 (0.601) a**	**0.0174**
BFI	13.167 (0.638)	15.833 (1.043)	15.333 (0.94)	14.909 (1.132)	0.1170
Fagna					
S_bact	20.417 (2.061)	20.833 (1.065)	21.333 (1.519)	19.667 (0.948)	0.54000
J_bact	0.916 (0.017)	0.907 (0.014)	0.909 (0.014)	0.857 (0.036)	0.61000
H_bact	2.719 (0.096)	2.74 (0.053)	2.75 (0.052)	2.531 (0.092)	0.24800
D_bact	0.078 (0.008)	0.073 (0.003)	0.073 (0.003)	0.093 (0.01)	0.39700
MBC	121.223 (18.565)	107.782 (16.607)	112.69 (20.649)	125.139 (13.111)	0.91900
BasR	7.404 (0.486)	8.24 (0.941)	9.27 (0.871)	8.714 (1.124)	0.39000
CumR	**311.009 (20.675) b**	**444.192 (32.708) a**	**505.605 (25.206) a**	**467.46 (40.924) a**	**0.00089**
qCO_2	3.699 (0.875)	4.172 (0.777)	5.7 (1.372)	3.726 (0.99)	0.57000
qM	**3.343 (0.236) b**	**3.914 (0.363) ab**	**4.702 (0.276) a**	**4.163 (0.36) ab**	**0.02320**
BFI	**14.25 (0.411) b**	**16.25 (0.579) a**	**16.75 (0.411) a**	**16.333 (0.711) a**	**0.01530**
Metaponto					
S_bact	21.167 (1.926)	22.75 (2.038)	23.083 (2.083)	22.75 (1.958)	0.9170
J_bact	0.905 (0.012)	0.915 (0.016)	0.931 (0.008)	0.897 (0.02)	0.3240
H_bact	2.712 (0.075)	2.81 (0.08)	2.872 (0.075)	2.751 (0.059)	0.5180
D_bact	0.081 (0.006)	0.072 (0.005)	0.068 (0.005)	0.076 (0.003)	0.4280
MBC	166.347 (24.441)	172.777 (14.53)	189.383 (16.723)	189.263 (16.984)	0.6130
BasR	**6.953 (0.468) b**	**11.313 (1.122) a**	**11.035 (1.65) a**	**10.421 (1.147) ab**	**0.0390**
CumR	**316.54 (25.375) b**	**535.25 (50.505) a**	**615.392 (80.04) a**	**529.869 (42.988) a**	**0.0063**
qCO_2	2.343 (0.487)	2.955 (0.384)	2.752 (0.545)	2.505 (0.343)	0.5920
qM	2.682 (0.208)	3.768 (0.35)	3.972 (0.467)	3.695 (0.285)	0.0524
BFI	**14.833 (0.441) b**	**18.333 (0.732) a**	**18.667 (1.025) a**	**18.25 (0.73) a**	**0.0091**

At the Metaponto site (Table 2), microbial respiration differed significantly among management systems for both basal ($p < 0.05$) and cumulative respiration ($p < 0.01$). BasR ranged from a minimum of 6.953 mg C–CO_2 mg^{-1} soil in CR soils to a maximum of 11.313 mg C–CO_2 mg^{-1} soil under MJ management (+62.7% compared to CR). Similarly, MM and NM soils exhibited significantly higher BasR values than CR soils (+58.7% and +49.8%, respectively). The lowest CumR value was again observed in CR soils, whereas

https://doi.org/10.3390/su18052575

significantly higher values were recorded under MJ (+69.1%), MM (+94.4%), and NM (+67.4%) management. Accordingly, the lowest BFI value was found in CR soils (14.833), while it increased significantly under MJ (+23.6%), MM (+25.8%), and NM (+23.0%) management ($p < 0.01$).

3.2. *Microarthropod Structure Community and Biodiversity Indices*

Twenty-one different microarthropod taxa were identified in soil samples collected in the three sites. MDS analysis of microarthropod abundance showed a spatial separation between set-aside management systems (MM, MJ, and NM) and traditional crop rotation in all sites (Figure S1). One-way ANOSIM analysis of microarthropod abundance confirmed these results (Table S2). Specifically, R values were highest in Caorle, ranging from 0.43 ($p < 0.0012$) in the MJ-CR pairwise comparison to 0.63 in the MM-CR ($p < 0.0006$) one. In the Fagna site, they ranged from 0.27 ($p < 0.0012$) for the pairwise comparison of MM-CR and NM-CR to 0.48 ($p < 0.0006$) for the MJ-CR one. Finally, R values were lowest in the Metaponto site, ranging from 0.32 ($p < 0.0048$) in the MJ-CR pairwise comparison to 0.35 ($p < 0.0012$) in the NM-CR one. Total microarthropod abundance was higher in all set-aside management systems than in conventional crop rotation in all sites, although only the Caorle and Fagna sites showed statistically significant differences. Furthermore, the highest abundances were found in the Caorle and Fagna sites, while the lowest abundances were reported in the Metaponto site.

Analysis of taxa abundance relative to the mean Bray–Curtis dissimilarity among these managements using SIMPER showed the highest overall dissimilarity values in the Caorle and Metaponto sites, with 48.26% and 46.25%, respectively (Tables S3–S5). The lowest total dissimilarity value was found in the Fagna site, corresponding to 36.05%. Although the taxa-wise breakdown of similarity showed that a similar number of taxa accounted for 95% of this similarity in each site (Caorle, 12 taxa; Fagna, 14 taxa; Metaponto, 14 taxa), a marked dominance of a few taxa was evident in the Caorle and Metaponto sites. Specifically, Collembola and Hymenoptera were prominent, representing over 60% of the cumulative contribution in the Caorle site, while, in the Metaponto site, Acarina and Collembola, accounted for almost 70% of the entire cumulative contribution. In contrast, in the Fagna site, Acarina and Collembola taxa represented only 45% of the entire cumulative contribution.

The abundance of microarthropod taxa showed some significant differences between the different management systems. In the Caorle site, the abundance of Collembola was higher in MM and NM than in CR; the individuals of Acarina, Isopoda, and Araneae were greater in MM than in the other management systems; the abundance of Diplura and Coleoptera was greater in MJ than in the other management systems. Furthermore, a high number of ants belonging to the species *Solenopsis fugax* Latreille (Hymenoptera: Formicidae) were found in all set-aside management systems even though they were non-significantly different compared to the CR. Some samples were discarded due to the presence of ant nests. At the Fagna site, the abundance of Acarina was higher in MJ than in CR, while the individuals of Collembola were greater in MM than in CR. Furthermore, the abundance of Isopoda was higher in MM and MJ than in the other management systems. In the Metaponto site, the abundance of Acarina was higher in MM than in CR.

The mean values of the richness (S_meso), Evenness (J_meso), Shannon diversity (H_meso), evenness (J), Simpson Dominance (D_meso), and QBS-ar (Soil Biological Quality—arthropods) indices for each management and related statistical analyses are reported in Table 3. Overall, indices showed similar results across the three sites. S_meso was significantly higher in MM and MJ compared to CR in the Caorle site and also for NM in the other two sites. J_meso was significantly higher in CR and MJ compared to MM and NM in the Caorle site, while it was only significantly higher in CR in the Metaponto site. No significant differences were found in the Fagna site for J_meso. D_meso evidenced significantly higher values in CR than in the other managements and in the Caorle and Metaponto sites; no significant differences were found in the Fagna site for it. On the contrary, H_meso showed contrasting evidence. In fact, in the Caorle and Metaponto sites, their highest values were recorded in CR. In contrast, while in the Fagna site, the highest values were reported in MM and MJ. QBS-ar was always significantly lower in CR than in all the set-aside management systems. Moreover, these values never reached 100, which is the generally accepted value, considered the numeric threshold indicating healthy soil in terms of biodiversity. On the other hand, QBS-ar indicated good soil quality in the other managements, especially in the mowing management systems (MM and MJ).

Table 3. Microarthropod community indices in soils managed with different practices (CR, crop rotation; MJ, mowing July; MM, mowing May; NM, no mowing) in the three sites. Table reports mean (and standard error in brackets) of each variable–management combination in the three different sites. The effect of the different management practices on the indices was assessed using the Kruskal–Wallis test and *p*-values were reported in table (column KW *p*-value). For significant results (evidenced in bold), a post hoc comparison between different management practices was performed using the Dunn test with the Benjamini–Hochberg correction for multiple testing. Results of the Dunn test are reported as letter notations.

	CR	MJ	MM	NM	KW *p*-Value
Caorle					
S_meso	6.25 (0.538) b	8.75 (0.617) a	9.08 (0.556) a	7.000 (0.492) b	**0.00292**
J_meso	0.727 (0.023) a	0.685 (0.045) a	0.429 (0.035) b	0.519 (0.033) b	**0.00003**
H_meso	1.388 (0.125) a	1.467 (0.149) a	1.065 (0.092) c	1.214 (0.116) b	**0.00723**
D_meso	0.744 (0.015) a	0.744 (0.024) a	0.71 (0.089) ab	0.642 (0.023) b	**0.0149**
QBS-ar	61.917 (6.803) b	104.917 (9.141) a	96.388 (5.73) a	80.333 (7.731) ab	**0.0187**
Fagna					
S_meso	8.25 (0.446) b	11.5 (0.485) a	10.833 (0.649) a	10.000 (0.492) a	**0.00111**
J_meso	0.404 (0.031)	0.349 (0.029)	0.373 (0.022)	0.341 (0.023)	0.399
H_meso	1.248 (0.047) c	1.323 (0.071) b	1.467 (0.061) a	1.264 (0.071) c	**0.0574**
D_meso	0.606 (0.027)	0.594 (0.003)	0.645 (0.021)	0.585 (0.029)	0.639
QBS-ar	102.167 (8.565) b	138.917 (7.607) a	132.33 (8.768) a	119.917 (10.067) a	**0.0197**
Metaponto					
S_meso	5.25 (0.552) b	8.083 (0.609) a	7.083 (0.514) a	8.667 (0.958) a	**0.00201**
J_meso	0.757 (0.033) a	0.363 (0.020) b	0.381 (0.032) b	0.313 (0.018) b	**0.000003**
H_meso	1.339 (0.079) a	1.109 (0.076) b	0.931 (0.056) b	0.993 (0.062) b	**0.00668**
D_meso	0.681 (0.022) a	0.486 (0.033) b	0.437 (0.033) b	0.442 (0.030) b	**0.0000705**
QBS-ar	56.083 (9.272) b	100.25 (12.601) a	91.75 (8.926) a	107.333 (10.407) a	**0.00515**

3.3. Relationship Between Environmental Variables and Biological Indicators

A non-metric multidimensional scaling (NMDS) ordination analysis was performed on all the obtained indices (microbial activity, bacterial and microarthropods diversity) and soil physicochemical properties to obtain a general overview of multivariate relationships between different samples (Figure 1). The sampling site strongly influenced the ordination. Caorle samples were positioned on the negative side of the first ordination axis, whereas samples from Metaponto and Fagna clustered on the positive side (Figure 1A). In addition, a difference was also observed based on the different sampling year. The year effect on the indices results were even more evident in panels B–D of Figure 1, reporting the ordination on the subsets of samples from Caorle, Fagna, and Metaponto, respectively. Samples were always clearly clustered according to the year (shape of point) but also based on management, particularly regarding conventional crop rotation (CR) which appeared separated in respect to the different set-aside management systems (MM, MJ, NM), especially for the Metaponto samples (Figure 1D).

https://doi.org/10.3390/su18052575

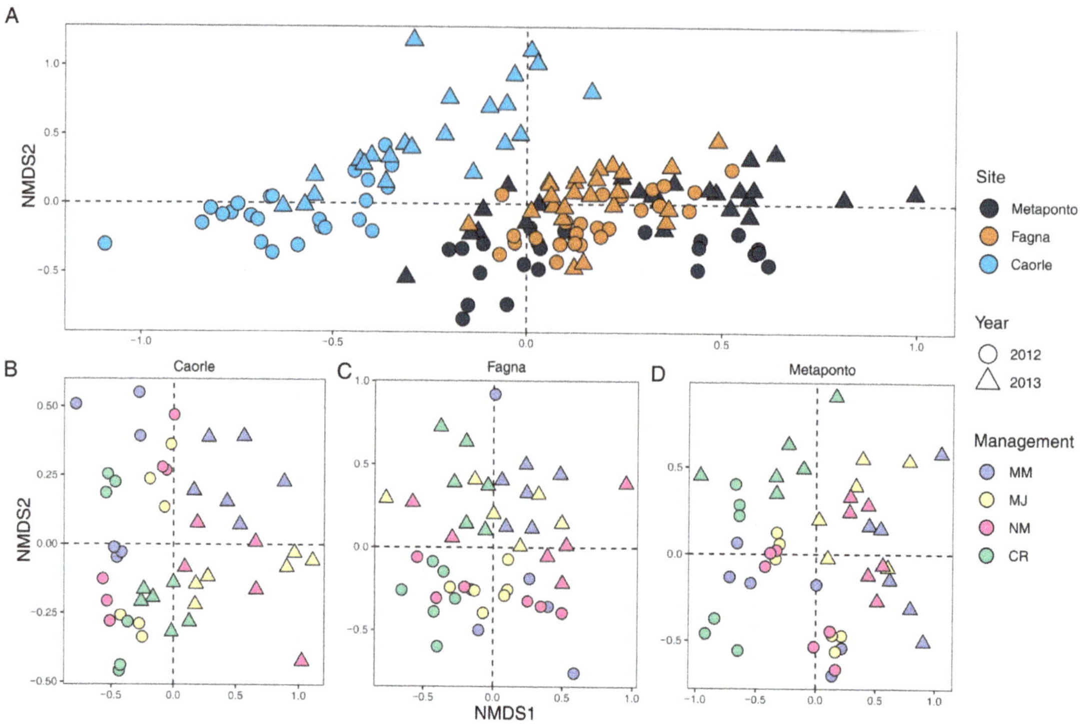

Figure 1. Non-metric multidimensional scaling ordination (NMDS) of all considered biodiversity indices and soil physicochemical properties. Shape of the points represents sampling year, while color of the points represents the sampling site in panel (**A**), or the management in the site-specific ordinations in panel (**B**) (Caorle), (**C**) (Fagna), and (**D**) (Metaponto).

The PERMANOVA analysis (Table 4) confirmed these observations. All the tested experimental factors (sampling site, season, year, management) were highly significant on the whole samples group (p-value < 0.001). The sampling site was the most determining factor (R^2 = 0.331) while sampling season the least (R^2 = 0.038). Also in this case, analysis was additionally performed on the subsets of samples from Caorle, Fagna, and Metaponto. All tested factors were highly significant in Caorle and Metaponto, while the sampling season was not significant in Fagna. Interestingly, the importance of management and year factors results were similar within the different sampling sites (R^2 approximately 0.25, 0.16, 0.24 for Caorle, Fagna, and Metaponto, respectively).

Table 4. Results of permutational multivariate ANOVA (PERMANOVA) analysis for testing the effect of different experimental factors (site, management, season and year) on all considered indices (using the Euclidean distance matrix). Both overall and within-site testing are reported. Reported numbers are R^2. For the p-value, the following notation applies: n.s. = p-value > 0.05, *** = p-value < 0.001).

PERMANOVA	Total	Caorle	Fagna	Metaponto
Site	0.33104 ***			
Management	0.09170 ***	0.25149 ***	0.16339 ***	0.24425 ***
Season	0.03791 ***	0.10066 ***	0.02023 n.s.	0.19014 ***
Year	0.11004 ***	0.25330 ***	0.15783 ***	0.23925 ***

The multiple factor analysis (MFA) confirmed the separation of samples on the first ordination axis based on the different sampling sites (Figure 2A), distinguishing Caorle

https://doi.org/10.3390/su18052575

samples from the others (Figure S2) in accordance with the NMDS results. This separation was mostly driven by soil texture (clay and sand content), together with microarthropod community evenness determined by the strong increment of Acarina, Collembola, and Hymenoptera taxa, especially in the Metaponto site (Figure 2C). Specifically, Caorle samples results are characterized by higher values of sand and microarthropods evenness, while Fagna and Metaponto have more clay soils, as evidenced by the plotting of the variables on the ordination plot (Figure 2D).

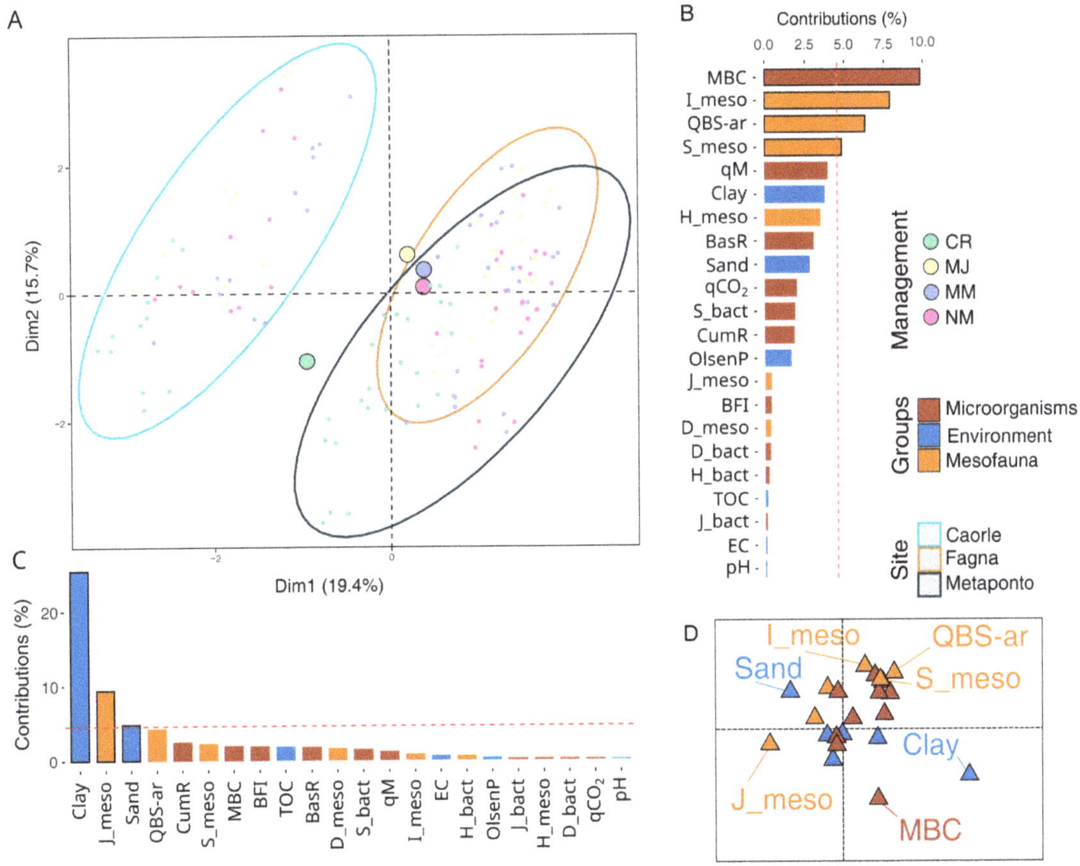

Figure 2. Multiple factor analysis (MFA) of all considered indices. (**A**) Ordination of the samples, where each point represents a sample, colored based on the management. The large dots with a black border represent the centroid of the management group. Finally, a gray ellipse is shown, representing site groups (the color of the ellipse border maps to site). (**B**,**C**) Bar plots reporting the contribution of each numerical variable on the first (**C**) and second (**B**) ordination axis. During MFA, each numerical variable was assigned to a variable group (i.e., bacteria, mesofauna and environment variables set), each represented with a different color. (**A**) Red dotted line is reported to represent the contribution threshold for identifying the most important variables (variables with contribution above this threshold were highlighted with a black outline for the bar plot). (**D**) Positioning of each variable (i.e., end of the vector) in the ordination plane. Only the most influencing variables (i.e., variables with contribution above the red dotted line in panels (**B**,**C**)) were labeled.

The MFA also highlighted a clear separation of the samples subjected to the different soil managements, as highlighted by the group centroids position (Figure 2A). Specifically, set-aside managements (MM, MJ, and NM) centroids clustered together, separated from the conventional crop rotation (CR) one. This separation was evident on both the ordination axes, with results mostly guided by both soil texture on the first axis (Figure 2C), soil community indices such as microarthropods evenness on the first axis (Figure 2C) and MBC and microarthropods indicators (number of individuals, richness and QBS-ar index) on the second axis (Figure 2B). The plotting of the variables on the ordination plot evidenced that the set-aside samples are those characterized by higher values of microarthropods indicators, compared to CR samples (Figure 2D).

Canonical Correspondence Analysis (CCA) was performed within each site (Supplementary Table S6, Figure S3), to evaluate the site-specific association of biological indicators and experimental/environmental parameters. Results were largely in accordance with the evaluation given by MFA. Interestingly, the CR management showed the highest canonical regression coefficient (to either CCA1 or CCA2) in respect to the other managements for all sites, differentiating them from the set-aside managements. Indeed, in both Caorle and Fagna sites, most of the community diversity indices, especially the mesofauna number of individuals, richness and QBS-ar index, tended to map in the quadrant diametrically opposite to CR, in accordance with MFA. In Fagna and Metaponto, Olsen P and TOC showed vectors pointing to the same direction and associated with set-aside management (MJ and MM); in Caorle, on the contrary, Olsen P and TOC vectors pointed in the opposite direction, with TOC associated with set-aside management (primarily MM) and Olsen P associated with conventional crop rotation (CR). In Metaponto CCA, the distribution of the indices' variables was less grouped at the origin of the axes, with a more homogeneous distribution within the plot.

4. Discussion

The selected sites, representing different Italian latitudes, allowed a broad evaluation of the efficacy of the set-aside management system on a national scale. These sites were, in fact, characterized by different climatic conditions, soil properties, and, consequently, by different crop rotation systems. As already reported by Landi et al. [22], after five years of application, the set-aside management strategies improved soil fertility in these sites, particularly after intensive crop rotations. However, even though the organic carbon content increased, the humification remained low. The best advisable set-aside practice involved mowing in May, which produced the highest organic matter increments. After a thorough investigation of the set-aside impact on the soil nematode community [22], in this work, microbial activity, microarthropod community structure, and their biodiversity indicators were assessed in the same sites.

4.1. *Effects of Set-Aside on Soil Microbial Community Structure and Activity*

The results of bacterial diversity of the three sites showed that no differences were observed within each of the sites under different managements. In fact, although the Shannon index (S_bact) showed increasing values across the three sites, ranging from an average of 18.253 in Caorle to 20.562 in Fagna, and to 22.437 in Metaponto, such differences did not cause the results to be significant. Similar variations were observed with J_bact, H_bact and D_bact indices. Overall, such a lack of difference in terms of the bacterial diversity within each site may likely be due to the low resolution of the DGGE technique as compared to the DNA metabarcoding combined with high-throughput sequencing approach, detecting only microorganisms present in high concentrations which may cause an underestimation of rare species [33,34]. Nevertheless, DGGE is yet a molecular technique successfully used for microbial community profiling and is still recognized as a powerful and cost-effective approach to address soil microbial diversity shifts [35–37]. In this context,

https://doi.org/10.3390/su18052575

we used this technique as a comparative analysis aiming at evaluating potential major changes in the dominant bacterial taxa. Thus, we rather suppose that such a lack of bacterial diversity indices under different management across the three sites may be likely related to the poor effects of soil management on the bacterial diversity of these soils, as reported in other works [38–40]. On the other hand, in this study, MBC showed just minor and not significant variation under different management, thus suggesting that local nutrient levels, total organic carbon and other chemical–physical soil properties were quite stable within each site. Nevertheless, in some occasions, this indicator was shown to have poor sensitivity to differences in soil management, as previously reported by Dos Santos et al. [41]. However, to some extent, what could explain this result is that the microbial biomass has stabilized in the soil. And if the soil is under stable conditions for a long period, the microbial biomass can also achieve a stable equilibrium with the soil substrate, mainly related to the organic carbon supply [42]. Moreover, considering the poor seasonal changes in organic carbon availability of these soils, which was shown in a previous work [22], some of the microorganisms in these soils would probably be in a latent state [43]. However, MBC is not an estimate of microbial activity, it only reflects the total living mass of microorganisms, whether in the latent state or not [27], and it is significantly correlated to climatic variables [35,44]. Thus, it is not surprising that MBC values may differ across the three sites but not within each of them. In this context, microbial respiration might play a more sensitive role to differentiating the management systems. For example, in the Fagna site the CumR values in CR indicated a lower microbial activity under conventional crop rotation, whereas BasR did not significantly change across the different managements, suggesting that the composition of the available organic matter was similar in all the samples. Nevertheless, the highest microbial respiration values have been measured in MM soils, where the label fraction of the organic residues after mowing in May was likely better mineralized under the favorable climatic conditions of the spring season. On the other hand, the lower CumR values of CR soils may reflect the lower organic carbon content due to the conventional agricultural intensity, as previously reported for these sites [22]. Consequently, it may explain the low qM values in CR soils as compared to the other managements, which were influenced by the higher proportion of organic material. In fact, qM values have been shown to significantly vary with soil fertilization [28,45]. Consistently, the BFI values in CR soils are classified as soils with "intermediate fertility". It is in line with previous validation of BFI which showed its reliability as an indicator to discriminate the soil biological fertility under different fertilization treatments [46,47]. Although MJ, MM and NM soils were also assigned to the same "intermediate fertility" class, they showed higher BFI scores (>16) than CR, displaying a similar trend as what was observed in the Metaponto site, thus confirming that the higher content of organic carbon played a key role. Moreover, the high content of organic carbon of MM soils observed across the three sites is reflected in the high value of microbial activity of MJ, MM, and NM soils expressed as soil respiration.

4.2. *Effects of Set-Aside on Soil Microarthropod Community Structure*

On the whole, a high variability in the soil microarthropod community structure was found across the three sites, which were characterized by different climatic zones, soil properties, and cropping histories. However, in all cases, the MDS analysis showed a separation between set-aside and crop-rotation management systems, demonstrating that this practice altered the soil microarthropod community. It is known that the composition and diversity of soil microarthropod communities depend more on their use than on geographic, climatic, and site-specific soil conditions [20]. In particular, deep soil tillage, along with soil moisture and compaction, represents the most important conditions influencing

https://doi.org/10.3390/su18052575

soil life [48]. Furthermore, SIMPER analysis demonstrated that set-aside management systems modified the composition of the microarthropod community primarily when the fields had been previously managed with an intensive crop rotation, such as in the Caorle and Metaponto sites. The overall mean dissimilarity was indeed not significantly affected by the conversion from conventional to set-aside management in Fagna, a site managed with a conservative rotation. This result is consistent with what was previously reported by Landi et al. [22] for the soil nematode community in the same trial. In line with what already reported by several authors, mites and springtails, the numerically dominant taxa, increased their abundance in all sites in the set-aside management systems when compared to the conventional crop rotation [20,21,49]. Specifically, mites were represented mainly by Oribatids, which strongly increased in all set-aside managements as already reported by Dubie et al. [50]. The feeding habits were favored by either the organic matter increase or the absence of tillage [51] leading to the return of grass biomass to the soil. The springtail community, characterized mainly by Entomobryomorpha and Poduromorpha forms, maintained all the ecological functions and increased in abundance in all set-aside managements [21]. In addition, in accordance with Landi et al. [20], the populations of the euedaphic group, mainly Diplura, and the hemi-edaphic group, primarily Coleoptera, Isopoda, Aranea, and Hymenoptera, also increased, especially in the mowing set-aside (MM and MJ). In particular, the beetle community, characterized mainly by predators belonging to Carabids, and to a lesser extent Staphylinids, increased in all set-aside managements due to the augmented food availability represented by micro- and mesofauna [22]. It is worth noting that the most relevant increments were reported in sites located in North and Central Italy, primarily because of the more suitable environmental conditions. On the contrary, phytophagous beetles, mainly represented by the Elateridae family in Fagna and Curculionidae in Metaponto, were not particularly abundant, being less than 20% of the entire Coleoptera community. Moreover, the NM highlighted the dominance of several groups, particularly the Formicidae in the Caorle site.

Among the indicators selected for this study, biodiversity indices appear to have limited potential as a metric for the evaluation of healthy/unhealthy systems, as microarthropods were only identified at high taxonomic ranks and results of this trial were not always consistent. However, some general considerations can be highlighted. The increase in richness in all sites managed with the set-aside practice, compared to the CR, indicated the improvement in biodiversity resulting from more natural, no-till management systems. Only in the Caorle set-aside system without mowing, the richness remained similar to crop rotation due to the dominance of certain taxa. The J index showed the highest CR values in all sites, indicating a high level of uniformity in the distribution of community individuals among the different species in a management system applied for many years. This observation highlighted how the application of set-aside represented a medium-term disturbance factor. The D index showed a reduction in dominance in all set-aside management systems compared to conventional management systems; however, the differences were negligible, except for the Metaponto site. The extremely warm climatic conditions recorded in the Metaponto site likely emphasized in this area the positive impact of set-aside in limiting the dominance of a few taxa. The H index showed very inconsistent results. On the contrary, the QBS-ar performed very well as an ecological indicator. In this trial, in accordance with what was previously reported for natural areas and set-aside land use, the measured threshold ranged approximately from 90 to 100 EMI (ecomorphological index); such an interval is generally accepted as an adequate soil quality standard [18,19,23,46,52,53]. Moreover, the intensive cereal-based crop rotation applied in the Caorle and Metaponto sites produced QBS-ar lower than 90 EMI, showing a disturbance in the soil microarthropod community. On the other hand, the more sustainable crop rotation based on cereal and alfalfa applied in the Fagna site achieved good soil quality.

4.3. Holistic/Combined Effects of Set-Aside Management

This study was designed to evaluate the effect of medium-term set-aside management in different climatic and pedological settings. Indeed, results confirmed that the three selected sampling sites were significantly different for several aspects of the microbial and microarthropod communities, as well as in respect to soil physicochemical properties. More specifically, Caorle clearly differentiated from the other two sampling sites. This separation was driven, among the considered variables, by soil texture, with Caorle samplings characterized by sandy soils. These experimental settings, statistically supported by the above-mentioned results, allow the following considerations on the effects of set-aside management to be generalizable to different pedoclimatic contexts. Samples subjected to conventional crop rotation were distinguished from set-aside managed ones. As already observed by Landi et al. [20], this separation result was guided by both soil physicochemical properties (specifically, soil texture) and soil biota indices, mainly microarthropod indicators. Set-aside managements were, in fact, characterized by higher numbers of microarthropod individuals, richness, and QBS-ar index. Their highest values in set-aside managed soils confirmed a general improvement in microarthropod biodiversity resulting from more natural management systems, as previously discussed [20,54]. Conversely, microbial biomass carbon (MBC) was related to CR, demonstrating a higher potential activity in conventional management.

In addition, the effect of the different management systems on soil biological and physicochemical indicators results were more evident when the three different sampling sites were separately considered. In the CCA, CR vectors were always very long and pointed in the opposite direction in respect to those from set-aside managements. This suggested that CR management differed most in terms of the observed biodiversity variables, compared to the others. Furthermore, CR vectors consistently showed a negative correlation with TOC and a positive correlation with all set-aside practices across all sites. This suggests that increasing the organic carbon content was the primary driver for improving both microbial parameters and microarthropod indicators. However, the response differed between the two, with microarthropods showing a rapid and strong reaction compared to a weaker response from the microbial community. This is well-evidenced by the greater distance from the origin of microarthropod individuals. Therefore, set-aside practices mainly positively influenced this indicator, demonstrating that certain taxa found optimal conditions for growth, with Acarina and Collembola being primarily responsible for this trend. On the other hand, although microbial parameters were positively affected by TOC, they were located near the origin, indicating the low impact of organic matter increment for the microbial community. Microbial community diversity was never among the contributing variables in the system under consideration, whereas the indices of microbial activity, particularly microbial biomass carbon (MBC), played a more relevant role [20]. As already observed by Hamer et al. [55], this may indicate that set-aside managements did not induce shifts in the composition of microbial communities compared with the conventional rotation, but rather modified the activity of the community, leading mainly to a greater microbial population growth.

Comparing the different managements, in the CCA, the vector of MM management is longer respect to that of MJ management, suggesting that MM differentiate more to CR, in respect to MJ, especially in Caorle. Similarly to what has already been reported by Landi et al. [22] on the soil nematode community in the same experimental sites, CCA in Fagna highlighted a wider distribution of indices' variables around the origin of the axes. This

https://doi.org/10.3390/su18052575

suggested that Fagna had a higher intra-site variability in respect to the other sites due to its previously discussed, more conservative management strategy prior to the set-aside application. Conversely, Caorle and Metaponto highlighted a stronger positive correlation between soil microarthropod abundance and TOC than Fagna, demonstrating the positive impact of set-aside management after more intensive rotations, as was already reported for the soil nematode community.

In general, as is widely reported in the literature, microarthropod indicators and their abundance are also influenced by microbial activity [54]. However, the low activity of the microbial community reported in this context probably did not allow us to highlight a strong relation. Indeed, as reported by Landi et al. [20], in a long-term condition this relation was more evident.

5. Conclusions

The national scale evaluation of three different set-aside managements compared to conventional crop rotation allowed the assessment of their effects on soil biological fertility and the diversity of the soil microbial and microarthropod communities. The set-aside practice improved soil biological fertility and microarthropod diversity. Nonetheless, the set-aside efficacy depended on the previously applied crop rotation, with major advantages particularly evident after intensive agricultural managements. Moreover, this study carried out on soil biota at a multitrophic level highlighted a network complexity strongly related to soil management. Among the indices selected for this work, microbial respiration and QBS-ar appeared to have the greatest potential as quality indicators of the management systems. The QBS-ar index indicated the achievement of an adequate soil quality standard in each set-aside management favoring the euedaphic microarthropods, while soil microbial respiration underlined a higher microbial activity in the same experimental conditions.

Our results suggest that the set-aside management is a medium-term beneficial practice for improving soil biodiversity and biological fertility.

Supplementary Materials: The following supporting information can be downloaded at: https://www.mdpi.com/article/10.3390/su18052575/s1, Figure S1: MDS analysis based on Bray–Curtis index of microarthropod community abundance from soil samples in Caorle, Fagna, and Metaponto sites; Figure S2: Multiple factor analysis (MFA) of all considered indices; Figure S3: Results of CCA; Table S1: Geographic parameters, climate, soil type, and management in the three selected sites; Table S2: One-way ANOSIM analysis based on Bray–Curtis index of microarthropod community abundance from soil samples; Table S3: Percentage contribution to the Bray–Curtis dissimilarity in soil microarthropod abundance (SIMPER analysis) in Caorle; Table S4: Percentage contribution to the Bray–Curtis dissimilarity in soil microarthropod abundance (SIMPER analysis) in Fagna; Table S5: Percentage contribution to the Bray–Curtis dissimilarity in soil microarthropod abundance (SIMPER analysis) in Metaponto; Table S6: Result of analysis of variance (ANOVA) on the terms of CCA model in different sites.

Author Contributions: Conceptualization, S.M. and S.L.; methodology, S.M. and S.L.; formal analysis, F.V. and S.D.D.; investigation, A.F., F.B., F.V. and S.D.D.; data curation, S.L. and S.M.; writing—original draft preparation, A.F.; writing—review and editing, A.F., F.V., S.D.D. and S.L.; supervision, S.M.; funding acquisition, S.M. All authors have read and agreed to the published version of the manuscript.

Funding: This research was conducted as part of the MO.NA.CO project (National Network for monitoring the Environmental Effectiveness of Cross Compliance and the Competitiveness Differential Affecting Agricultural Enterprises), funded on 18 February 2011 by the Italian Ministry of Agriculture, Food Sovereignty, and Forests (MASAF), under grant number 'COSVIR II Prot.0004010', within Action 1.2.2, 'Interregional Workshops for Development', of the 'Operational Programme National Rural Network 2007–2013'.

https://doi.org/10.3390/su18052575

Institutional Review Board Statement: Not applicable.

Informed Consent Statement: Not applicable.

Data Availability Statement: The data presented in this study are available upon request from the corresponding author.

Acknowledgments: We wish to thank Giorgio Moretti, Lorenzo Furlan, Francesco Fracasso, Francesco Fagotto, Francesco Montemurro, Stefano Fiore and their co-workers for the setup and the management of the monitoring fields, making this work possible.

Conflicts of Interest: The authors declare no conflicts of interest.

References

1. Matson, P.A.; Parton, W.J.; Power, A.G.; Swift, M.J. Agricultural intensification and ecosystem properties. *Science* **1997**, *277*, 504–509. [CrossRef] [PubMed]
2. Tsiafouli, M.A.; Thébault, E.; Sgardelis, S.P.; de Ruiter, P.C.; van der Putten, W.H.; Birkhofer, K.; Hemerik, L.; De Vries, F.T.; Bardgett, R.D.; Brady, M.V.; et al. Intensive agriculture reduces soil biodiversity across Europe. *Glob. Change Biol.* **2015**, *21*, 973–985. [CrossRef] [PubMed]
3. Geisseler, D.; Scow, K.M. Long-term effects of mineral fertilizers on soil microorganisms—A review. *Soil Biol. Biochem.* **2014**, *75*, 54–63. [CrossRef]
4. Sánchez-Bayo, F.; Wyckhuys, K.A.G. Worldwide decline of the entomofauna: A review of its drivers. *Biol. Conserv.* **2019**, *232*, 8–27. [CrossRef]
5. Torsvik, V.; Øvreås, L. Microbial diversity and function in soil: From genes to ecosystems. *Curr. Opin. Microbiol.* **2002**, *5*, 240–245. [CrossRef]
6. van der Heijden, M.G.A.; Bardgett, R.D.; van Straalen, N.M. The unseen majority: Soil microbes as drivers of plant diversity and productivity in terrestrial ecosystems. *Ecol. Lett.* **2008**, *11*, 296–310. [CrossRef]
7. Philippot, L.; Chenu, C.; Kappler, A.; Rillig, M.C.; Fierer, N. The interplay between microbial communities and soil properties. *Nat. Rev. Microbiol.* **2024**, *22*, 226–239. [CrossRef]
8. Falkowski, P.G.; Fenchel, T.; Delong, E.F. The microbial engines that drive Earth's biogeochemical cycles. *Science* **2008**, *320*, 1034–1039. [CrossRef]
9. Lavelle, P.; Decaëns, T.; Aubert, M.; Barot, S.; Blouin, M.; Bureau, F.; Margerie, P.; Mora, P.; Rossi, J.P. Soil invertebrates and ecosystem services. *Eur. J. Soil Biol.* **2006**, *42*, S3–S15. [CrossRef]
10. Bardgett, R.D.; van der Putten, W.H. Belowground biodiversity and ecosystem functioning. *Nature* **2014**, *515*, 505–511. [CrossRef]
11. Moore, J.C.; McCann, K.; de Ruiter, P.C. Modeling trophic pathways, nutrient cycling, and dynamic stability in soils. *Pedobiologia* **2004**, *47*, 507–519. [CrossRef]
12. Maaß, S.; Caruso, T.; Rillig, M.C. Functional role of microarthropods in soil aggregation. *Pedobiologia* **2015**, *58*, 59–63. [CrossRef]
13. Tilman, D.; Cassman, K.G.; Matson, P.A.; Naylor, R.; Polasky, S. Agricultural sustainability and intensive production practices. *Nature* **2002**, *418*, 671–677. [CrossRef] [PubMed]
14. Lal, R. Restoring soil quality to mitigate soil degradation. *Sustainability* **2015**, *7*, 5875–5895. [CrossRef]
15. Delgado-Baquerizo, M.; Maestre, F.T.; Reich, P.B.; Jeffries, T.C.; Gaitan, J.J.; Encinar, D.; Berdugo, M.; Campbell, C.D.; Singh, B.K. Microbial diversity drives multifunctionality in terrestrial ecosystems. *Nat. Commun.* **2016**, *7*, 10541. [CrossRef]
16. Bardgett, R.D.; Bullock, J.M.; Lavorel, S.; Manning, P.; Schaffner, U.; Ostle, N.; Chomel, M.; Durigan, G.; Fry, E.L.; Johnson, D.; et al. Combatting global grassland degradation. *Nat. Rev. Earth Environ.* **2021**, *2*, 720–735. [CrossRef]
17. Tóth, Z.; Hornung, E.; Báldi, A. Effects of set-aside management on certain elements of soil biota and early stage organic matter decomposition in a high nature value area, Hungary. *Nat. Conserv.* **2018**, *29*, 1–26. [CrossRef]
18. Biaggini, M.; Bazzoffi, P.; Gentile, R.; Corti, C. Effectiveness of the GAEC cross compliance standards rational management of set aside, grass strips to control soil erosion and vegetation buffers along watercourses on surface animal diversity and biological quality of soil. *Ital. J. Agron.* **2011**, *6*, 100–106. [CrossRef]
19. Raglione, M.; Lorenzoni, P.; Bonifazi, A.; Nannelli, R.; Gagnarli, E.; Bazzoffi, P.; Gentile, R. Effectiveness of the GAEC cross compliance standard rational management of set aside (weed control through mowing) in assuring a minimum level of habitats maintenance. *Ital. J. Agron.* **2011**, *6*, 92–99. [CrossRef]
20. Landi, S.; Binazzi, F.; Papini, R.; d'Errico, G.; Del Duca, S.; Fabiani, A.; Mocali, S. Effect of long-term set-aside management system on soil health by evaluation of biodiversity indicators. *Diversity* **2025**, *17*, 240. [CrossRef]
21. Mocali, S.; Landi, S.; Fabiani, A.; Piccolo, R.; Agnelli, A.; d'Errico, G.; Mazza, G.; Fedrizzi, M.; Sperandio, G.; Guerrieri, M.; et al. Environmental effectiveness of GAEC cross-compliance Standard 4.2 on biodiversity in set-aside management and economic evaluation of the competitiveness gap for farmers, part I. *Ital. J. Agron.* **2015**, *10*, 711. [CrossRef]

https://doi.org/10.3390/su18052575

22. Landi, S.; Papini, R.; d'Errico, G.; Brandi, G.; Rocchini, A.; Roversi, P.F.; Bazzoffi, P.; Mocali, M. Effect of different set aside management systems on soil nematode community and soil fertility in North, Central and South Italy. *Agric. Ecosyst. Environ.* **2018**, *261*, 251–260. [CrossRef]
23. Parisi, V.; Menta, C.; Gardi, C.; Jacomini, C.; Mozzanica, E. Microarthropod communities as a tool to assess soil quality and biodiversity: A new approach in Italy. *Agric. Ecosyst. Environ.* **2005**, *105*, 323–333. [CrossRef]
24. Felske, A.; Wolterink, A.; Van Lis, R.; Akkermans, A.D.L. Phylogeny of the main bacterial 16S rDNA sequences in Drentse A grassland soils (The Netherlands). *Appl. Environ. Microbiol.* **1998**, *64*, 871–879. [CrossRef] [PubMed]
25. Baiano, S.; Fabiani, A.; Fornasier, F.; Ferrarini, A.; Innangi, M.; Mocali, S.; Morra, L. Biowaste compost amendment modifies soil biogeochemical cycles and microbial community according to aggregate classes. *Appl. Soil Ecol.* **2021**, *168*, 104132. [CrossRef]
26. Isermeyer, H. Eine einfache Methode zur Bestimmung der Bodenatmung und der Karbonate im Boden. *Z. Pflanzenernähr. Bodenkd.* **1952**, *56*, 26–38. [CrossRef]
27. Vance, E.D.; Brookes, P.C.; Jenkinson, D.S. An extraction method for measuring microbial biomass carbon. *Soil Biol. Biochem.* **1987**, *19*, 703–707. [CrossRef]
28. Mocali, S.; Paffetti, D.; Emiliani, G.; Benedetti, A.; Fani, R. Diversity of heterotrophic aerobic cultivable microbial communities of soils treated with fumigants and dynamics of metabolic, microbial, and mineralization quotients. *Biol. Fertil. Soils* **2008**, *44*, 557–569. [CrossRef]
29. Renzi, G.; Canfora, L.; Salvati, L.; Benedetti, A. Validation of the soil biological fertility index (BFI) using a multidimensional statistical approach: A country-scale exercise. *Catena* **2017**, *149*, 294–299. [CrossRef]
30. Oksanen, J.; Simpson, G.; Blanchet, F.; Kindt, R.; Legendre, P.; Minchin, P.; O'Hara, R.; Solymos, P.; Stevens, M.; Szoecs, E.; et al. Vegan: Community Ecology Package. R Package Version 2.6-10, 2025. Available online: https://CRAN.R-project.org/package=vegan (accessed on 5 February 2026).
31. R Core Team. *R: A Language and Environment for Statistical Computing*; R Foundation for Statistical Computing: Vienna, Austria, 2024. Available online: https://www.R-project.org/ (accessed on 5 February 2026).
32. Le, S.; Josse, J.; Husson, F. FactoMineR: An R package for multivariate analysis. *J. Stat. Softw.* **2008**, *25*, 1–18. [CrossRef]
33. Vasselon, V.; Domaizon, I.; Rimet, F.; Kahlert, M.; Bouchez, A. Application of high-throughput sequencing metabarcoding to diatom biomonitoring: Do DNA extraction methods matter? *Freshw. Sci.* **2017**, *36*, 162–177. [CrossRef]
34. Al-Mailem, D.M.; Kansour, M.K.; Radwan, S.S. Capabilities and limitations of DGGE for the analysis of hydrocarbonoclastic prokaryotic communities directly in environmental samples. *MicrobiologyOpen* **2017**, *6*, e00495. [CrossRef] [PubMed]
35. Mocali, S.; Gelsomino, A.; Nannipieri, P.; Pastorelli, R.; Giagnoni, L.; Petrovicova, B.; Renella, G. Short-term resilience of soil microbial communities and functions following severe environmental changes. *Agriculture* **2022**, *12*, 268. [CrossRef]
36. Wu, T.; Sabula, M.; Milner, H.; Strickland, G.; Liu, G. Agricultural practice negatively affects soil bacterial diversity and nitrogen functional genes compared to adjacent native forest soils. *Appl. Soil Ecol.* **2023**, *186*, 104856. [CrossRef]
37. Karna, R.R.; Kumara, S.T.; McCracken, V.J.; Fowler, T.J.; Lin, Z.Q. Enhancement of selenium volatilization for phytoremediation: Role of plant and soil microbe interaction. *Front. Plant Sci.* **2024**, *15*, 1504528. [CrossRef]
38. Rachwał, K.; Gustaw, K.; Kazimierczak, W.; Waśko, A. Is soil management system really important? Comparison of microbial community diversity and structure in soils managed under organic and conventional regimes with some view on soil properties. *PLoS ONE* **2021**, *16*, e0256969. [CrossRef]
39. Lupatini, M.; Suleiman, A.K.A.; Jacques, R.J.S.; Antoniolli, Z.I.; Kuramae, E.E.; Roesch, L.F.W. Soil-borne bacterial structure and diversity does not reflect community activity in Pampa biome. *PLoS ONE* **2013**, *8*, e76465. [CrossRef]
40. Wu, T.; Chellemi, D.O.; Graham, J.H.; Martin, K.J.; Rosskopf, E.N. Comparison of soil bacterial communities under diverse agricultural land management and crop production practices. *Microb. Ecol.* **2008**, *55*, 293–310. [CrossRef]
41. dos Santos, J.V.; Bento, L.R.; Bresolin, J.D.; Mitsuyuki, M.C.; Oliveira, P.P.A.; Pezzopane, J.R.M.; Martin-Neto, L. Long-term effects of intensive grazing and silvopastoral systems on soil physicochemical properties, enzymatic activity, and microbial biomass. *Catena* **2022**, *219*, 106619. [CrossRef]
42. Wang, W.J.; Dalal, R.C.; Moody, P.W.; Smith, C.J. Relationships of soil respiration to microbial biomass, substrate availability and clay content. *Soil Biol. Biochem.* **2003**, *35*, 273–284. [CrossRef]
43. Blagodatskaya, E.; Kuzyakov, Y. Active microorganisms in soil: Critical review of estimation criteria and approaches. *Soil Biol. Biochem.* **2013**, *67*, 192–211. [CrossRef]
44. Insam, H. Are the soil microbial biomass and basal respiration governed by the climatic regime? *Soil Biol. Biochem.* **1990**, *22*, 525–532. [CrossRef]
45. Kaur, S.; Dheri, G.S.; Benbi, D.K. Effect of long-term fertilization in maize–wheat cropping system on carbon mineralization in soil. *Carbon Manag.* **2019**, *10*, 523–532. [CrossRef]
46. Vittori Antisari, L.; Ferronato, C.; De Feudis, M.; Natali, C.; Bianchini, G.; Falsone, G. Soil biochemical indicators and biological fertility in agricultural soils: A case study from northern Italy. *Minerals* **2021**, *11*, 219. [CrossRef]

https://doi.org/10.3390/su18052575

47. Francaviglia, R.; Renzi, G.; Ledda, L.; Benedetti, A. Organic carbon pools and soil biological fertility are affected by land use intensity in Mediterranean ecosystems of Sardinia, Italy. *Sci. Total Environ.* **2017**, *599*, 789–796. [CrossRef] [PubMed]
48. Jacomini, C.; Nappi, P.; Sbrilli, G.; Mancini, L. *Indicatori Ed Indici Eco Tossicologici E Biologici Applicati Al Suolo: Stato Dell'Arte*; Agenzia Nazionale per la Protezione dell'Ambiente (ANPA): Rome, Italy, 2000.
49. Menta, C.; Conti, F.D.; Pinto, S.P.; Bodini, A. Soil Biological Quality index (QBS-ar): 15 years of application at global scale. *Ecol. Indic.* **2018**, *85*, 773–780. [CrossRef]
50. Dubie, T.R.; Greenwood, C.M.; Godsey, C.; Payton, M.E. Effects of Tillage on Soil Microarthropods in Winter Wheat. *Southwest. Entomol.* **2011**, *36*, 11–20. [CrossRef]
51. Betancur-Corredor, B.; Lang, B.; Russell, D.J. Reducing tillage benefits the soil micro- and mesofauna in a global meta-analysis. *Eur. J. Soil Sci.* **2022**, *73*, e13321. [CrossRef]
52. Menta, C.; Leoni, A.; Bardini, M.; Gardi, C.; Gatti, F. Nematode and microarthropod communities: Comparative use of soil quality bioindicators in covered dump and natural soils. *Environ. Bioind.* **2008**, *3*, 35–46. [CrossRef]
53. Ferrazzi, P.; Berger, F.; Verti, F. Biomonitoraggio della Qualità dei Suoli nei Comuni di Agenda 21 Laghi. In Proceedings of the 4th Forum di Agenda 21 Laghi, Torino, Italy, 16 June 2007.
54. Del Duca, S.; Mocali, S.; Vitali, F.; Fabiani, A.; Cucu, M.A.; Valboa, G.; d'Errico, G.; Binazzi, F.; Storchi, P.; Perria, R.; et al. Impacts of soil management and sustainable plant protection strategies on soil biodiversity in a Sangiovese vineyard. *Land* **2024**, *13*, 599. [CrossRef]
55. Hamer, U.; Makeschin, F.; Stadler, J.; Klotz, S. Soil organic matter and microbial community structure in set-aside and intensively managed arable soils in NE-Saxony, Germany. *Appl. Soil Ecol.* **2008**, *40*, 465–475. [CrossRef]

https://doi.org/10.3390/su18052575

sustainability

MDPI

Article

Yield Formation and Stability of Maize Under Monoculture in Response to Biological Amendments, Weather Variability and Cultivar Maturity

Katarzyna Rymuza *, Elżbieta Radzka, Krzysztof Kapela and Marek Gugała

Institute of Agriculture and Horticulture, University of Siedlce, Prusa 14, 08-110 Siedlce, Poland; elzbieta.radzka@uws.edu.pl (E.R.); krzysztof.kapela@uws.edu.pl (K.K.); marek.gugala@uws.edu.pl (M.G.)
* Correspondence: katarzyna.rymuza@uws.edu.pl; Tel.: +48-(25)-6431246

Abstract

Contemporary agriculture faces the challenge of sustaining crop productivity amid increasing climatic pressures and simplified agronomic practices, such as monoculture. A field experiment conducted from 2022 to 2024 aimed to determine the effects of meteorological conditions and biological amendments on grain yield and yield structure in three maturity groups of continuous maize (*Zea mays* L.; FAO 200, 230 and 260). The split-plot experiment included applications of the biological amendments Neosol, Bactim Gleba and UGmax. Deteriorating agrometeorological conditions over the years studied led to a progressive decline in mean grain yield, reaching the lowest value in 2024 (5.06 Mg ha^{-1}). The cultivar belonging to the FAO 260 maturity group exhibited the highest yield potential. Application of all biological amendments resulted in a significant increase in grain yield and thousand-grain weight compared with the untreated control. The most effective treatment was UGmax which increased mean grain yield by approximately 14% and thousand-grain weight by 19% compared with the control. Path analysis revealed hierarchical relationships among components of ear structure and grain yield. The primary direct effect on yield increase was the number of kernels per ear, with thousand-grain weight also contributing significantly depending on maturity group. In later-maturing cultivars, kernel number per ear played the dominant role, whereas thousand-grain weight was more influential in earlier-maturing ones. The economic analysis demonstrated that all of the applied biological amendments generated a positive net profit, with the highest additional revenue obtained following the application of UGmax (160 USD·ha^{-1}). These results confirm that biostimulant application affected grain yield formation, and reduced yield losses under stress conditions.

Keywords: maize; monoculture; biological amendments; drought tolerance; sustainable agriculture; yield structure

1. Introduction

Maize (*Zea mays* L.) ranks among the most important crop species grown worldwide [1]. Its versatile uses make it a promising cereal [2,3]. Maize plays a key role in global food and energy security, a significance that has become even more pronounced in 2026 amid escalating climatic pressures, geopolitical instability, and environmental constraints. In many European regions practising intensive agriculture, maize has increased its share in cropping systems, with a growing tendency towards monoculture. This shift is driven

 https://doi.org/10.3390/su18052542

by economic viability, farm specialisation, and simplified production organisation [4]. The results of numerous studies conducted under diverse soil and climatic conditions indicate that simplifications in the cultivation system, including long-term monoculture, may lead to reduced yield stability, increased susceptibility to abiotic stresses, and a gradual decline in agroecosystem productivity [5–7]. Moreover, the observed increased sensitivity of plants to abiotic stresses, including drought, may in the long term lead to yield reduction [8]. These trends underscore the need for sustainable agronomic practices that integrate biological and technical approaches to mitigate the drawbacks of monoculture, improve soil health, and enhance production system resilience to water deficits [9]. It has been demonstrated that the use of biological amendments can enhance plant productivity [10,11]. Moreover, their application may strengthen plant resistance to environmental stress, which in turn can improve both the quality and quantity of the yield [12]. Contemporary agriculture must maintain maize productivity under increasingly unpredictable weather patterns and simplified agronomic systems, including repeated cultivation on the same site—a practice not uncommon for maize [13]. The use of biological amendments therefore becomes essential, as they support plants during stress periods and help ensure satisfactory and relatively stable yields across successive years. It is also important to understand how maize cultivars of differing maturity respond to such interventions. Identifying these differences and the mechanisms underlying yield formation in altered environments can guide the development of cropping technologies that are both economically viable and environmentally sustainable. In maize research, information on grain yield alone is often insufficient. To achieve meaningful improvements in productivity, it is necessary to identify the specific traits contributing to yield and their interactions [14,15]. Path analysis is particularly valuable in this context, as it quantifies the direct and indirect effects of traits on yield under varying cropping conditions [16,17].

Because path analysis enables the identification of traits that genuinely determine grain yield [18,19] rather than merely correlate with it, this study applies Wright's path coefficient method to address a significant gap in the existing literature. To date, research on biological amendments in maize has largely focused on yield magnitude, with limited attention given to the hierarchical structure of yield components and the causal relationships underlying productivity formation under monoculture conditions. The novelty of the present study lies in the integrated analytical framework combining classical analysis of variance (ANOVA) with Wright's path analysis in a long-term field monoculture experiment. Such a methodological combination remains rarely applied in agronomic studies evaluating biological amendments. While ANOVA identifies statistically significant treatment effects, path analysis simultaneously clarifies biological cause–effect relationships among yield components. This dual approach allows not only quantification of productivity responses, but also explanation of the mechanisms driving those responses. Moreover, the study jointly evaluates three interacting determinants of maize productivity: interannual meteorological variability, cultivar maturity group (FAO 200, 230, 260), and biological amendment application. By integrating these factors within a single hierarchical model, the research provides a more comprehensive understanding of yield-formation mechanisms under monoculture stress. The theoretical contribution of this work therefore lies in identifying cultivar-specific yield-formation strategies and amendment-induced modifications in causal yield structure, which may support the design of climate-adaptive and resource-efficient cropping systems.

This research focused on the productive effects of biostimulant application and their influence on yield components, rather than on the dynamics of physicochemical soil parameter changes occurring over time. This study aimed to evaluate the effects of varying meteorological conditions and biological amendment applications on maize yield and yield-

https://doi.org/10.3390/su18052542

structure traits under monoculture conditions, and to determine hierarchical relationships among yield components using path analysis.

2. Materials and Methods

The field experiment was set up as a split-plot arrangement with three replicates and conducted from 2022 to 2024 at the Agricultural Experimental Farm in Zawady (52°03′ N, 22°33′ E), which belongs to the University of Siedlce. Each year, the experiment was established on the same site to achieve the effects of continuous monoculture cropping. The harvest area of each plot was 30 m^2. The experiment examined the following factors:

Factor A—Maize cultivars:

A1—Early-maturing hybrid, FAO 200;
A2—Medium–early-maturing hybrid, FAO 230;
A3—Medium–late-maturing hybrid, FAO 260.

Factor B—Biological amendments:

B1—Untreated control;
B2—Neosol;
B3—Bactim Gleba;
B4—UGmax.

The soil was classified as a proper lessive soil (podzolic group) with a sandy granulometric composition, a very good rye complex, and a quality class IVa [20]. Prior to maize sowing, the soil nutrient status was analysed. The soil reaction was slightly acidic (pH ranged from 5.67 to 5.70). The soil showed low available phosphorus and medium available potassium and magnesium levels (Table 1).

Table 1. Soil content of available nutrients before establishing the experiment.

Parameter	Unit	Value
pH in KCL	–	5.8
Phosphorus (P)	$mg \cdot kg^{-1}$	40.5
Potassium (K)	$mg \cdot kg^{-1}$	102.6
Magnesium (Mg)	$mg \cdot kg^{-1}$	41.4

Soil pH was determined using the potentiometric method. The content of the available forms of phosphorus and potassium was determined using the Egner–Riehm method (DL), while magnesium content was determined using the Schachtschabel method.

The physio-chemical soil parameters were determined prior to the establishment of the experiment. Detailed monitoring of the biological, chemical, and physical changes in the soil properties during the three-year study period was not included in the experimental design. The investigation focused primarily on the productive response of plants (yield and yield components) as an indicator of soil–plant system functioning. Therefore, interpretations concerning soil improvement refer to functional effects observed in plant responses rather than to directly measured changes in soil parameters.

Maize was grown in monoculture. Following harvest of the preceding crop, post-harvest tillage operations were performed, followed by winter ploughing to a depth of 25 cm. In early spring, agronomic practices were conducted to level the soil surface and promote warming. Mineral NPK was applied before planting. Potassium was supplied as potash salt at 250 kg ha^{-1}, equivalent to 124.5 kg K ha^{-1}. Nitrogen was applied at 100 kg N ha^{-1}, corresponding to 291 kg ha^{-1} of ammonium nitrate (34.4% N). Fertilisers were incorporated into the soil using a combined cultivator. Each spring, pre-plant applications of biological amendments were made, and they were mixed into the soil to a depth of

https://doi.org/10.3390/su18052542

5–10 cm. Bactim Gleba (Intermag sp. z o.o., Olkusz, Poland) is a bioproduct that accelerates revitalisation of 'fatigued' soils, i.e., those degraded to varying degrees by intensive cropping and unbalanced fertilisation. It contains *Bacillus* spp. bacteria isolated from natural soil environments, occurring in the form of spores, and a humic co-formulator supporting microbial development. It was applied as a soil spray at 1.5 L ha^{-1}. Granular Neosol (PRP Polska sp. z o.o., Warsaw, Poland) acts as an activator of soil microbial biomass, stimulates humic acid production, promotes humus formation through enhanced biological activity, and restores soil fertility. The active substance MIP SOIL in Neosol supplies minerals (iron, zinc, boron, and manganese compounds on a calcium and magnesium carbonate matrix) essential for soil microflora and for enzymes involved in organic matter transformations. It was applied at 150 kg granules ha^{-1}. UGmax (Agrowitan sp. z o.o., Skarszewy, Poland) contains microorganisms along with key macro- and microelements. It improves soil physical, chemical, and biological properties, increases organic matter content, and enhances soil structure. Its advanced formulation, including lactic acid bacteria (*Lactobacillus*) and an organic matter content of at least 40%, accelerates the natural decomposition of organic matter. UGmax was applied as a soil spray at 2 L ha^{-1}.

Maize was sown using a pneumatic precision drill with simultaneous row fertilisation using triple superphosphate (46% P_2O_5) at 71.8 kg ha^{-1}, equivalent to 33 kg P ha^{-1}. Seeds were placed in rows spaced 75 cm apart.

During the growing season, the experiment was conducted under real rainfall conditions and no additional irrigation was applied at the test sites.

Measurements were taken at full maturity (BBCH 89) on plants harvested from a 1 m^2 area within each plot, excluding border rows. The following biometric traits were assessed:

- Number of kernel rows per ear;
- Number of kernels per row;
- Number of kernels per ear;
- Thousand-grain weight (TGW, g).

Grain yield was determined from mechanical harvest of the plot area and recalculated to Mg ha^{-1} at 14% grain moisture.

2.1. Meteorological Conditions

Data on the air temperature and atmospheric precipitation sums during the maize growing season were obtained from an automatic meteorological station located at the Agricultural Experimental Farm (AEF) in Zawady, belonging to the University of Siedlce, equipped with sensors and software manufactured by LAB-EL Laboratory Electronics sp. z.o.o, Reguły, Poland and regularly calibrated under laboratory conditions. Due to its local, institutional status, the station is not included in the Global Historical Climate Network Daily (GHCN-D) database and does not have an RSM identification number.

These data served as the basis for evaluating the thermal and moisture conditions in the years studied and their influence on maize growth and development. The Sielianinov hydrothermal coefficient (K) was calculated, and the hydrothermic conditions were classified according to Skowera and Puła [21]. Nine moisture classes were distinguished. Values of $K \leq 0.4$ indicate extremely dry conditions. The range $0.4 < K \leq 0.7$ corresponds to very dry conditions, while $0.7 < K \leq 1.0$ indicates dry conditions. The interval $1.0 < K \leq 1.3$ denotes fairly dry conditions. Values of $1.3 < K \leq 1.6$ are considered optimal from the standpoint of moisture supply. The range $1.6 < K \leq 2.0$ characterises fairly moist conditions, $2.0 < K \leq 2.5$ moist conditions, $2.5 < K \leq 3.0$ very moist conditions, and $K > 3.0$ extremely moist conditions.

Mean air temperature during the study period (2022–2024) ranged from 8.2 °C in April to 21.0 °C in August (Table 2). The coolest month in the analysed half-year was April,

https://doi.org/10.3390/su18052542

while July and August were the warmest, favouring vigorous plant growth during the growing season. Interannual temperature variability was moderate, with 2024 showing the highest values in most months. Monthly precipitation totals exhibited substantial variability, both seasonally and between years. Over the 2022–2024 period, the highest average precipitation occurred in July (62.3 mm) and June (48.5 mm), consistent with the typical summer convective rainfall maximum. The lowest average totals were recorded in May (27.6 mm) and April (22.4 mm). Notably high precipitation was observed in July 2022 (95.7 mm), which may have caused temporary soil water excess. In contrast, precipitation in May 2024 was very low (5.2 mm), suggesting the potential for short-term dry conditions.

Table 2. Meteorological conditions at the Zawady AEF station in 2022–2024.

Month/Year	April		May		June		July		August		September	
	t	P	t	P	t	P	t	P	t	P	t	P
2022	5.2	31.5	13.6	31.1	19.9	26.5	19.3	95.7	21.0	39.3	11.7	64.9
2023	8.7	12.4	13.4	46.5	18.0	53.6	20.3	31.4	21.3	25.0	18.0	16.6
2024	10.8	23.2	16.6	5.2	19.0	65.4	21.6	59.7	20.7	55.5	18.1	25.5
Mean 2022–2024	8.2	22.4	14.5	27.6	19.0	48.5	20.4	62.3	21.0	39.9	15.9	35.7

t—Mean monthly air temperature, P—monthly total precipitation.

Moisture conditions during the maize growing season from 2022 to 2024 showed considerable variation, with a predominance of dry and very dry periods, particularly during the critical developmental stages of the plants. In 2022, favourable conditions occurred only in April (wet) and July (moist), whereas May (dry), June (very dry), and August (very dry) exhibited moisture deficits that likely restricted plant growth and grain filling (Table 3). In 2023, dry and very dry conditions prevailed for most of the growing season, encompassing key phases of intensive growth, flowering, and grain filling. In 2024, moisture conditions were more variable, yet dry conditions still dominated during critical phases, especially in May (very dry) and during flowering and grain filling.

Table 3. Values of Sielianinov hydrothermal coefficient (K).

Year/Month	2022	2023	2024
April	2.01 w.	0.48 v.d.	0.72 d.
May	0.74 d.	1.12 r.d.	0.10 ex.d.
June	0.44 v.d.	0.99 d.	1.15 r.d.
July	1.60 op.	0.50 v.d.	0.89 d.
August	0.60 v.d.	0.38 ex.d	0.86 d.
September	1.84 f.m.	0.31 ex.d	0.47 v.d.

Month: Extremely dry—ex.d., very dry—v.d., dry—d., rather dry—r.d., optimal—op., rather wet—r.w., wet—w., fairly moist—f.m.

2.2. Statistical Analysis

Because the experiment was conducted on the same site across successive years, appropriate statistical methods were required to account for the repeated use of the location. The blocks and sub-blocks randomly selected in the first year remained fixed in subsequent years. In such cases, following standard principles for field experiments, the two-factor split-plot design was subjected to a combined (three-year) analysis according to a split-split-plot model, in which years were treated as main plots (factor A levels).

https://doi.org/10.3390/su18052542

Analysis of variance for each trait was performed using the following mathematical model [22]:

$$y_{ijlp} = m + a_i + g_j + e_{ij}{}^{/1/} + b_l + ab_{il} + e_{ijl}{}^{/2/} + c_p + ac_{ip} + bc_{lp} + abc_{ilp} + e_{ijlp}{}^{/3/},$$

where

y_{ijlp} is the value of the trait for the *i*-th level of factor A (year), *l*-th level of factor B (cultivar), and *p*-th level of factor C (biological amendment) in the *j*-th block (replicate). Other terms are defined as follows:

m is the overall experimental mean;
a_i, b_l, and c_p are the main effects of the respective factors;
g_j is the effect of the *j*-th block;
$e_{ij}{}^{/1/}$ is the whole-plot error;
ab_{il}, ac_{ip}, and bc_{lp} are the two-factor interaction effects;
$e_{ijl}{}^{/2/}$ is the sub-plot error;
abc_{ilp} is the three-factor interaction effect;
$e_{ijlp}{}^{/3/}$ is the sub-sub-plot error.

Means were compared and interactions evaluated using the Tukey's Honestly Significant Difference (HSD) test [22].

In addition to analysis of variance, Wright's path analysis was applied to determine the direct and indirect relationships between the yield components and final grain yield [14,23,24]. The analysis was based on a biologically justified hierarchical model that accounted for the sequential development of generative organs and the grain-filling process. The model adopted a sequential structure in which morphological traits (number of kernel rows and number of kernels per row) determine the number of kernels per ear, which in turn, together with thousand-grain weight (TGW), shapes the final yield.

Analysis of variance was conducted using Statistica 13.3 software (with the *Zestaw Przyrodnika* add-on). Direct and indirect path effects according to Wright were calculated using structural equation modelling (SEM) available in Statistica 13.3. All calculations were performed at a significance level of $\alpha = 0.05$.

2.3. Economic Analysis

To assess the practical applicability of the examined technology, a simplified economic analysis was conducted using the gross margin method, which is commonly employed in agronomic research to evaluate the profitability of a single technological factor [25]. The analysis considered only the differences resulting from the application of soil conditioners, assuming identical costs for all other technological components. Such an approach makes it possible to determine the economic efficiency of the examined factor independently of production cost variability across farms. The economic assessment used the long-term average market price of maize grain, calculated on the basis of stock exchange quotations from the end of December in 2022, 2023, and 2024 [26]. Using the average grain price enabled an objective evaluation of the economic performance of the tested products by eliminating the influence of substantial year-to-year market fluctuations. The average purchase price of maize grain, converted to USD, amounted to 193 USD per Mg. The estimated purchase costs of the products, depending on the supplier, were as follows: UGmax (rate 2 $L \cdot ha^{-1}$), approximately 24–29 $USD \cdot ha^{-1}$; Bactim Gleba (rate 1.5 $L \cdot ha^{-1}$), approximately 22–27 $USD \cdot ha^{-1}$; and Neosol (rate 150 $kg \cdot ha^{-1}$), approximately 44–53 $USD \cdot ha^{-1}$ [27]. For data standardisation purposes, all costs and revenues were converted from PLN to USD according to the average exchange rate of the National Bank

https://doi.org/10.3390/su18052542

of Poland on the last working day of December for each of the analysed years (1 USD = 4.14 PLN) [28].

3. Results

Statistical analysis revealed significant effects of the meteorological conditions across the years studied, the genetic characteristics of the cultivars, and the applied biological amendments on maize grain yield.

Deteriorating agrometeorological conditions contributed to a progressive decline in plant productivity. The mean grain yield across all cultivars decreased by 2.25 Mg ha^{-1} over the study period (from 7.31 to 5.06 Mg ha^{-1}) (Table 3). The primary factors limiting yield in the final year were precipitation deficits during the critical growth phase (May), combined with elevated mean air temperatures, which intensified abiotic stress on the plants.

The cultivar with the highest FAO number (260) displayed the greatest stability and yield potential, with a three-year mean grain yield significantly higher than those of the other cultivars. Cultivars with shorter growing seasons produced significantly lower yields, confirming that, under monoculture conditions, cultivars with greater biological potential make more effective use of available environmental resources.

Cultivar responses to the prevailing conditions varied across the years studied. In 2022, the longest-season cultivar (FAO 260) and cv. FAO 230 yielded similarly. In 2023, grain yield increased significantly with increasing length of the growing season. Under the drought conditions of 2024, cv. FAO 200 and FAO 230 produced similar yields.

All applied biological amendments (Neosol, Bactim Gleba, UGmax) resulted in a significant increase in grain yield compared with the untreated control. UGmax proved the most effective, increasing mean grain yield by approximately 14% compared with the control. The statistically significant differences among the products confirmed that the applied biological amendments improved plant performance under monoculture conditions.

The year × biological amendment interaction indicates that the efficacy of the amendments depended on the meteorological conditions during the growing season. Only in 2022 were significant differences observed in the yield-promoting effects of each product. In 2023 and 2024, yields obtained after the application of Neosol and Bactim Gleba were similar and significantly higher than the control. UGmax exhibited a stabilising effect on production, consistently producing the highest yields in every year (Table 4).

Table 4. Grain yield of maize (Mg ha^{-1}) grown in monoculture by year, cultivar, and biological amendment.

		Year		
Cultivar	2022	2023	2024	Mean
FAO 200	6.69 B	6.02 C	4.83 B	5.84 C
FAO 230	7.54 A	6.57 B	4.85 B	6.32 B
FAO 260	7.72 A	6.89 A	5.49 A	6.70 A
Mean	7.31 a	6.49 b	5.06 c	
Biological amendment	2022	2023	2024	
Control	6.78 D	6.13 C	4.76 C	5.90 D
Neosol	7.19 C	6.38 B	5.02 B	6.20 C
Bactim Gleba	7.44 B	6.58 B	5.06 B	6.39 B
UGmax	7.84 A	6.86 A	5.42 A	6.73 A

Means in rows followed by different lowercase letters (a, b, c) differ significantly at $p \leq 0.05$ (for years). Means in columns followed by different uppercase letters (A, B, C, D) differ significantly at $p \leq 0.05$ (for cultivars, biological amendments, year × cultivar interactions, and year × biological amendment interactions).

Analysis of the results demonstrated a significant effect of cultivar and year-to-year weather conditions on the thousand-grain weight (TGW) of maize (Table 5). The highest

https://doi.org/10.3390/su18052542

three-year mean TGW was recorded for cv. FAO 260 (300.28 g), a value significantly greater than those of cv. FAO 230 and FAO 200. The cultivar with the shortest growing season produced the lowest grain mass (252.26 g).

Table 5. TGW of maize (g) grown in monoculture by year, cultivar, and biological amendment.

	Year			
Cultivar	2022	2023	2024	Mean
FAO 200	266.06 a	249.19 b	241.54 c	252.26 C
FAO 230	297.22 a	286.82 b	274.44 c	286.16 B
FAO 260	306.63 a	309.67 ab	284.54 b	300.28 A
Mean	289.97 a	281.89 a	266.84 b	
Biological amendment	2022	2023	2024	
Control	268.2 a	255.3 b	242.0 c	255.16 D
Neosol	273.6 b	284.4 a	268.9 b	275.65 C
Bactim Gleba	299.8 a	289.9 a	264.4 b	284.70 B
UGmax	318.3 a	298.0 b	292.0 b	302.77 A
Biological amendment	Cultivar			
	FAO 200	FAO 230	FAO 260	
Control	237.16 B	260.84 C	267.48 D	
Neosol	243.63 A	284.09 B	299.23 C	
Bactim Gleba	257.82 A	288.29 B	307.99 B	
UGmax	270.44 A	311.43 A	326.43 A	

Means in rows followed by different lowercase letters (a, b, c) differ significantly at $p \leq 0.05$ (for years, year × cultivar interactions, and year × biological amendment interactions). Means in columns followed by different uppercase letters (A, B, C, D) differ significantly at $p \leq 0.05$ (for cultivars, biological amendments, and year × biological amendment interactions).

Weather conditions during the 2022 growing season favoured the accumulation of assimilates in the grain, resulting in the highest mean TGW (289.97 g). In subsequent years, a decline in this parameter was observed, although it was statistically significant only in 2024.

The application of biological amendments affected maize TGW. All of the tested products increased grain mass compared with the untreated control (255.16 g). The most beneficial effect was observed for UGmax, which increased mean TGW by approximately 19% compared with the control. Significant positive effects were also recorded for Bactim Gleba (284.70 g) and Neosol (275.65 g).

Analysis of the interaction between biological amendments and years showed that only in the control treatment did TGW decrease significantly from year to year (by 5% and 6%, respectively). The weather conditions prevailing in 2023 favoured the highest TGW in plants treated with Neosol and Bactim Gleba. UGmax contributed to maintaining TGW at a relatively stable level; after three years of monoculture (and under the least favourable climatic conditions), TGW did not decline markedly compared with 2023.

Cultivar responses to the applied biological amendments varied. The cultivar with the shortest growing season (FAO 200) produced grains of similar weight across all amendments, with values significantly higher than in the control. In cv. FAO 230, the TGW obtained with Bactim Gleba (288.29 g) and Neosol (284.09 g) was statistically similar, yet significantly lower than after the application of UGmax. In the longest-season cultivar, each amendment significantly differentiated the TGW, with values ranging from 299.23 g (Neosol) to 326.43 g (UGmax). In all maturity groups, the lowest TGW values were recorded in the control treatments.

Based on the presented data, both the year studied and the cultivars exerted a significant influence on the number of kernel rows per ear in maize. The overall mean number of kernel rows across the experiment ranged from 13.32 to 14.81. The most favourable year

https://doi.org/10.3390/su18052542

for ear structure development was 2022, when the highest mean number of kernel rows (14.64) was recorded. This value was statistically significantly higher than those achieved in the other years. In 2023 and 2024, a decline in the number of kernel rows was observed, although the difference between these two years was not statistically significant. This indicates that conditions in 2023 and 2024 were less conducive to ear development than those in 2022.

Analysis of the cultivar means revealed clear genetic differences. The later-maturing cultivars (FAO 230 and FAO 260) developed significantly higher numbers of kernel rows per ear (14.12 and 14.17, respectively). The difference between these two cultivars was not statistically significant (Table 6).

Table 6. Number of kernel rows per ear in continuous maize by year and cultivar.

	Year			
Cultivar	2022	2023	2024	Mean
FAO 200	14.49	13.58	13.32	13.79 B
FAO 230	14.81	13.83	13.72	14.12 A
FAO 260	14.63	13.92	13.97	14.17 A
Mean	14.64 a	13.77 b	13.67 b	

Means in rows followed by different lowercase letters (a, b) differ significantly at $p \leq 0.05$ (for years). Means in columns followed by different uppercase letters (A, B) differ significantly at $p \leq 0.05$ (for cultivars).

The number of kernels per row was significantly influenced by all examined factors: the weather conditions during the years studied, the cultivars, and the biological amendments (Table 7). The highest mean number of kernels per row was recorded in 2022 (37.15), a value significantly greater than that observed in 2024 (29.78). The year 2023 represented an intermediate period and did not differ statistically from the other years. The later-maturing FAO 260 cultivar produced the highest number of kernels per row (mean 36.53), a statistically significant difference compared with cv. FAO 200 and FAO 230 (approximately 31–32).

Table 7. Number of kernels per row of maize grown in monoculture by year, cultivar, and biological amendment.

	Year			
Cultivar	2022	2023	2024	Mean
FAO 200	35.94	31.32	28.12	31.79 B
FAO 230	36.19	31.30	28.56	32.02 B
FAO 260	39.32	37.60	32.66	36.53 A
Mean	37.147 a	33.406 ab	29.78 b	
Biological amendment	2022	2023	2024	
Control	34.11 a	33.64 a	26.60 b	31.45 C
Neosol	37.29 a	33.38 b	30.02 c	33.56 B
Bactim Gleba	38.43 a	34.04 b	31.31 c	34.59 A
UGmax	38.77 a	32.56 b	31.20 b	34.17 AB
Biological amendment		Cultivar		
	FAO 200	FAO 230	FAO 260	
Control	30.30 B	30.02 B	34.03 B	
Neosol	31.98 A	32.06 A	36.65 A	
Bactim Gleba	33.93 A	32.54 A	37.31 A	
UGmax	30.96 B	33.44 A	38.12 A	

Means in rows followed by different lowercase letters (a, b, c) differ significantly at $p \leq 0.05$ (for years, year × biological amendment interactions and cultivar × biological amendment interactions). Means in columns followed by different uppercase letters (A, B, C) differ significantly at $p \leq 0.05$ (for cultivars and biological amendments).

https://doi.org/10.3390/su18052542

All applied biological amendments had a beneficial effect on kernel number compared with the control (31.45). The most favourable results were obtained with Bactim Gleba (34.59) and UGmax (34.17). Neosol also significantly improved this trait compared with the control, although it was less effective than Bactim Gleba.

Statistical analysis revealed significant interactions between the biological amendments and year, as well as between the cultivars and biological amendments. In the untreated control, a drastic reduction in kernel number occurred in 2024 compared with the previous years. Products such as Bactim Gleba and Neosol helped maintain higher yield-structure parameters even in less-favourable years, although a downward trend was evident across all combinations. The analysis also showed that cultivar responses to biological amendments varied depending on maturity group. In the earliest cultivar (FAO 200), Neosol (31.98) and Bactim Gleba (33.93) significantly improved kernel number compared with the control. For the cultivars with longer growing seasons (FAO 230 and FAO 260), all tested amendments produced values significantly higher than the control. The highest absolute value was recorded in FAO 260 following application of UGmax (38.12). Across all maturity groups, the lowest kernel numbers per row were observed in the control.

The total number of kernels per ear showed trends similar to those observed for kernels per row (Table 8). The year 2022 was clearly the most favourable for kernel set. In 2023 and 2024, significant reductions were recorded—approximately 15% and 25%, respectively, compared with 2022. Cv. FAO 260 produced an average of 517.74 kernels per ear, a value significantly higher (by approximately 14–17%) than those of the other cultivars. Biological amendments significantly increased the number of kernels per ear. Bactim Gleba (493.75) was the most effective, while Neosol (469.24) also produced a significant improvement compared with the control but was less effective than Bactim Gleba. The untreated control had the lowest kernel number (435.43). Interaction analysis indicates that in the most challenging year, 2024, the use of biological amendments, particularly Bactim Gleba and UGmax, resulted in approximately 18–21% more kernels per ear compared with the control, suggesting their role in mitigating environmental stress in maize monoculture.

Table 8. Number of kernels per ear of maize grown in monoculture by year, cultivar, and biological amendment.

	Year			
Cultivar	2022	2023	2024	Mean
FAO 200	520.79	424.75	375.07	440.20 B
FAO 230	535.76	432.54	391.92	453.41 B
FAO 260	575.09	523.07	455.06	517.74 A
Mean	543.88 a	460.12 b	407.35 b	
Biological amendment	2022	2023	2024	
Control	494.33 a	452.50 b	359.47 c	435.43 C
Neosol	540.39 a	460.03 b	407.31 c	469.24 B
Bactim Gleba	568.29 a	476.07 b	436.89 c	493.75 A
UGmax	572.51 a	451.88 b	425.72 b	483.37 AB

Means in rows followed by different lowercase letters (a, b, c) differ significantly at $p \leq 0.05$ (for years and year × biological amendment interactions). Means in columns followed by different uppercase letters (A, B, C) differ significantly at $p \leq 0.05$ (for cultivars and biological amendments).

The Wright's path analysis conducted for the period 2022–2024 revealed varying contributions of individual components to the formation of kernel number per ear and final kernel yield. Over the study period, the number of kernels per ear was determined by the number of kernel rows and the number of kernels per row (R^2 = 0.99). The number of kernels per row exerted the strongest direct effect (0.818), whereas the number of kernel

https://doi.org/10.3390/su18052542

rows had a significant but weaker direct influence (0.352). For kernel yield, the analysis showed that the number of kernels per ear was the primary determinant, displaying a strong direct effect (0.634). Thousand-grain weight (TGW), although significantly correlated with yield, had a stronger indirect rather than direct effect on the yield through kernel number (0.385 vs. 0.274). In 2022, a year characterised by a high coefficient of determination for yield (R^2 = 0.80), TGW played a dominant role. Its direct effect on yield (0.827) was considerably stronger than that of kernel number per ear (0.094), which acted mainly indirectly. In 2023, the structure of yield formation changed markedly. TGW almost entirely determined yield through a direct effect (0.975), whereas the influence of kernel number per ear was negative and operated solely through indirect mechanisms. This suggests that, in this season, factors affecting grain filling were of paramount importance. In 2024, the most balanced contribution from both components was observed. Both kernel number per ear (0.537) and TGW (0.314) exhibited statistically significant direct effects, with kernel number remaining the principal yield determinant (Table 9).

Table 9. Path coefficients for maize yield components in 2022–2024.

Year	Outcome Variable	Causal Variable	Total Correlation	Direct Effect	Indirect Effect via Variable:	
					Value	Variable Name
2022–2024	Number of kernels per ear (R^2 = 0.99)	Number of rows per ear Number of kernels per row	0.644 * 0.943 *	0.352 * 0.818 *	0.292 0.126	Number of kernels per row Number of rows
	Grain yield (R^2 = 0.68)	TGW Number of kernels per ear	0.659 * 0.800 *	0.274 * 0.634 *	0.385 0.166	Number of kernels per row TGW
2022	Number of kernels per ear (R^2 = 0.99)	Number of rows per ear Number of kernels per row	0.483 * 0.870 *	0.491 * 0.875 *	−0.008 −0.04	Number of kernels per row Number of rows
	Grain yield (R^2 = 0.80)	TGW Number of kernels per ear	0.894 * 0.686 *	0.827 * 0.094	0.067 0.593	Number of kernels per row TGW
2023	Number of kernels per ear (R^2 = 0.99)	Number of rows per ear Number of kernels per row	0.361 * 0.940 *	0.340 * 0.932 *	0.022 0.008	Number of kernels per row Number of rows
	Grain yield (R^2 = 0.88)	TGW Number of kernels per ear	0.933 * 0.356 *	0.975 * −0.092	−0.042 0.488	Number of kernels per ear TGW
2024	Number of kernels per ear (R^2 = 0.99)	Number of kernels per row	0.570 * 0.894 *	0.449 * 0.829 *	0.120 0.065	Number of kernels per row Number of rows
	Grain yield (R^2 = 0.61)	TGW Number of kernels per ear	0.669 * 0.745 *	0.314 * 0.537 *	0.365 0.208	Number of kernels per ear TGW

*—significant relationships at $p \leq 0.05$.

The application of different biological amendments modified both the strength and the nature of relationships among maize yield components (Table 10). Combining correlation analysis with path analysis made it possible to determine how individual products altered the yield-forming structure relative to the control treatment. In the control, a very strong correlation was observed between the number of kernels per ear and the grain yield (r = 0.885). This association arose predominantly from the direct effect of that trait (0.723).

https://doi.org/10.3390/su18052542

Thousand-grain weight (TGW) was significantly correlated with yield (r = 0.733); however, path analysis indicated that its direct effect was not significant (0.236). The high correlation stemmed mainly from an indirect effect mediated through the number of kernels per ear (0.497). Biological amendments altered these relationships. Following Neosol application, the number of kernels per ear retained its dominant role as the main yield determinant (direct effect 0.694). In this treatment, TGW exhibited the lowest correlation with yield among all variants studied (r = 0.446), and its influence on yield was distributed between direct and indirect pathways. With Bactim Gleba, the highest coefficient of determination for yield was recorded (R^2 = 0.88), indicating that the adopted model explained the yield structure very precisely. Although the correlation between TGW and yield remained high and significant (r = 0.715), path analysis showed that the number of kernels per ear continued to act as a stable direct yield determinant (0.570), whereas TGW influenced yield largely indirectly by increasing the number of kernels per ear (indirect effect 0.390). As with the other amendments, UGmax application resulted in the number of kernels per ear strongly and directly shaping yield (0.599). In this variant, however, a clear indirect effect of TGW on final yield was evident (0.403), suggesting that the product enhances interdependence between kernel size and kernel number per ear. Analysis of the ear structural parameters revealed that, across all variants receiving biological amendments (Neosol, Bactim Gleba, UGmax), the primary direct determinant of total kernel number per ear was the number of kernels per row. These products strengthened the correlation between the number of rows and total kernel number compared with the control, indicating improved exploitation of the ear's biological potential following soil biostimulation.

Table 10. Path coefficients for maize yield components across biological amendments applied.

Biological Amendment	Outcome Variable	Causal Variable	Total Correlation	Direct Effect	Indirect Effect via Variable	
					Value	Variable Name
Control	Number of kernels per ear (R^2 = 0.99)	Number of rows per ear Number of kernels per row	0.539 * 0.899 *	0.437 * 0.846 *	0.101 0.05	Number of kernels per row Number of rows
	Grain yield (R^2 = 0.82)	TGW Number of kernels per ear	0.733 * 0.885 *	0.236 0.723 *	0.497 0.162	Number of kernels per ear TGW
Neosol	Number of kernels per ear (R^2 = 0.99)	Number of rows per ear Number of kernels per row	0.650 * 0.395 *	0.375 * 0.807 *	0.275 0.128	Number of kernels per row Number of rows
	Grain yield (R^2 = 0.62)	TGW Number of kernels per ear	0.446 * 0.765 *	0.208 0.694 *	0.238 0.071	Number of kernels per ear TGW
Bactim gleba	Number of kernels per ear (R^2 = 0.99)	Number of rows per ear Number of kernels per row	0.707 * 0.946 *	0.360 0.787 *	0.346 0.159	Number of kernels per row Number of rows
	Grain yield (R^2 = 0.68)	TGW Number of kernels per ear	0.715 * 0.793 *	0.325 0.570 *	0.390 0.222	Number of kernels per ear TGW
UGmax	Number of kernels per ear (R^2 = 0.99)	Number of rows per ear Number of kernels per row	0.692 * 0.969 *	0.282 0.830 *	0.410 0.139	Number of kernels per row Number of rows
	Grain yield (R^2 = 0.63)	TGW Number of kernels per ear	0.661 * 0.772 *	0.257 0.599 *	0.403 0.173	Number of kernels per ear TGW

*—significant relationships at $p \leq 0.05$.

Path analysis revealed differences in the yield-forming strategies among maize cultivars differing in maturity group (Table 11). In all FAO groups, the model for the variable

https://doi.org/10.3390/su18052542

'number of kernels per ear' showed an almost perfect fit ($R^2 = 0.99$). In early-maturing cultivars (FAO 200), grain yield was determined to a nearly equal degree by thousand-grain weight (TGW, direct effect 0.513) and by the number of kernels per ear (0.466). This points to a balanced contribution from both yield components. In this maturity group, the number of kernels per ear depended predominantly on the number of kernels per row (direct effect 0.781), whereas the influence of the number of rows was exerted largely through indirect pathways. In the medium–early-maturing group (FAO 230), the relationships shifted towards ear structure. Here, the number of kernels per ear exerted a markedly stronger direct effect on yield (0.677) than did the TGW (0.261). Kernel weight played a secondary role, while the primary yield determinant was the number of kernels per row (direct effect 0.776). The most pronounced differences appeared in late-maturing cultivars (FAO 260). Path analysis indicated that, in this group, the sole key determinant of yield was the number of kernels per ear (very strong direct effect 0.727). It should be emphasised that TGW showed no significant direct influence in this FAO group (0.149); its correlation with yield (0.569) arose almost entirely from an indirect effect mediated through the number of kernels per ear (0.420). This indicates that, in late cultivars grown in monoculture, ear architecture, particularly the number of kernels per row, rather than individual kernel weight, governs yield success.

Table 11. Path coefficients for maize yield components according to cultivar earliness.

Cultivar	Outcome Variable	Causal Variable	Total Correlation	Direct Effect	Indirect Effect via Variable: Value	Indirect Effect via Variable: Variable Name
FAO 2000	Number of kernels per ear ($R^2 = 0.99$)	Number of rows per ear	0.724 *	0.353 *	0.371 *	Number of kernels per row
		Number of kernels per row	0.949 *	0.781 *	0.168	Number of rows
	Grain yield ($R^2 = 0.76$)	TGW	0.785 *	0.513 *	0.271	Number of kernels per ear
		Number of kernels per ear	0.765 *	0.466 *	0.300	TGW
FAO 230	Number of kernels per ear ($R^2 = 0.99$)	Number of rows per ear	0.734 *	0.356 *	0.378	Number of kernels per row
		Number of kernels per row	0.950 *	0.776 *	0.173	Number of rows
	Grain yield ($R^2 = 0.71$)	TGW	0.618 *	0.261 *	0.357	Number of kernels per ear
		Number of kernels per ear	0.814 *	0.677 *	0.138	TGW
FAO 260	Number of kernels per ear ($R^2 = 0.99$)	Number of rows per ear	0.508 *	0.456 *	0.052	Number of kernels per row
		Number of kernels per row	0.889 *	0.862 *	0.028	Number of rows
	Grain yield ($R^2 = 0.67$)	TGW	0.569 *	0.149	0.420 *	Number of kernels per ear
		Number of kernels per ear	0.813 *	0.727 *	0.086	TGW

*—significant relationships at $p \leq 0.05$.

The economic analysis showed that all tested amendments generated a positive economic return (Table 12). The highest additional revenue was obtained following the application of UGmax (160 USD·ha^{-1}), followed by Bactim Gleba (95 USD·ha^{-1}). UGmax also achieved the highest profitability ratio (5.50–6.60). In contrast, Neosol exhibited the lowest economic efficiency due to its relatively high cost in relation to the yield increase obtained, with a profitability ratio close to the break-even threshold (1.09–1.31).

https://doi.org/10.3390/su18052542

Table 12. Economic efficiency of using biological amendments in maize cultivation (means for 2022–2024).

Biological Amendment	Yield Increase vs. Control ($Mg{\cdot}ha^{-1}$)	Additional Revenue ($USD{\cdot}ha^{-1}$)	Amendment Cost ($USD{\cdot}ha^{-1}$)	Net Economic Gain ($USD{\cdot}ha^{-1}$)	Profitability Index (BCR)	Break-Even Grain Price ($USD{\cdot}Mg^{-1}$)
Neosol	0.3	58	44–53	+14 to +5	1.31–1.09	146–176
Bactim Gleba	0.49	95	22–27	+73 to +68	4.3–3.5	45–55
UG Max	0.83	160	24–29	+136 to +131	6.5–5.5	29–35

4. Discussion

The United Nations Sustainable Development Goals highlight the crucial role of the agricultural sector in ensuring food security, protecting natural resources, and supporting adaptation to climate change. In countries with a strong reliance on crop production, such as Poland, technologies that stabilise yields under increasing environmental pressure, particularly drought and soil degradation in monoculture systems, are gaining strategic importance. Numerous long-term experiments have demonstrated that such systems lead to reduced yield stability and the deterioration of soil quality [29–33]. The study focused primarily on the productive response of plants (yield and yield components) as an indicator of soil–plant system functioning. The combined use of ANOVA and path analysis made it possible to assess the influence of the meteorological conditions, cultivar earliness, and the action of the biological amendments on the mechanism of yield formation through its individual components. ANOVA results obtained in this study showed that, under monoculture conditions, maize grain yield was strongly determined by meteorological conditions. Path coefficient analysis, in turn, enabled identification of the direct and indirect effects of individual traits on the final yield variable [16,34], providing deeper insight into the complex relationships between yield components and maize grain yield in monoculture.

Our findings demonstrated that the traits with the strongest direct influence on kernel yield were the number of kernels per ear and the thousand-grain weight (TGW). These results are consistent with observations by Huda et al. [35] and Matin et al. [16], who indicated that these parameters constitute the primary determinants of maize productivity. Path analysis further revealed that the relative importance of yield components varied depending on weather conditions. In the year with favourable weather (2022), TGW played the dominant role in shaping yield, while the effect of the number of kernels per ear was indirect. This indicates that kernel-filling efficiency was the decisive factor. Under water stress, the structure of relationships shifted: TGW exhibited a very strong direct effect on yield, whereas the direct effect of the number of kernels per ear was negative and insignificant. This suggests the presence of compensatory mechanisms, whereby plants partially offset reductions in kernel number by increasing individual kernel weight. At the same time, the application of biological amendments significantly mitigated the decline in monoculture productivity, which is particularly important from a sustainability perspective, as it contributes to yield stabilisation. The use of UGmax and Neosol led to a substantial reduction in yield losses in years characterised by unfavourable hydrothermal conditions. The observed response may be associated with soil-functioning improvement mechanisms described in the literature [3,21].

This is supported by numerous studies showing that organic or biological amendments can effectively mitigate the negative effects of monoculture [36–39]. UGmax produced the highest mean yield across the three-year cycle (6.73 $Mg{\cdot}ha^{-1}$), exceeding the control by more than 14%, while Bactim Gleba and Neosol increased yields to 6.39 and 6.20 $Mg{\cdot}ha^{-1}$, respectively. These production effects align with earlier research on UGmax in maize and potato, where yield increases and improved plant health were reported [40–43]. At the same time, Piotrowska et al. [44] demonstrated that this product increased soil microbiological activity and the content of plant-available nutrients. Novák et al. [45] noted that Neosol affected the physical properties of heavy soils as well as the physiological status of plants. The stabilisation of yield components observed in the present study is consistent with such findings. The plant response recorded in the present study may also be associated with mechanisms described in the literature, including those related to water management and soil aggregate structure [5,6,8].

From a physio-biological perspective, the plant responses observed here are consistent with studies describing the role of conditioners and organic amendments in shaping the rhizosphere microbiome and nutrient uptake efficiency [11,12,34].

ANOVA results further indicated that cv. FAO 260 achieved significantly higher mean yield (6.70 $Mg \cdot ha^{-1}$) than FAO 230 (6.32 $Mg \cdot ha^{-1}$) and FAO 200 (5.84 $Mg \cdot ha^{-1}$). Cv. FAO 260 also exhibited the highest TGW (300.3 g), suggesting a greater capacity for assimilate accumulation even under declining water availability. This is consistent with findings by Berzsenyi et al. [30] and Księżak et al. [31], who reported that, in long-term monoculture, high-yielding cultivars utilise available resources more effectively than earlier-maturing forms.

Path analysis revealed clear differences in yield-forming strategies between maturity groups and biological amendment variants, confirming the usefulness of this method in agronomic research [14,24]. The differences observed between maturity groups correspond to the findings of Matin et al. [16] and Munawar et al. [46], who emphasised the variability of yield-formation mechanisms depending on genotype and environmental conditions. In cv. FAO 200, yield was shaped almost equally by TGW (direct effect 0.513) and the number of kernels per ear (0.466), indicating a balanced compensatory strategy. Similar mechanisms were reported by Rafiq et al. [15] and Soumya and Kamatar [47] who noted that early cultivars often compensate limitations in one component through another. In cv. FAO 230, the dominant role was played by the number of kernels per plant, with the primary determinant being the number of kernels per row. A similar pattern was described by Pavlov et al. [48] and Matin et al. [16], highlighting the importance of generative development in shaping yield in medium–early-maturing cultivars. The higher productivity of late-maturing cultivars, confirmed by ANOVA, resulted from the strong direct effect of the number of kernels per ear as the main yield-forming trait. This relationship aligns with the findings of Matin et al. [16] and Huda et al. [35], who emphasised that kernel number is the trait with the highest potential to determine yield in maize. This suggests that, under monoculture and water stress, cultivars which mature later in the season maintain productivity primarily by preserving kernel number, which may be crucial for yield stability under climate change.

Path analysis results for biological amendments are particularly noteworthy from a sustainability perspective. In the control, yield was mainly affected by the number of kernels per ear (direct effect 0.723), while TGW acted predominantly through indirect pathways (0.497), indicating limited kernel-filling capacity under declining soil fertility. After Bactim Gleba application, the number of kernels per ear remained the main and stable determinant of yield (0.570), with strong indirect effects of TGW (0.390).

For UGmax, the interdependence between grain number and grain weight was additionally strengthened (indirect effect of TGW = 0.403), which may indicate more favourable conditions for grain filling. This observation is consistent with the mechanisms described by Piotrowska et al. [44], Keteku et al. [49], and Odugbenro et al. [50], who reported links between biological amendments, soil structural properties, and nutrient uptake efficiency. From an agronomic perspective, this may imply the potential to stabilise yields without increasing fertilisation intensity, which is particularly relevant in the context of sustainable agriculture.

https://doi.org/10.3390/su18052542

From a practical perspective, the profitability of using biological amendments is of key importance, as it depends on the balance between their cost and the resulting yield increase. Our results confirm that microbial-based products (UGmax and Bactim Gleba) are cost-effective solutions for mitigating the negative effects of maize monoculture. For Neosol, the break-even price is higher, indicating that its use is more sensitive to market volatility.

Detailed economic analyses that would account for the full production costs and the long-term effects of biological amendments on maize yield were beyond the scope of the present study. However, incorporating such analyses in future research is essential to fully assess the feasibility of implementing these solutions at the farm scale. The need for continued investigation in this direction is further justified by the inherent limitations of the current work. The most important of these include the relatively short, three-year duration of the experiment, which may not fully capture the long-term consequences of biostimulant use in maize monoculture. Moreover, this study focused on plant production parameters without direct monitoring of physicochemical and biological changes occurring in the soil. It should also be noted that the obtained results may have been influenced by specific local conditions and interannual meteorological variability.

5. Conclusions

Maize productivity under continuous monoculture was strongly determined by interannual meteorological variability. Increasing water deficit and heat stress during the generative development stage led to a progressive decline in yield, primarily through a reduction in the number of kernels per ear. This trait proved to be the key structural determinant of yield stability under abiotic stress.

The integration of ANOVA with Wright's path analysis provided new insight into the biological hierarchy of yield formation in monoculture systems. The results demonstrated that maturity class determines the mechanism of productivity maintenance: early cultivars relied on a balanced contribution of kernel number and kernel weight, whereas medium- and late-maturing cultivars depended mainly on preserving kernel number, with thousand-kernel weight acting largely indirectly. These findings highlight the central role of reproductive structure preservation in sustaining yield under simplified cropping systems.

From a practical perspective, the targeted use of biological amendments combined with appropriate cultivar selection (in the present conditions particularly FAO 260) may improve yield stability in simplified cropping systems. Microbial-based products showed the most stable economic performance, with UGmax providing the highest profitability due to both yield increase and a favourable cost–benefit relation, whereas other amendments were more dependent on market conditions.

The study was limited by the relatively short three-year monoculture period and by the lack of direct measurements of the soil physical, chemical and biological properties. Therefore, long-term experiments integrating crop productivity with direct soil quality indicators are necessary to fully explain the mechanisms underlying yield stabilisation in monoculture maize systems.

Author Contributions: Conceptualization: M.G. Methodology: K.R., E.R., K.K. and M.G.; Formal analysis: K.R. and E.R.; Original draft preparation: K.R., E.R., K.K. and M.G.; Writing, review and editing: K.K. and M.G. All authors have read and agreed to the published version of the manuscript.

Funding: This research received no external funding.

Institutional Review Board Statement: Not applicable.

Informed Consent Statement: Not applicable.

Data Availability Statement: The data are contained within the article.

https://doi.org/10.3390/su18052542

Conflicts of Interest: The authors declare no conflicts of interest.

References

1. Morris, M.L. *Impacts of International Maize Breeding Research in Developing Countries, 1966–1998*; CIMMYT: Texcoco, Mexico, 2002.
2. Zhang, W.; Xiong, Y.; Li, Y.; Qiu, Y.; Huang, G. Effects of organic amendment incorporation on maize (*Zea mays* L.) growth, yield and water-fertilizer productivity under arid conditions. *Agric. Water Manag.* **2022**, *269*, 107663. [CrossRef]
3. Kouame, A.K.; Bindraban, P.S.; Kissiedu, I.N.; Atakora, W.K.; El Mejahed, K. Identifying drivers for variability in maize (*Zea mays* L.) yield in Ghana: A meta-regression approach. *Agric. Syst.* **2023**, *209*, 103667. [CrossRef]
4. Wezel, A.; Casagrande, M.; Celette, F.; Vian, J.F.; Ferrer, A.; Peigné, J. Agroecological practices for sustainable agriculture. *Agron. Sustain. Dev.* **2014**, *34*, 1–20. [CrossRef]
5. Six, J.; Elliott, E.T.; Paustian, K. Soil macroaggregate turnover and microaggregate formation. *Soil Biol. Biochem.* **2000**, *32*, 2099–2103. [CrossRef]
6. Bronick, C.J.; Lal, R. Soil structure and management: A review. *Geoderma* **2005**, *124*, 3–22. [CrossRef]
7. Franzluebbers, A.J. Soil organic matter stratification ratio as an indicator of soil quality. *Soil Tillage Res.* **2002**, *66*, 95–106. [CrossRef]
8. Kibblewhite, M.G.; Ritz, K.; Swift, M.J. Soil health in agricultural systems. *Philos. Trans. R. Soc. B* **2008**, *363*, 685–701. [CrossRef]
9. Guilherme, M.R.; Aouada, F.A.; Fajardo, A.R.; Martins, A.F.; Paulino, A.T.; Davi, M.F.T.; Rubira, A.F.; Muniz, E.C. Superabsorbent hydrogels based on polysaccharides for application in agriculture as soil conditioner and nutrient carrier: A review. *Eur. Polym. J.* **2015**, *72*, 365–385. [CrossRef]
10. Medina, E.; Paredes, C.; Bustamante, M.A.; Moral, R.; Moreno-Caselles, J. Relationships between soil physico-chemical, chemical and biological properties in a soil amended with spent mushroom substrate. *Geoderma* **2012**, *173–174*, 152–161. [CrossRef]
11. Qiao, C.; Penton, C.R.; Xiong, W.; Liu, C.; Wang, R.; Liu, Z.; Xu, X.; Li, R.; Shen, Q. Reshaping the rhizosphere microbiome by bio-organic amendment to enhance crop yield in a maize-cabbage rotation system. *Appl. Soil Ecol.* **2019**, *142*, 136–146. [CrossRef]
12. Liu, Q.; Cui, H.; Yang, W.; Wang, F.; Liao, H.; Zhu, Q.; Qin, S.; Lu, P. Soil conditioner improves soil properties, regulates microbial communities, and increases yield and quality of *Uncaria rhynchophylla*. *Sci. Rep.* **2024**, *14*, 13398. [CrossRef] [PubMed]
13. Koynarska, K. Study the influence of mineral fertilization on the yield and its structural elements of maize grown as a short-lived monoculture. *J. Mt. Agric. Balk.* **2023**, *26*, 162–180.
14. Gozdowski, D.; Mądry, W.; Wyszyński, Z. Analysis of correlation and path coefficients in evaluation of relationships between grain yield and its components of two spring barley cultivars. *Bull. Plant Breed. Acclim. Inst.* **2008**, *248*, 23–31. (In Polish)
15. Rafiq, C.M.; Rafique, M.; Hussain, A.; Altaf, M. Studies on heritability, correlation and path analysis in maize (*Zea mays* L.). *J. Agric. Res.* **2010**, *48*, 35–38.
16. Matin, M.Q.I.; Uddin, M.S.; Rohman, M.M.; Amiruzzaman, M.; Azad, A.K.; Banik, B.R. Genetic variability and path analysis studies in hybrid maize (*Zea mays* L.). *Am. J. Plant Sci.* **2017**, *8*, 3101–3109. [CrossRef]
17. Prasad, B.V.V.V.; Shivani, D. Correlation and path analysis in maize (*Zea mays* L.). *J. Genet. Genom. Plant Breed.* **2017**, *1*, 1–7.
18. Nemati, A.; Sedghi, M.; Sharifi, R.S.; Seiedi, M.N. Investigation of correlation between traits and path analysis of corn (*Zea mays* L.) grain yield in Ardabil region. *Not. Bot. Horti Agrobot. Cluj-Napoca* **2009**, *37*, 194–198.
19. Toebe, M.; Cargnelutti Filho, A. Multicollinearity in path analysis of maize (*Zea mays* L.). *J. Cereal Sci.* **2013**, *57*, 453–462. [CrossRef]
20. World Reference Base for Soil Resources. International soil classification system for naming soils and creating legends for soil. In *World Soil Resources Reports 106; Field Experiment*; Food and Agriculture Organization: Rome, Italy, 2014.
21. Skowera, B.; Pula, J. Skrajne warunki pluwiotermiczne w okresie wiosennym na obszarze Polski w latach 1971–2000. *Acta Agrophys.* **2004**, *3*, 171–177. (In Polish)
22. Trętowski, J.; Wójcik, A.R. *Metodyka Doświadczeń Rolniczych*; WSRP: Siedlce, Poland, 1991; p. 538. (In Polish)
23. Wright, S. The method of path coefficients. *Ann. Math. Stat.* **1934**, *5*, 161–215. [CrossRef]
24. Kozak, M. Analiza związków przyczynowo-skutkowych w agronomii i hodowli roślin. *Bull. Plant Breed. Acclim. Inst.* **2011**, *259*, 3–21. (In Polish) [CrossRef]
25. Produkcja Koszty i Nadwyżka Bezpośrednia Wybranych Produktów Rolniczych w 2024 Roku. Available online: https://ierigz.waw.pl/publikacje/poza-seria/26028,61,3,0,produkcja-koszty-i-nadwyzka-bezposrednia-wybranych-produktow-rolniczych-w-2024-roku.html (accessed on 18 February 2026).
26. Notowania cen Kukurydzy. Available online: https://www.agrolok.pl/notowania/notowania-cen-kukurydzy.htm (accessed on 17 February 2026).
27. Bactim Gleba. Available online: https://esklepagro.pl/pl/p/BACTIM-GLEBA-/717 (accessed on 19 February 2024).
28. Archiwum Kursów Średnich. Available online: https://nbp.pl/archiwum-kursowo (accessed on 18 February 2025).
29. Berzsenyi, Z.; Győrffy, B. Effect of crop rotation and fertilization on maize and wheat yields and yield stability in long-term Experiments. *Agrokémia Talajt.* **1997**, *46*, 77–98.
30. Berzsenyi, Z.; Győrffy, B.; Lap, D. Effect of crop rotation and fertilisation on maize and wheat yields and yield stability. *Eur. J. Agron.* **2000**, *13*, 225–244. [CrossRef]

https://doi.org/10.3390/su18052542

31. Księżak, J.; Bojarszczuk, J.; Staniak, M. Comparison of maize yield and soil chemical properties under monoculture and crop rotation. *J. Elem.* **2018**, *23*, 659–672. [CrossRef]
32. Fuchs, A.; Berger, V.; Steinbauer, K.; Köstl, T.; Wuttej, D.; Jungmeier, M. The long-term effects of monoculture maize cultivation on plant diversity. *Phytocoenologia* **2021**, *50*, 397–408. [CrossRef]
33. Wolińska, A.; Kruczyńska, A.; Podlewski, J.; Słomczewski, A.; Grządziel, J.; Gałązka, A.; Kuźniar, A. Does intercropping improve monocultural soils? *Agronomy* **2022**, *12*, 613. [CrossRef]
34. Kote, U.B.; Kumar, P.R.; Ahamed, M.L.; Rani, Y.A.; Adilakshmi, V.S.R. Correlation and path analyses in maize (*Zea mays* L.). *Electron. J. Plant Breed.* **2014**, *5*, 538–544.
35. Huda, M.N.; Hossain, M.S.; Sonom, M. Genetic variability, character association and path analysis in maize. *Bangladesh J. Plant Breed. Genet.* **2016**, *29*, 21–30. [CrossRef]
36. Rusinamhodzi, L.; Corbeels, M.; Van Wijk, M.T.; Rufino, M.C.; Nyamangara, J.; Giller, K.E. Long-term effects of conservation agriculture on maize yield. *Agron. Sustain. Dev.* **2011**, *31*, 657–673. [CrossRef]
37. Adeux, G.; Giuliano, S.; Cordeau, S.; Savoie, J.-M.; Alletto, L. Low-input maize-based cropping systems. *Agriculture* **2017**, *7*, 74. [CrossRef]
38. Zhang, L.; Yuan, J.; Zhang, M.; Zhang, Y.; Wang, L.; Li, J. Long term effects of crop rotation and fertilization on crop yield stability in southeast China. *Sci. Rep.* **2022**, *12*, 14234. [CrossRef]
39. Bockstaller, C.; Blatz, A.; Rapp, O.; Koller, R.; Slezack, S.; Schaub, A. Sustainable alternative to irrigated maize monoculture in a maize-dominated cropped area: Lessons learned from a system experiment. *Heliyon* **2024**, *10*, e30400. [CrossRef]
40. Sulewska, H.; Szymańska, G.; Pecio, A. Evaluation of UGmax soil additive in maize cultivation. *J. Res. Appl. Agric. Eng.* **2009**, *54*, 120–124.
41. Zarzecka, K.; Gugała, M.; Milewska, A. Oddziaływanie użyźniacza glebowego UGmax na plonowanie ziemniaka. *Prog. Plant Prot.* **2011**, *51*, 1–7. (In Polish)
42. Zarzecka, K.; Gugała, M. Plonotwórcze działanie użyźniacza glebowego UGmax. *Inżynieria Ekol.* **2012**, *28*, 144–148. (In Polish)
43. Frąckowiak-Pawlak, K. Wyniki wieloletnich doświadczeń z UGmax. *Wiadomości Rol.* 2011, 2. Available online: https://agro.icm.edu.pl/agro/element/bwmeta1.element.agro-ed67dfb8-f7ae-4bcc-9377-ddb1cf95ea67 (accessed on 19 February 2024). (In Polish)
44. Piotrowska, A.; Długosz, J.; Zamorski, R.; Bogdanowicz, P. Changes in soil properties after UGmax application. *Pol. J. Environ. Stud.* **2012**, *21*, 455–463.
45. Novák, V.; Šařec, P.; Křížová, K.; Novák, P.; Látal, O. Impact of biostimulator NeOsol and manure on soil properties. *Sustainability* **2021**, *14*, 438. [CrossRef]
46. Munawar, M.; Shahbaz, M.; Hammad, G.; Yasir, M. Correlation And Path Analysis Of Grain Yield Components In Exotic Maize (*Zea mays* L.) Hybrids. *Int. J. Sci. Basic Appl. Res.* **2013**, *12*, 22–27.
47. Soumya, H.; Kamatar, M. Correlation and path analysis in maize hybrids. *J. Farm Sci.* **2017**, *30*, 153–156.
48. Pavlov, J.; Delić, N.; Marković, K.; Crevar, M.; Čamdžija, Z.; Stevanović, M. Path analysis for morphological traits in maize. *Genetika* **2015**, *47*, 295–301. [CrossRef]
49. Keteku, A.K.; Intanon, P.; Terapongtanakorn, S. Impact of soil amendments and fertilizers on maize (*Zea mays* L.) growth and yield and on physical, chemical, and biological soil properties. *Songklanakarin J. Sci. Technol.* **2021**, *43*, 1078–1085.
50. Odugbenro, G.O.; Liu, Z.; Oluwasemire, K.O.; Erinle, K.O.; Sun, Y. Soil quality alteration and maize (*Zea mays* L.) yield after organic amendments application to a Pellic Vertisol in China. *Soil Sci. Ann.* **2023**, *74*, 177042. [CrossRef]

https://doi.org/10.3390/su18052542

sustainability

Article

Balanced Fertilization of Winter Wheat with Potassium and Magnesium—An Effective Way to Manage Fertilizer Nitrogen Sustainably

Agnieszka Andrzejewska, Katarzyna Przygocka-Cyna and Witold Grzebisz *

Department of Agricultural Chemistry and Environmental Biogeochemistry, Poznan University of Life Science, Wojska Polskiego 28, 60-637 Poznan, Poland; agnieszka.andrzejewska@up.poznan.pl (A.A.); katarzyna.przygocka-cyna@up.poznan.pl (K.P.-C.)

* Correspondence: witold.grzebisz@up.poznan.pl; Tel.: +48-618-77-88

Abstract

In agricultural practice, in addition to determining the nitrogen (N_f) dose, it is necessary to effectively control its effect on currently grown crops. Meeting these conditions requires not only the use of phosphorus (P) and potassium (K), but also nutrients such as magnesium (Mg) and sulfur (S). This hypothesis was verified in a single-factor field experiment with winter wheat (WW) carried out in the 2015/2016, 2016/2017, and 2017/2018 growing seasons. The experiment consisted of seven variants: absolute control (AC), NP, NPK-MOP (K as Muriate of Potash), NPK-MOP+Ki (Kieserite), NPK-KK (K as Korn–Kali), NPK-KK+Ki, and NPK-KK+Ki+ES (Epsom Salt). The use of K as MOP increased grain yield (GY) by 6.3% compared to NP. In the NPK-KK variant, GY was 13% (+0.84 t ha^{-1}) higher compared to NP. Moreover, GYs in this fertilization variant (FV) were stable over the years (coefficient of variation, CV = 9.4%). In NPK-KK+Ki+ES, the yield increase was the highest and mounted to 17.2% compared to NP, but the variability over the years was also the highest (CV $\approx$ 20%). The amount of N in grain N (GN) increased progressively from 4% for NPK-MOP to 15% for NPK-KK and 25% for NPK-KK+Ki+ES in comparison to NP. The nitrogen harvest index was highly stable, achieving 72.6 $\pm$ 3.1%. All analyzed NUE indices showed a significant response to FVs. The PFP-N_f (partial factor productivity of N_f) indices increased on NPK-MOP by 5.8%, NPK-KK by 12.9%, and NPK-KK+Ki+ES by 17.9% compared to NP. The corresponding N_f recovery of N_f in wheat grain was 47.2%, 55.9%, and 64.4%, but its total recovery by wheat (grain + straw) was 67%, 74.5%, and 87.2%, respectively. In terms of the theoretical and practical value of the tested indexes, two indices, namely, NUP (nitrogen unit productivity) and NUA (nitrogen unit accumulation), proved to be the most useful. From the farmer's production strategy, FV with K applied in the form of Korn–Kali proved to be the most stable option due to high and stable yield, regardless of weather conditions. The increase in the number of nutritional factors optimizing the action of nitrogen in winter wheat caused the phenomenon known as the "scissors effect". This phenomenon manifested itself in a progressive increase in nitrogen unit productivity (NUP) combined with a regressive trend in unit nitrogen accumulation (NUA) in the grain versus the balance of soil available Mg (Mg_b). The studies clearly showed that obtaining grain that met the milling requirements was recorded only for NUA above 22 kg N t^{-1} grain. This was possible only with the most intensive Mg treatment (NPK-KK+Ki and NPK-KK+Ki+ES). The study clearly showed that three of the six FVs fully met the three basic conditions for sustainable crop production: (i) stabilization and even an increase in grain yield; (ii) a decrease in the mass of inorganic N in the soil at harvest, potentially susceptible to leaching; and (iii) stabilization of the soil fertility of P, K, and Mg.

Sustainability **2025**, *17*, 6705 https://doi.org/10.3390/su17156705

Keywords: grain yield; yield components; nitrogen accumulation; crop nutrient balance; soil nutrient balance; indicators of nitrogen use efficiency

1. Introduction

Nitrogen, provided that there is sufficient water availability, is a decisive factor for crop plant growth and the formation of yield components [1,2]. Moreover, the nitrate nitrogen ion ($N\text{-}NO_3$) is considered as crucial for plant morphogenesis. This inorganic N form, acting as a plant morphogen, affects both roots and shoots growth. Its deficiency in the growth medium changes the morphology of the root system, which forces it to grow in areas rich in nitrates [3,4]. This specificity of crop plants; response to soil nitrate availability is considered a key factor in breeding new varieties. The steep, cheap, and deep breeding concept focuses on exploring nitrate resources in the soil [5]. The critical stages of wheat yield formation and demand for N are well defined [6]. Despite this knowledge, the nitrogen use efficiency of nitrogen fertilizers (NUE) is low. As thousands of field experiments have shown, the recovery of applied N_f is in the range of 30–50% [7]. The remaining N_f is dispersed in the environment, thus posing a threat to the stability of natural ecosystems [8].

The consequence of inefficient N_f use by crops is the yield gap, or more precisely, the nitrogen gap (N gap) [9]. In the case of wheat, it is significant and results not only from the impact of environmental factors [10]. The N gap amelioration strategy requires a holistic view of the plant's demand for nutrients during the formation of the yield structure [11]. The key reason of the N gap is not only, as it is generally believed, an inappropriate N dose, N form, or N_f application date (N term) [12]. The key reason for low NUE in the second stage/level of growth factor validation is the imbalance between the amount of inorganic N in the soil/plant system during the growing season and the availability of nutrients that condition N uptake and its use by the plant [13]. Nutritional factors must, therefore, be an important factor determining sustainable N management in the agroecosystem [9,13].

The first nutrient in the above approach is P [14,15]. However, in intensive agricultural practice, P is excessively applied, and, therefore, its recovery is low [2,16]. For example, in nutritionally well-nourished winter oilseed rape, P remobilization from vegetative parts during the seed filling period has remained unchanged for over 50 years, but yields have doubled [17–19]. The nutrient taken up by high-yielding crop plants in the largest amount is potassium (K). This rule does not apply only to root and tuber crops, which is documented and widely recognized. In the case of potatoes, it has been recently documented that the narrower the K:N ratio, the higher the NUE obtained, and, thus, the higher the tuber yield [20,21]. There are only a few examples of higher K uptake by cereals compared to N during the pre-flowering growth [22,23]. Projections of global K consumption for the future (2050) do not take into account these relationships [24]. Finally, it should be clearly emphasized that even in some academic textbooks, the recommended K:N ratio used in fertilizers is below 1.0 [25]. As a result, the use of K fertilizers by farmers in crop production, despite this well-documented knowledge, is far below the realistic needs of high-yielding seed crops [26]. A consequence of current practice is an imbalance of N and K, resulting in N_f inefficiency, leading to the formation of the N gap [2,9,20,21].

K has many functions in the plant, starting from photosynthesis, through ions transport in the xylem, plant growth, and finally, yield [27]. The role of K in the uptake and transport of inorganic N ions is well known. This cation dominates, among others, in the process of NO_3^- uptake, according to the rule ($2K^+ \times NO_3^-$) [14,28]. In the case of crop plants, the role of K concerns the physiological and molecular processes responsible for the daily cycle of stomata [29,30], hence the important role of K in alleviating, at least partly, water stress in crop plants [31,32]. No less important is K for the quantity and quality of the yield. K ions are involved in the transport of assimilates from leaves to storage organs (root and tuberous plants) or to generative organs (seeds and grains). N management during the vegetation of seed plants is confirmed by the dynamics of K uptake, as well as the dynamics of plant growth during this period [22,33]. On this basis, it is justified to state that K is a key nutrient determining N use efficiency (NUE), regardless of its source (i.e., soil, atmospheric N_2 fixation, N dose in the applied fertilizer).

The sustainable strategy of high-yielding crops fertilization with N cannot be limited to P and K. A good N_f management strategy requires the use of nutrients, referred to in agricultural practice as secondary nutrients [34]. In fields with regulated soil pH, these requirements are met by magnesium (Mg) and sulfur (S). Both nutrients play a very important role in the metabolism of crop plants and in the yield formation [35,36]. Both nutrients are present in magnesium sulfate, which is easily soluble in water (i.e., Kieserite, $MgSO_4 \times H_2O$ and Epsom Salt (ES), $MgSO_4 \times 7H_2O$). Kieserite is usually applied to the soil, and ES as an aqueous solution on plant leaves in the form of a spray [37]. The key function of Mg in the plant is its presence as a component of chlorophyll. Its necessity for all energy conversion processes, including N uptake and transport in the plant, has been well documented [38]. In turn, S, being in a stable relationship with N, is necessary for the synthesis of sulfur amino acids. In this way, it determines the quality of plant proteins [39,40].

The above presented considerations on the role of nutrients controlling yield-forming effect of N_f are consistent with the assumptions of the sustainable intensification of agriculture (SIA) concept. Its key assumption is the effective use of critical means of production, such as fertilizer nitrogen (N_f) [9,41]. Achieving this goal is only possible if N_f is used efficiently and its utilization rate does not lead to yield reduction. Furthermore, the N_{min} content after harvest of the currently cultivated crop should be at a level that does not pose a threat to the environment [2,8,12]. The second assumption of rational nutrient management is to maintain a stable level of soil fertility, in accordance with the Law of Return [13,42]. The degree of crop plants' response to nutritional growth factors can best be assessed by analyzing the sensitivity of N productivity indices to the effects of other nutrients. Classic N efficiency indices can be generated using field experiments with so-called absolute control. Operational indices, however, do not require this database because they are based on the direct measurement of the yield and N accumulation [1,2,43,44].

It was assumed that in naturally fertile K soil, a moderate dose of K fertilizer, but supported by an appropriate amount of Mg and S, guarantees a significant increase in the GY of winter wheat. The expected increase in winter wheat yields results from increased N_f productivity, which is the effect of the active exploration of nutrients from both applied fertilizers and soil resources. Based on this hypothesis, the following set of objectives was established. The primary objective was to assess the effect of balanced K–MgS fertilization on grain yield, its components, and N accumulation in winter wheat at harvest. The second objective was to determine the best set of K and MgS fertilization regimes to maximize winter wheat grain yield. The third objective was to assess the effect of K and Mg application in different fertilizer variants on the nutrient balance in wheat and soil. The fourth objective was to assess the usefulness of differently developed NUE indices for a reliable and practical assessment of N economics in winter wheat.

2. Materials and Methods

2.1. Experimental Site

Studies on the effect of balanced fertilization of winter wheat were carried out in the growing seasons 2015/2016, 2016/2017, and 2017/2018 at the Brody Experimental Far, Poznan University of Life Sciences (52°26′ N, 16°18′ E, 92 m a.s.l.), Poznań, Poland. The field experiment was conducted on Albic Luvisol formed from loamy sand underlined by light loam. The soil pH was slightly acidic, being the lowest in the growing season 2017/2018. The organic carbon (C_{org}) content was low (<20 g kg^{-1}). The content of available nutrients, measured before applying fertilizers, was very high in all growing season and stable for phosphorus (P). The content of potassium (K) was on the border of the low and medium availability classes. The content of available Mg was in the high class in the first growing season and in the medium class in the other two growing seasons. The content of calcium (Ca), despite the year-to-year variability was in the low class. Of the four micronutrients examined, the most stable content in the years of the study was found for iron (Fe) and manganese (Mn), which were in the medium class. The content of available copper (Cu) was in the low class in the first and the third growing season, and in the medium class in the second growing season. The content of zinc (Zn) showed the same trend as for Cu but was a class higher. The amount of mineral N (N_{min}), measured just before the spring regrowth of winter wheat in the 0.0–0.6 m layer, was generally moderate, ranging from 48 to 66 kg ha^{-1} (Table 1).

Table 1. Soil agrochemical characteristics of the topsoil in consecutive growing seasons [3].

Soil Characteristics	Growing Seasons		
	2015/2015	2016/2017	2017/2018
pH [1]	6.1 ± 0.2	6.1 ± 0.3	5.7 ± 0.2
C_{org} [2], g kg^{-1}	15.8 ± 3.3	16.0 ± 2.6	15.9 ± 2.8
N mineral, kg ha^{-1}	66 ± 16	64 ± 18	48 ± 14
P [5], mg kg^{-1}	210 ± 15VH [4]	212 ± 20VH	286 ± 32VH
K [5], mg kg^{-1}	145 ± 15L	152 ± 12L	155 ± 16M
Mg [5], mg kg^{-1}	120 ± 11H	97 ± 10M	91 ± 12M
Ca [6], mg kg^{-1}	740 ± 125L	1195 ± 130L	840 ± 95L
Cu [7], mg kg^{-1}	0.9 ± 0.4L	2.2 ± 0.5M	1.4 ± 0.6L
Zn [7], mg kg^{-1}	2.4 ± 0.9M	4.1 ± 1.2H	4.8 ± 1.4M
Mn [7], mg kg^{-1}	36 ± 12M	57 ± 16M	54 ± 14M
Fe [7], mg kg^{-1}	299 ± 66M	285 ± 58M	365 ± 84M

[1] 1.0 M KCl soil/solution ratio 1:2.5 m/v; [2] loss on ignition; [3] Mehlich 3 [45]; [4] availability classes: L—low; M—medium; H—high; VH—very high [46–48] [5–7].

2.2. Weather Conditions

The local climate in the geographical area of the field experiment is classified as a moderate climatic zone (transitional between Atlantic and Continental). In the summer months, continental conditions dominate, with high temperatures, thunderstorms, and locally distributed rains. The climate of the study area in the period 1964–2015 was characterized by an average annual temperature of 8.1 °C (in the range from 6.6 to 10 °C) and total precipitation of about 580 mm per year (in the range from 310 to 840 mm). Weather conditions were very variable in subsequent growing seasons (Table 2).

Table 2. Main meteorological characteristics during the winter wheat growing season relative to long-term averages at Brody Experimental Farm during the study.

Months	2015/2016		2016/2017		2017/2018		Long-Term 1961–2015	
	T [1]	P [2]	T	P	T	P	T	P
September	17.9	57.9	17.4	9.9	13.8	60.7	13.3	67.3
October	5.8	40.6	8.4	76.6	11.1	79.4	8.5	48.8
November	8.1	21.7	3.3	31.8	5.4	51.7	3.7	40.7
December	5.8	45.1	2.2	37.6	2.7	31.3	0.0	45.5
January	5.9	38.0	−2.4	12.5	2.1	57.6	−1.6	47.7
February	−1.7	28.5	1.2	30.4	−2.5	3.7	−0.6	40.2
March	3.7	33.9	7.1	40.5	0.7	20.6	2.9	32.3
April	4.1	31.2	7.9	25.7	12.9	65.3	8.1	39.2
May	8.8	29.7	14.0	49.2	17.1	19.2	13.2	37.4
Juni	15.3	76.1	17.6	106.0	19.1	31.5	16.6	57.4
July	18.0	94.8	18.4	160.8	20.7	134.9	18.3	63.8

[1] Temperature, °C; [2] precipitation, mm.

In the 2015/2016 growing season, the most unfavorable conditions were revealed in early January, when temperatures were below −10 °C for several consecutive days. From early February, temperatures remained in the range of 0–5 °C, which was high enough to break the winter dormancy of wheat plants. As a result, the crop reached the full tillering phase at the end of winter. Suitable temperatures for growth of wheat dominated throughout the spring part of the wheat growing season. Extreme temperatures occurred in June.

In general, the meteorological conditions in the 2016/2017 growing season were very favorable for winter wheat growth. Sufficiently high temperature and high precipitation in October resulted in an accelerated rate of plant growth. Temperatures in February and especially in March were high enough to break plant dormancy and initiate intensive regrowth of plants. The total amount of precipitation was favorable for plant growth in the critical stages of yield formation (i.e., stem elongation, germination and flowering, grain filling).

In general, meteorological conditions in the 2017/2018 growing season were unfavorable for winter wheat. Extremely high temperatures, which significantly exceeded long-term averages, were first recorded at the end of March. This trend was observed throughout most of the spring part of the growing season. In the third decade of July, the difference reached almost 3 °C, which accelerated the rate of plant ripening. The amount and distribution of rainfall was unfavorable for wheat. There was a deficit of precipitation throughout the whole spring vegetation. Total rainfall from February to the end of June was 140 mm, but the long-term average is 207 mm. In May, the amount of rainfall was 19.2 mm against the long-term average of 37.4 mm.

2.3. Experimental Design

The data used in this study were based on a one-factor field experiment replicated four times. Details of the experimental design (fertilization variants, and FVs) are shown in Table 3. The total area of a single plot was 30 m^2 (2 × 15 m). The winter wheat was sown annually after winter oilseed rape in the first week of October at 300 grains m^{-2}. The crop was harvested the following year at the end of July from an area of 19.5 m^{-2} (1.5 × 13 m). Nitrogen fertilizer was applied to the top using a fertilizer spreader. It was applied in the form of ammonia nitrate (34:0:0; producer Grupa Azoty Puławy, Puławy, Poland) in line with the following experimental schedule:

(1) The first N dose of 80 kg N ha^{-1} was applied at the end of winter, just before the start (regrowth) of winter wheat vegetation.
(2) The second N dose of 70 kg N ha^{-1} was applied at the beginning of the shoot elongation phase (BBCH 31/32).

Table 3. The experimental design and nutrient doses.

Treatment (Acronyms)	Fertilization Variants	N	P_2O_5	K_2O [kg ha^{-1}]	MgO
AC	Absolute Control—AC	0	0	0	0
NP	NP	150	40	0	0
NPK-MOP	NPK–MOP	150	40	80	0
NPK-MOP+Ki	NPK–MOP+Kieserite	150	40	80	30
NPK-KK	NPK–Korn–Kali	150	40	80	12
NPK-KK+Ki	NPK–Korn–Kali+Kieserite	150	40	80	12 + 18
NPK-KK+Ki+ES	NPK–Korn–Kali+Kieserite+EPSO Top	150	40	80	12 + 18 + 6.4

Phosphorus was applied in the form of triple superphosphate (46% P_2O_5; producer: Grupa Azoty Fosfory, Gdańsk, Poland). Potassium was applied in accordance with experimental schedule in two forms: (i) muriate of potash (MOP, 60% K_2O; producer: K+S, Kassel, Germany) and (ii) Korn–Kali (K-MgO-Na_2O-SO_3 $\rightarrow$ 40-6-3-12.5; producer: K+S, Kassel, Germany). Magnesium was applied as ESTA Kieserite (25% MgO, 50% SO_3 producer: K+S, Kassel, Germany). All these fertilizers were applied to the soil two weeks before wheat sowing. EPSO Top (16% MgO, 32% SO_3) was applied to the wheat foliage in accordance with the schedule presented in Table 4. Plant protection was conducted in accordance with the principles of the Code of Good Practice.

Table 4. Schedule winter wheat foliar fertilization during the growing season using EPSO Top.

Application	Maximum Concentration	Rate [kg ha^{-1}]	BBCH Code	Description
1st	5%	10	19	Nine leaves unfolded (autumn)
2nd	(5 kg in 100 dm^3 of	15	30	Beginning of stem elongation
3rd	water)	15	55	Early ear emergence

2.4. Plant Material Sampling and Analysis

The plant material used to determine dry matter and the N content in grain and straw of winter wheat was collected at the full ripening stage (BBCH 90). The N content was determined in plant material using the standard macro Kjeldahl method [49]. The results were expressed on a dry matter basis. The mass of N in winter wheat parts was calculated based on the grain and straw biomass and the N content. For mineral nutrients, the collected plant sample was dried at 65 °C and then mineralized at 550 °C. The obtained ash was dissolved in 33% HNO_3. The P concentration was measured using the vanadium–molybdenum method with a Specord 2XX/40 (Analytik Jena, Jena, Germany) at a wavelength of 436 nm. The concentrations of K and Mg were determined using flame-type atomic absorption spectrometry. The results were expressed on a dry matter basis. The mass of the nutrients in winter wheat at harvest was calculated based on the grain and crop residues biomass and the content of a given nutrient.

2.5. Calculated Parameters

Equations for calculating the amount of N in grain or total biomass of winter wheat, as well as algorithms for N management indicators and N use efficiency (NUE) indicators, are presented below [50].

1. Grain density, GD

$$GD = NE \times GE,\ number\ grains\ \mathrm{m}^{-2} \times 1000$$

2. Harvest index

$$HI = \frac{GY}{TB} \times 100\%$$

3. Nitrogen accumulation in grain, GN

$$GN = GY \times N,\ \ \mathrm{kg\ N\ ha}^{-1}$$

4. Nitrogen accumulation in straw, SN

$$SN = SY \times N,\ \ \mathrm{kg\ N\ ha}^{-1}$$

5. Total accumulation of nitrogen in wheat biomass, TN

$$TN = GN + SN,\ \ \mathrm{kg\ ha}^{-1}$$

6. Nitrogen harvest index, NHI

$$NHI = \frac{GN}{TN} \times 100\%$$

7. Nitrogen unit accumulation in grain, $NUA\text{-}GN$

$$NUA\text{-}GN = \frac{GN}{GY}\ \ \mathrm{kg\ N} \times \mathrm{t}^{-1}$$

8. Nitrogen unit accumulation in total wheat biomass, $NUA\text{-}TN$

$$NUA\text{-}TN = \frac{TN}{GY}\ \ \mathrm{kg\ N} \times \mathrm{t}^{-1}$$

9. Nitrogen unit productivity—grain, $NUP\text{-}GN$

$$NUP\text{-}GN = \frac{GY \times 1000}{GN}\ \ \mathrm{kg\ grain} \times \mathrm{kg}^{-1}\ \mathrm{N}$$

10. Nitrogen unit productivity—total N, $NUP\text{-}TN$

$$NUP\text{-}TN = \frac{GY \times 1000}{TN}\ \ \mathrm{kg\ grain} \times \mathrm{kg}^{-1}\ \mathrm{N}$$

11. Partial factor productivity of fertilizer N, $PFP\text{-}N_f$

$$PFP\text{-}N_f = \frac{GY \times 1000}{N_f}\ \ \mathrm{kg\ grain\ kg}^{-1}\ N_f$$

12. Agronomic nitrogen efficiency, ANE

$$ANE = \frac{GY_N \times 1000 - GY_{AC} \times 1000}{N_f}\ \ \mathrm{kg\ grain} \times \mathrm{kg}^{-1}\ N_f$$

13. Nitrogen recovery, $R\text{-}GN$

$$R\text{-}GN = \frac{GN - GN_{AC}}{N_f} \times 100\%$$

14. Nitrogen recovery, $R\text{-}TN$

$$R\text{-}GN = \frac{TN - TN_{AC}}{N_f} \times 100\%$$

15. Physiological nitrogen efficiency, *PhNE-GN*

$$PhNE\text{-}GN = \frac{GY_N \times 1000 - GY_{AC} \times 1000}{GN_{AC} - GN_{AC}} \text{ kg grain kg}^{-1}\text{ N}$$

16. Physiological nitrogen efficiency, *PhNE-TN*

$$PhNE\text{-}TN = \frac{GY_N \times 1000 - GY_{AC} \times 1000}{TN_{AC} - TN_{AC}} \text{ kg grain kg}^{-1}\text{ N}$$

where

GY—grain yield, t ha^{-1};
N—N content in grain or crop residues;
g kg^{-1} DW, N_{AC}—nitrogen control;
N_f—treatments fertilized with nitrogen.

The crop nutrient balance (CNB) is the difference between the amount of nutrients introduced in fertilizers and accumulated in the biomass of winter wheat during harvest. The crop residue replacement ratio (CRRR) was calculated as the ratio of a given nutrient in harvest residues and CNB. Both indices were calculated using algorithms:

$$CNB = Nu_{in} - Nu_{out}, \text{ kg ha}^{-1}$$

$$CRRR = \frac{Nu}{CNB}$$

where

Nu—nutrients: N, P, K, Mg, kg ha^{-1};
in—nutrient input in applied fertilizers, kg ha^{-1};
out—nutrient output in winter wheat biomass, kg ha^{-1}.

The net nitrogen balance (NNb) during the winter wheat growing season was calculated using the formula

$$NNb = (N_{mins} + N_f) - (TN_h + N_{minh}), \text{ kg N ha}^{-1}$$

where

N_{min}—the amount of mineral N in soil, kg N ha^{-1};
s, h—sampling time, spring, harvest;
TN—the amount of N in winter wheat biomass at harvest, kg N ha^{-1}.

The nitrogen mass balance during the spring vegetation of winter wheat and the balance of available P, K, and Mg forms were calculated using algorithms:

$$N_b = N_s - N_h, \text{ kg ha}^{-1}$$

$$Mg_b = Mg_{bf} - Mg_h, \text{ mg kg}^{-1}$$

where

N—mass of N_{min}, kg ha^{-1};
Mg (P, K)—content of available Mg, mg kg^{-1};
b—nutrient balance, kg ha^{-1} or mg kg^{-1};
s, h, bf—soil sampling dates: spring, harvest, and before winter wheat sowing.

2.6. Statistical Analysis

The effects of the experimental factor (fertilization variants, FVs) in subsequent years of the study were assessed for primary or aggregate winter wheat characteristics and developed N management and NUE indicators. The examined variables were evaluated based on the coefficient of variation (CV) using ranges proposed by Wilding and Drees [51]. According to the proposed ranges, CV <15%, 15% < CV < 35%, and >35% indicate low, moderate, and high variability in the obtained data, respectively. Means were separated by honest significant difference (HSD) using Tukey's method when the F-test indicated significant factorial effects at the level of $p < 0.05$. In the second step of the diagnostic procedure, stepwise regression was applied to discriminate the set of plant traits determining the grain yield. In the computational procedure, a consecutive variable was removed from the multiple linear regressions in a step-by-step manner. The best regression model was chosen based on the highest F-value [52]. Pearson correlation and linear regression STATISTICA 12 software was used for all statistical analyses (StatSoft Inc., Tulsa, OK, USA, 2013).

3. Results

3.1. Grain Yield and Yield Components

The grain yield (GY) of winter wheat significantly depended on fertilization variants (FVs), but it was variable in subsequent years of the study (Table 5). The average GY was 6.6 ± 1.5 t ha^{-1}, showing moderate variability in response to the Y × FVs interaction (CV = 22.9%). The key reason of the observed year-to-year variability was weather in 2018. In this particular year, GY was 22% lower compared to 2017, which was the highest. GY at the absolute control (AC) was twice as low compared to the most intensive FVs, i.e., NPK-KK+Ki+ES. At the same time, no significant differences were found between GYs in the AC variant in subsequent years (Figure 1). The same level of GY stability, as indicated by CV <10%, was achieved for NP and NPK-KK (Table S1). However, GY for the latter FV was 13.1% greater compared to NP. The highest GY combined with the highest year-to-year variability (CV = 19.8%) was the NPK-KK+Ki+ES attribute. It was 18% (+1.14 t ha^{-1}) higher compared to NP. GY was significantly, except for thousand grain weight (TGW), correlated with primary yield components (Table A1). The strongest direct effect was exerted by the number of grain per ear (GE). The relationship obtained was best described by a quadratic regression model (Figure S1):

$$GY = -0.011GE^2 + 0.65GE - 6.31 \; for \; n = 21, \; R^2 = 0.57, \; p \le 0.01$$

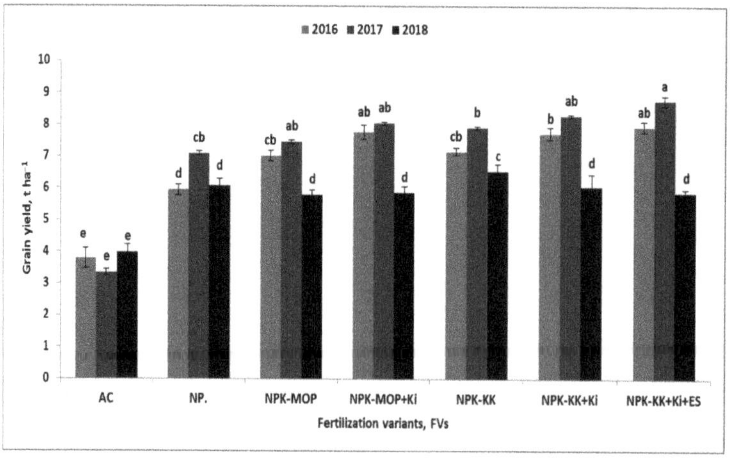

Figure 1. The effect of fertilization systems on the grain yield of winter wheat in subsequent years of the study. Means followed by the same letter within a column indicate the lack of a significant difference between the treatments. The vertical bar in the column is the standard error of the mean.

Table 5. Basic characteristics of winter wheat at harvest: grain yield, yield components, and biomass.

Factor	Level of Factor	GY $t\ ha^{-1}$	NE No. Ears m^{-2}	GE No. Grains $Ears^{-1}$	GD No. Grains $m^{-2} \times 1000$	TGW g	SY $t\ ha^{-1}$	TB $t\ ha^{-1}$	HI %
Year (Y)	2015/2016	6.8 b	524.8 a	27.6 b	14.2 b	48.1 a	8.4 a	15.2 a	44 b
	2016/2017	7.3 a	475.9 a	32.7 b	15.4 a	47.4 ab	8.1 a	15.4 a	47 a
	2017/2018	5.7 c	405.9 b	34.3 b	13.5 b	42.7 b	6.4 b	12.1 a	48 a
Fc, p		125 ***	15.8 ***	10.7 ***	20.2 ***	36.0 ***	18.5 ***	45.4 ***	5.6 **
Fertilization variants (FVs)	AC	3.7 d	432.0	19.0 b	7.8 a	47.9 a	5.2 b	8.9 b	43 b
	NP	6.4 c	459.7	32.8 a	14.1 c	45.1 b	8.1 a	14.4 a	45 ab
	NPK-MOP	6.8 bc	513.0	29.4 a	14.6 bc	46.3 b	7.8 a	14.6 a	47 ab
	NPK-MOP+Ki	7.2 ab	465.5	35.6 a	16.1 a	44.9 b	8.0 a	15.2 a	47 ab
	NPK-KK	7.2 ab	454.1	35.3 a	15.5 ab	46.6 b	8.1 a	15.3 a	47 ab
	NPK-KK+Ki	7.3 a	453.4	34.9 a	15.7 ab	46.8 b	7.9 a	15.2 a	48 a
	NPK-KK+Ki+ES	7.5 a	504.4	33.8 a	16.6 a	45.0 b	8.5 a	16.0 a	47 ab
Fc, p		154 ***	1.4 ns	13.2 ***	80.8 ***	2.4 *	8.5 ***	34.9 ***	2.6 *
Source variation for the studied interaction									
Y × FV		***	ns	ns	***	**	*	***	2.7 **
Average		6.6	468.9	31.5	14.4	46.1	7.7	14.2	0.46
Standard deviation		1.5	66.8	6.9	3.1	3.1	1.6	3.0	0.03
Coefficient of variation, %		22.9	14.3	21.8	21.9	6.8	20.6	20.8	7.4

Means followed by the same letter within a column indicate the lack of a significant difference between the treatments; ***, **, and * indicate significant differences at $p < 0.001$, $p < 0.01$, and $p < 0.05$, respectively; ns—non-significant. Legend: GY—grain yield; NE—number of ears per m^2; GE—number of grains per ear; GD—grain density, number of grains per m^2; TGW—thousand grain weight; SY—straw yield; TB—total biomass at harvest; HI—harvest index.

The maximum GY of 6.82 t ha^{-1} was found for GE of 35.4 grains ear^{-1}. Such values and ones higher were recorded only once in 2016, three times in 2017, and four times in 2018. The highest GE obtained in 2018 was associated with the lowest number of ears per m^2 (NE, Table 5). At the same time, the highest NE in 2016 was associated with the lowest GE. In 2018, NE was lower by 22.3% compared to 2016, whereas GE was higher by 24.3%. The variability in NE, resulting from the effect of FVs and years, expressed by coefficient of variation (CV) was in the low class (<14.3%), while that of GE was in the medium class (21.3) (Table 5). Both wheat traits did not show any response the Y × FVs interaction. The third primary GY component, i.e., thousand grain weight (TGW), responded significantly to the examined interaction (Table 5). The average TGW was 46.1 ± 3.1 g and resulted from a significant decrease in 2018 compared to 2016 down to (−10 and −9 g, respectively) (Figure A1). The TGW variability assessed by CV was the highest, exceeding 10%, for NPK-MOP and NPK-KK+Ki+ES. The most stable FV was NPK-KK+Ki, for which TGW reached 47 g, but CV was only 0.9% (Table S1). Although TGW did not show any significant relationship with GY, together with GD, it explained 99.9% of the variability in GY:

$$GY = -6.58 + 0.48GD + 0.14TGW \quad for\ n = 21,\ R^2 = 99.6,\ p \leq 0.001$$

Straw yield (SY), averaged across the tested factors, was 7.7 ± 1.6 t ha^{-1}, being at the same level of variability as found for GY (Table 5). It was significantly correlated with GY and especially with the total wheat biomass (TB) at maturity. At the same time, SY was significantly associated with NE and GD, but not with GE (Table A1). The yield structural index, which is the harvest index (HI), reached, on average, 46 ± 3%, being significantly driven by the Y × FVs interaction (Table 5). Both, SY and HI explained the 99.3% variability in GY:

$$GY = -10.8 + 0.86SY + 23.4HI \quad for\ n = 21,\ R^2 = 99.3,\ p \leq 0.001$$

3.2. Nitrogen Accumulation—N Management Indicators

Nitrogen accumulation in wheat grain (grain nitrogen, GN) was affected by FVs, but the values were significantly variable between years. The average GN was 125.2 ± 34.4 kg ha^{-1} (CV = 27.5%). However, no interaction between these two factors was found (Table 6). Significantly higher GN compared to both other years was recorded in 2017. The effect of the tested FVs was very specific. The application of NP doubled GN (+123%) compared to AC. A significant increase of 18.2%, compared to NP, was recorded the first time for NPK-KK+Ki. This upward trend continued up to NPK-KK+Ki+ES and amounted to 6.3% compared to the previous FV. To sum up, the increase in GN in this particular FV compared to NP was 25.5%. The highest GN stability over the study years was found for NPK-KK (CV = 6.3%) and the highest variability for NP (CV = 17.4%) (Table S1).

The average amount of N in straw (SN) was 46.7 ± 12.6 kg ha^{-1} (CV of 26.9%). Similarly to GN, the highest SN was recorded in 2017. The amount of N in the NP plot was almost 1.5 times higher compared to AC. However, no significant differences were found between FVs. The total N (TN) accumulated by winter wheat at maturity was 171.9 ± 46.1 kg ha^{-1} (CV of 26.8%). This wheat trait was almost completely dependent on GN (r = 0.99 ***), and only slightly weaker on SN (r = 0.95 ***) (Table A2). All these wheat traits showed a strong effect on GY, being slightly weaker for SN (r = 0.82 ***). No significant relationship was found between the studied N traits and yield components such as NE and TGW. At the same time, a strong response of GE was noted, which was the highest for GN (r = 0.82 ***). A significant HI response to discussed N traits was revealed only for GN and TN. The nitrogen harvest index (NHI) was the only winter wheat trait that did not show any significant response to the tested factors and their interaction. The average NHI was 72.6 ± 3.1%, and CV was only 4.3% (Table 6). This wheat trait did not correlate significantly with all the characteristics studied, except HI (r = 0.82 ***; Table A2).

Table 6. Nitrogen accumulation by winter wheat–nitrogen characteristics and nitrogen use efficiency indices.

Factor	Level of Factor	GN	SN	TN	NHI	PFP-N_f	NUP-GN	NUP-TN	NUA-GN	NUA-TN
			kg ha^{-1}		%	kg grain kg^{-1} N_{fk}	kg grain kg^{-1} GN	kg grain kg^{-1} TN	kg N t^{-1} grain	kg N t^{-1} TB
Year (Y)	2015/2016	117.3 b	43.0 b	160.3 b	71.5	45.1 b	63.0 a	43.6 a	17.1 b	23.6 c
	2016/2017	139.5 a	51.6 a	191.1 a	73.6	48.5 a	53.4 b	39.4 b	18.9 ab	25.8 b
	2017/2018	118.7 b	45.6 ab	164.3 b	72.7	38.2 c	50.2 b	36.5 b	20.4 a	28.3 a
Fc. p		10.5 ***	3.8 *	16.8 ***	0.7 ns	125 ***	9.4 ***	12.1 ***	10.1 ***	17.3 ***
Fertilization variants (FV)	AC	53.7 a	20.3 b	74.0 a	72.7	–	70.1 a	50.2 a	14.7 b	20.2 b
	NP	119.7 b	50.4 a	170.1 b	70.0	42.5 b	55.4 b	37.9 b	18.7 ab	26.6 ab
	NPK-MOP	124.5 bc	49.9 a	174.4 b	71.0	45.0 bc	56.2 ab	39.4 b	18.5 ab	26.0 ab
	NPK-MOP+Ki	141.5 a–c	50.3 a	191.8 ab	73.6	48.1 ab	52.3 b	38.2 b	19.8 a	26.9 ab
	NPK-KK	137.5 a–c	48.1 a	185.7 ab	74.0	48.0 a	54.0 b	39.4 b	19.2 ab	25.9 ab
	NPK-KK+Ki	149.0 ab	53.6 a	202.6 a	73.5	49.0 a	50.0 b	36.6 b	20.4 a	27.8 a
	NPK-KK+Ki+ES	150.3 a	54.4 a	204.7 a	73.5	50.1 a	50.7	37.2 b	20.3 a	27.7 a
Fc. p		32.7 ***	11.9 ***	52.1 ***	0.6 ns	154 ***	4.3 **	8.7 ***	5.8 ***	9.1 ***
Source variation for the studied interaction										
Y × FV		ns	ns	ns	ns	***	ns	ns	ns	ns
Average		125.2	46.7	171.9	72.6	47.1	55.5	39.8	18.8	25.9
Standard deviation		34.4	12.6	46.1	3.1	6.5	8.9	5.8	2.5	3.4
Coefficient of variation, %		27.5	26.9	26.8	4.3	13.7	16.0	14.5	13.1	13.2

Similar letters in the column mean no significant differences in the Tukey test; ***, **, * indicate significant differences at $p < 0.001$, $p < 0.01$, and $p < 0.05$, respectively; ns—non-significant. Legend: GN—amount of nitrogen in grain; SN—amount of nitrogen in straw; TN—total amount of nitrogen in winter wheat biomass; NHI—nitrogen harvest index; PFP-N_f—partial factor productivity of fertilizer nitrogen; NUP-GN—nitrogen unit productivity in grain N; NUP-TN—nitrogen unit productivity in total N in winter wheat biomass; NUA-GN—nitrogen unit accumulation for grain N; NUA-TN—nitrogen unit accumulation in total N in winter wheat biomass; TB—total winter wheat biomass.

The first of the discussed N management indices, i.e., PFP-N_f, showed the same trend as GY (Table 6). This state results directly from the method of calculation. However, CV of this index was slightly smaller compared to that of GY, reaching 22%. The basic operational indicators of N management by a crop plant are N unit productivity (NUP) and N unit accumulation (NUA). Both were calculated based on GN or TN. All responded to experimental factors, but not to the interaction between them. This method of calculating presupposes lower values for NUP-TN, which was noted. The average NUP-GN was 55.5 ± 8.9, and CV of 16%. The average NUP-TN was lower by $^1/_3$ compared to NUP-GN and amounted to 39.8 ± 5.8 with CV of 14.5%. The highest NUP was recorded in 2016. The highest values of this index, regardless of the tested FVs, were found for AC. For NUP-GN, it was higher by about 25% compared to NPK-MOP and NP. The lowest, amounting to 50 kg grain kg^{-1} GN, was recorded for the FVs with Mg. In the case of NUP-TN, its values compared to AC decreased by 22%, which was found for NPK-MOP and NPK-KK. The greatest decrease in NUP-TN relative to NUP-GN was noted just for AC and NP variants. The most stable in this respect were FVs intensively fertilized with Mg. The effect of the year factor on NUA indices was much more pronounced for NUA-TN than for NUA-GN. In the latter case, its value in 2017 was intermediate to the other years. In contrast, NUA-TN followed the order 2016 < 2017 < 2018. The largest difference between both indices revealed in 2018. The highest values of both indices were recorded for FVs with KK and Mg and the lowest for AC. This difference was about 37.5% for both indices. At the same time, the increase in NUA-TN compared to NUA-GN was 37% for AC and 42% for NP, which was the highest.

For NUP-type indices, which also include AC, higher R^2 values, expressing stronger correlation between both wheat traits, were obtained for TN as an independent variable. The obtained relationships are presented in the following set of equations:

(1) Grain nitrogen, GN:

$$GY = -0.01NUP^2 + 1.13NUP - 23.7 \; for \; n = 21, \; R^2 = 0.49, \; p \le 0.05$$

(2) Total nitrogen, TN:

$$GY = -0.03NUP^2 + 2.63NUP - 44.7 \; for \; n = 21, \; R^2 = 0.71, \; p \le 0.01$$

The cardinal values, i.e., the maximum grain yield (GY_{max}) and the optimal NUP (NUP_{op}) value, were 8.01 t ha^{-1} and 56.3 kg grain kg^{-1} accumulated N in grain and 7.63 t ha^{-1} and 40 kg grain kg^{-1} N in the total biomass of winter wheat during harvest. The FV that met both of these conditions turned out to be NPK-MOP.

3.3. *Nitrogen Use Efficiency Indicators*

Five classic nitrogen use efficiency (NUE) indices were calculated (Table 7). The agronomic nitrogen efficiency (ANE) indicator was the only one among those tested that showed a significant response to the Y × FVs interaction (Figure A2). It reached, averaged across the studied factors, the value of 22.5 ± 7.9 kg grain kg^{-1} N_f. However, at the same time, it showed a high variability, as indicated by the coefficient of variation (35.2%). The key driver of ANE was the course of the weather during the growing season (Table 2). In the first growing season, this index was the largest, slightly exceeding 30 kg grain kg^{-1} N_f. In this particular growing season, its values extended from 25 kg grain kg^{-1} N_f in the NP plot, and 27 kg grain kg^{-1} N_f in NPK-MOP, to 36 kg grain kg^{-1} N_f in NPK-KK+Ki+ES. The ANE index in 2017/2018 was 2.2 times smaller compared to the previous season. The variability of ANE was high. In four out of six FVs, CV exceeded 35% (the high CV class) (Table S1). The lowest CV of 28% was recorded in NPK-KK and was significantly higher, but still in the medium variability class (15–35%). The highest CV, approaching 50%, was recorded for the most intensive FV, i.e., NPK-KK+Ki+ES. ANE was significantly but weakly correlated with NE and with TGW (moderately) but not with GE. At the same time, it showed strong correlation with SY (r = 0.80 **) (Table A3).

Table 7. Nitrogen use efficiency indicators.

Factor	Level of Factor	ANE kg grain kg^{-1} N_f	R-GN $\%N_f$ in GN	R-TN $\%N_f$ in TN	PhNE-GN kg grain kg^{-1} GN	PhNE-TN kg grain kg^{-1} TN
Year (Y)	2015/2016	23.2 b	51.4 b	63.3 b	50.1 a	38.2 a
	2016/2017	30.5 a	67.8 a	96.6 a	45.4 a	32.0 b
	2017/2018	13.7 c	47.6 b	68.6 b	29.3 b	20.4 c
Fc, p		131 ***	14.5 ***	30.8 ***	25.0 ***	44.8 ***
Fertilization variants (FVs)	NP	17.8 c	44.0 c	64.1 c	42.2	28.1
	NPK-MOP	20.4 bc	47.2 bc	67.0 bc	45.2	31.7
	NPK-MOP+Ki	23.5 ab	58.4 a–c	78.5 a–c	40.7	30.5
	NPK-KK	23.4 ab	55.9 a–c	74.5 a–c	43.9	32.9
	NPK-KK+Ki	24.3 ab	63.6 ab	85.7 ab	38.5	28.7
	NPK-KK+Ki+ES	25.5 a	64.4 a	87.2 a	39.2	29.2
Fc, p		7.6 ***	4.4 **	4.4 **	0.7 ns	1.0 ns
Source variation for the studied interaction						
Y × FVs		**	ns	ns	ns	ns
Average		22.5	55.6	76.2	41.6	30.2
Standard deviation		7.9	12.8	18.0	9.9	8.0
Coefficient of variation, %		35.2	23.0	23.6	23.8	26.7

Similar letters in the column mean no significant differences in the Tukey test; *** and ** indicate significant differences.at $p < 0.001$ and $p < 0.01$, respectively; ns—non-significant. Legend: ANE—agronomic nitrogen efficiency; R-GN—nitrogen recovery in grain; R-TN—nitrogen recovery in total N in winter wheat biomass; PhNE—physiological efficiency of N in grain; PhTN—nitrogen efficiency of N in winter wheat biomass.

Two N_f recovery (R) indices were calculated. The first was based on N accumulated only in grain (GN) and the second was based to the total N in wheat biomass at harvest (TN). Logically, R-GN values were generally lower compared to R-TN (Table 7). The effect of the year factor on the R indices was high, but not as great as in the case of ANE. The highest N_f recovery, regardless of the type of R index, was recorded in the 2016/2017 growing season. In the case of R-TN, it approached 100%. In relation to the FVs studied, the highest R indicators, regardless of its type, were found for the NPK-KK+Ki+ES variant. For this plot, R-GN was higher than NP by 17.2 p.p. and higher than NPK-MOP by 6 p.p. In the case of R-TN, these differences were 20.2 and 8.7 p.ps., respectively. Both indices did not correlate with primary yield components. However, a strong correlation coefficient, especially for R-GN with GY, was found (Table A3).

Indices of physiological N use efficiency (PhNE) were calculated in the same way as R indices. The growing season (year) was revealed as the only factor differentiating the PhNE values. Therefore, almost the same seasonal trends were observed. The highest PhNE values were recorded in the 2015/2016 growing season. Only for TN did significant differences between the first two seasons occur. The lowest values, significantly, were recorded in the 2017/2018 growing season. Both indices correlated significantly and strongly with NE and TGW. At the same time, a negative correlation with GE was noted, and it was significant for PhNE-GN (Table A3). The obtained relationships with GY are presented in the following set of equations:

(1) Grain nitrogen, GN:

$$GY = -0.005PhPN^2 + 0.45PhPN - 3.02 \; for \; n = 18, \; R^2 = 0.77, \; p \leq 0.01$$

(2) Total nitrogen, TN:

$$GY = -0.008NPhPN^2 + 0.58PhPN - 2.15 \; for \; n = 18, \; R^2 = 0.79, \; p \leq 0.01$$

The cardinal values of the obtained equations were 7.80 t ha^{-1} and 48 kg grain kg^{-1} accumulated N in grain and 7.88 t ha^{-1} and 34.8 kg grain kg^{-1} N in the total biomass of wheat during harvest. The FV that met this condition for GN turned out to be NPK-KK in the first two growing seasons. With respect to TN as the predicator of PhPN, the closest to this condition was also NPK-KK, but only in 2016/2017 growing seasons.

3.4. Crop Nutrient Balance

The nutrient balance in winter wheat clearly showed the dependence of this trait for a given nutrient on the Y × FV interaction (Table 8). Negative balance values indicate an increase in nutrient mass in crop biomass at winter wheat harvest in response to its input, i.e., fertilizer rate. The N balance clearly indicates the dominance of the year factor. The average for 2017 was almost twice as high as those recorded in the other growing seasons. Nevertheless, the FV effect was highly specific (Figure 2). The most negative N_{cb} values, with little variation between years, were recorded for AC (CV = 8.4%). The lowest value of this trait, also showing the greatest variation between years, was recorded for NP (CV = 232%). Ncb doubled for NPK-MOP compared to NP. The presence of Mg in each FV increased the index value compared to NPK-MOP, which was twice as high for NPK-KK-KI+ES. At the same time, a decrease in the variability of this trait between years was noted (CV: 45% vs. 59%). The most stable N_{bc} in the growing seasons, but lower by 20 kg N ha^{-1}, was observed for NPK+KK (CV = 21%). The N replacement ratio (N_r) coefficient showed a dominance of the year factor. In the 2017/2018 season, it was twice as high as in 2016/2017 and three times higher than in 2015/2016 (Table 8). The imbalance in N_r was mainly revealed in AC, for which it was only 0.3. This coefficient for the remaining FVs, except for the highly unbalanced NP (CV = 341%), was above 1.0. The most stable FV turned out to be NPK-KK (Figure S2).

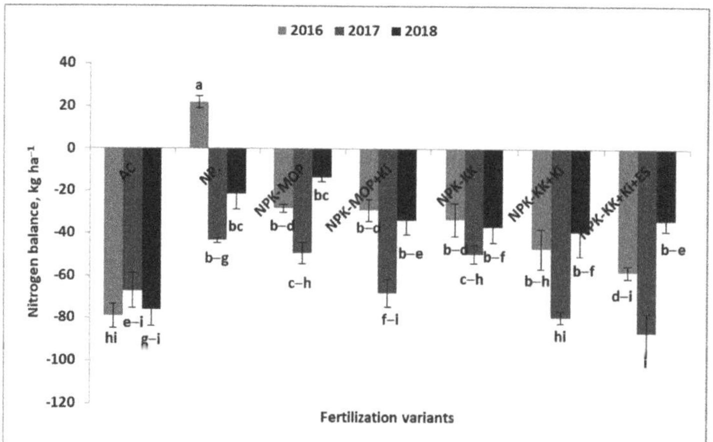

Figure 2. Nitrogen balance in winter wheat during the growing season in subsequent years of the study. Means followed by the same letter within a column indicate the lack of a significant difference between the treatments. The vertical bar in the column is the standard error of the mean.

Table 8. Crip nutrient balance and crop residue nutrient replacement ratio.

Factor	Level of Factor	Ncb	Pcb	Kcb	Mgcb	Nr	Pr	Kr	Mgr
			$kg\ ha^{-1}$					-	
Year (Y)	2015/2016	−36.0 a	−20.0 b	−65.0 b	−6.1 a	0.69 b	0.56 b	1.21 b	−1.68 b
	2016/2017	−63.2 b	−8.9 a	−65.5 b	−7.1 a	0.90 b	1.28 a	1.12 b	−0.65 ab
	2017/2018	−36.2 a	−20.4 b	−49.7 a	−11.8 b	1.91 a	0.50 b	1.79 a	0.32 a
Fc, p		42.3 ***	96.4 ***	9.1 ***	53.9 ***	53.9 ***	72.3 ***	52.4 ***	9.5 ***
Fertilization variants (FVs)	AC	−74.0 e	−18.4 b	−59.1 a	−10.4 c	0.27 c	0.37 b	0.65 c	0.59 a
	NP	−14.3 a	−13.0 a	−107.9 b	−17.9 e	0.77 bc	1.02 a	0.67 c	0.57 a
	NPK-MOP	−30.3 b	−15.1 ab	−40.8 b	−17.2 e	2.32 a	0.96 a	2.19 a	0.55 a
	NPK-MOP+Ki	−43.4 b–d	−16.2 ab	−55.1 b	−2.5 b	1.34 ab	0.80 a	1.42 b	1.50 a
	NPK-KK	−39.8 bc	−16.5 ab	−53.9 b	−13.2 d	1.33 ab	0.82 a	1.40 b	0.84 a
	NPK-KK+Ki	−55.1 cd	−16.7 ab	−56.5 b	−0.5 b	1.07 ab	0.76 a	1.41 b	−4.84 b
	NPK-KK+Ki+ES	−59.3 de	−19.1 b	−47.1 b	3.4 a	1.07 ab	0.76 a	1.87 a	−3.91 b
Fc, p		28.8 ***	3.9 ***	23.2 **	173 ***	21.1 ***	7.1 ***	54.4 ***	26.8 ***
Source variation for the studied interaction									
Y × FV		***	*	***	***	***	**	***	***
Average		−45.2	−16.4	−60.1	−8.3	1.17	0.78	1.37	−0.67
Standard deviation		25.8	6.1	25.1	8.6	1.16	0.45	0.71	4.16
Coefficient of variation. %		−57.1	−37.3	−41.7	−103.4	99.22	57.51	51.83	−619.86

Means followed by the same letter within a column indicate the lack of a significant difference between the treatments; ***, **, and * indicate significant differences at $p < 0.001$, $p < 0.01$, and $p < 0.05$, respectively; ns—non-significant. Legend: N, P, K, Mg—nutrients; Cb—nutrient balance; r—crop residue nutrient replacement ratio.

Phosphorus balance (P_{cb}) showed an inverse annual trend compared to N_{cb}. In the second growing season, it was more than twice as low as in the other years (Table 8). The effect of the year factor on the variability of P_{cb} (CV) between fertilizers ranged from 31% for NPK-KK to 48% for NPK-MOP (Figure S3). The P replacement ratio (Pr) clearly indicated that only in the second growing season, excluding AC, were winter wheat crop residues able to cover the P_{cb} (Figure S4).

Potassium balance (K_{cb}) was at the same level in the first two growing seasons, but significantly lower in the 2017/2018 (Table 8). The impact of FVs was variable in subsequent growing seasons (Figure A3). The largest, negative balance, exceeding 100 kg K ha^{-1}, was observed for NP, which also showed the greatest variability between growing seasons (CV = 51%). The impact of the remaining FVs was almost twice lower, ranging from −41 to −57 kg K ha^{-1} for NPK-MOP and NPK-KK+Ki, respectively. Again, NPK-KK proved to be the most stable FV (CV = 11%). The K replacement ratio (Kr), despite variability in subsequent growing seasons, was lower than 1.0 only for AC and NP. For the remaining FVs, its value significantly exceeded the threshold, showing the lowest variability for NPK-KK (CV = 7%) (Figure A4).

Magnesium balance (Mg_{cb}) showed an inverse trend compared to K_{cb}. In the 2017/2018 growing season, it was significantly higher compared to the first growing seasons (Table 8). The effect of FVs was variable in subsequent growing seasons (Figure 3). The deepest negative balance was recorded for NP and NPK-MOP, amounting to −17.5 kg M ha^{-1}. At the same time, it was very stable for these two FVs (CV, 16.9% and 6.2%, respectively). Negative Mg_{cb} values were also recorded for NPK-KK in all three seasons. In the 2017/2018 growing season, Mcb was at the same level as recorded for non-magnesium FVs. A completely opposite situation was observed for NPK-KK+Ki+ES, where net positive Mg_{cb} was detected. MOP and KK variants with Kieserite were characterized by low Mg_{cb} values, but at the same time, showed a high year-to-year variability (CV). The values of the Mg replacement ratio (Mg_r) were strongly dependent on a given FV (Figure S5). For magnesium-free FVs, it was at the level of 0.6. For NPK-KK, it increased to 0.8. However, for the magnesium-FVs, it was positive, i.e., the applied Mg was not fully utilized by winter wheat.

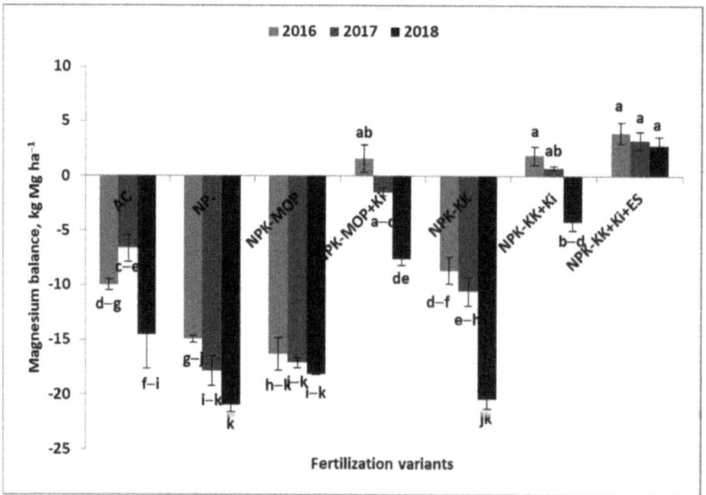

Figure 3. The effect of fertilization variants in consecutive years of the study on the magnesium winter wheat balance. Means followed by the same letter within a column indicate the lack of a significant difference between the treatments. The vertical bar in the column is the standard error of the mean.

3.5. Soil Nutrient Balance

The analysis of soil nutrient content, carried out at the harvest of winter wheat, included both the N_{min} mass and the content of available P, K, and M and their balance during the growing season. The key factor influencing the studied soil properties was the weather in the subsequent growing seasons (Table 9). The only exception was Mg balance (Mg_b). Among the studied nutrients, only for P indices was no response to FVs and to the interaction of Y x FV found. The most important factor was the growing season. In the first two consecutive years, the P content was in the high class, and in the third year, it was in the very high class. The P balance (P_b) was at the same level in the first and third years, and more than twice lower in the second year.

The amount of N_{min} at harvest depended on the Y × FV interaction. However, the dominant factor was weather in subsequent growing seasons (Table 9). The amount of N_{min} mass at harvest decreased in subsequent years of the study. In 2018, it was 2.5 times lower than in 2016. The average N_{min} balance—in fact, its decrease compared to the beginning of the spring vegetation of winter wheat—averaged for the experimental factors was approximately 41 kg N ha^{-1} and showed very low variation (CV < 10%) (Figure 4). Among the FVs studied, the most stable N_{min} balance was the attribute of the NP plot (<4%). Compared to NP, CV increased more than 4-fold on NPK-MOP and 10-fold on NPK-KK+Ki+ES (Table S1).

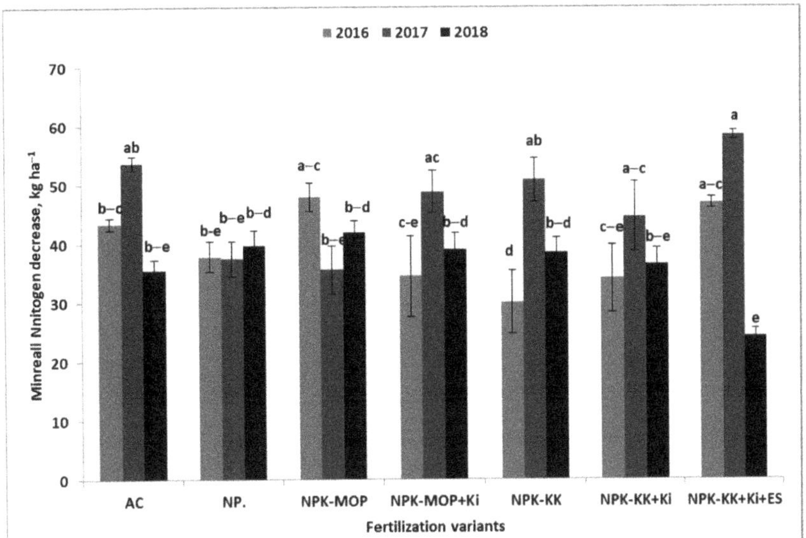

Figure 4. The effect of fertilization systems on the mineral N balance in subsequent years of the study. Means followed by the same letter within a column indicate the lack of a significant difference between the treatments. The vertical bar in the column is the standard error of the mean.

Table 9. Content of basic nutrients at winter wheat harvest and balance during the growing season mg kg^{-1} soil.

Factor	Level of Factor	N	N_b	N_{in}	N_{out}	NN_b	P	P_b	K	K_b	Mg	Mg_b
Year	2015/2016	26.8 a	39.2 b	194.6	187.1 b	7.49 a	169.2 c H [1]	40.8 a	87.5 c L	57.5 a	92.9 H	27.1 a
(Y)	2016/2017	17.0 b	47.0 a	192.6	208.1 a	−15.5 b	194.3 b H	17.7 b	116.2 b M	35.8 b	90.6 H	6.4 b
	2017/2018	11.6 c	36.4 b	176.6	175.9 b	0.69 a	250.2 a VH	35.8 a	149.7 a H	5.3 c	90.7 H	0.3 c
Fc. p		35.0 ***	17.6 ***	17.6 ***	15.4 ***	8.0 ***	348 ***	29.9 ***	186 ***	132 ***	2.0 ns	232 ***
Fertilization	AC	15.2	44.1	59.3	89.2 c	−29.8 c	198.5 H	37.5	119.2 ab H	31.4 a	90.2 H	12.4 a
variants	NP	21.0	38.3	209.3	191.1 b	18.2 a	201.1 H	34.9	104.9 b H	45.8 a	91.1 H	11.6 a
(FVs)	NPK-MOP	17.6	41.8	209.3	192.0 b	17.3 a	208.9 VH	27.1	121.7 a H	29.0 b	92.7 H	10.0 a
	NPK-MOP+Ki	18.6	40.7	209.3	210.4 ab	−1.1 ab	205.6 H	30.4	117.0 ab H	33.6	90.9 H	11.7 a
	NPK-KK	19.6	39.7	209.3	205,3 ab	4.0 ab	207.6 VH	28.4	124.3 a H	26.4 b	90.3 H	12.3 a
	NPK-KK+Ki	21.0	38.3	209.3	223.6 a	−14.3 bc	208.4 VH	27.6	120.6 ab H	30.0 b	90.2 H	12.5 a
	NPK-KK+Ki+ES	16.2	43.1	209.3	221.0 a	−11.6 bc	202.0 H	34.0	117.0 H	33.7 a	94.2 H	8.5 b
Sourece variation for the studied interaction												
Fc. p		1.3 ns	1.2 ns	1.3 ns	53.0 ***	7.5 ***	1.4 ns	1.5 ns	3.2 **	3.4 *	1.2 ns	1.4 *
Y × FV		***	***	***	ns	ns	ns	ns	***	***	*	*
Average		18.5	40.9	187.9	190.4	−2.5	204.6	31.4	117.8	32.9	91.4	11.3
Standard deviation		4.1	3.2	54.4	47.5	22.4	19.8	6.6	15.5	13.4	1.4	6.7
Coefficient of variation.%		22.1	7.8	29.0	24.9	911	9.7	21.1	13.2	40.7	1.5	59.7

[1] availability class: L—low; M—medium; H—high; VH—very high [43]; means followed by the same letter within a column indicate the lack of a significant difference between the treatments; ***, ** and * indicate significant differences at $p < 0.001$. $p < 0.01$. and $p < 0.05$ respectively; ns—non-significant. Legend: N, P, K, Mg—nutrient content in the soil at harvest; b—balance of nutrient during the growing season; N_{in}—the sum of N_f and the amount of N_{min} in spring; N_{out}—the sum of n accumulated in winter wheat biomass at harvest and the amount of N_{min} at harvest; NN_b—net nitrogen balance.

Net nitrogen balance (NNb) showed significant variability between years and in response to the experimental factor, but not to the interaction between both factors (Table 9). The values of the studied indicator were negative only in the second growing season. In the first one, they were positive and in the third, they approached zero. Fertilization treatments revealed the presence of four FVs groups, consisting of the following:

1. AC; NNb was negative, meaning that the relative N increase was 50%.
2. NPK-KK+Ki and NPK-KK+Ki+ES; NNb was negative, and the N increase was at least twice as low as AC.
3. NPK-MOP+Ki and NPK-KK; NNb fluctuated around zero.
4. NP+NPK-MOP; NNb was positive, which means that inorganic resources were not fully utilized by wheat.

The relationship between Nout and NNb was negative, as shown by the following equation:

$$NNb = -0.83Nout + 173.5\ for\ n = 21,\ R^2 = 0.87,\ p \leq 0.001$$

This means that an increase in Nout resulted in a linear decrease in NNb. The equilibrium value between the two traits was reached for Nout, which was 209.8 kg Nout ha^{-1}. Only above this critical value did NNb become negative. Among the N characteristics and indices of winter wheat, the strongest correlation with NNb was observed for GN and TN, as well as for R-GN and R-TN. The following set of equations is presented for GN and RGN:

1. GN:

$$GN = -0.84NNb + 138.9,\ for\ n = 18,\ R^2 = 0.88,\ p \leq 0.001$$

2. R-GN:

$$R\text{-}GN = -0.57NNb + 56.8\ for\ n = 18,\ R^2 = 0.83,\ p \leq 0.001$$

According to the obtained equations, the values of both wheat traits increased with decreasing NNb values.

The content of available K, unlike N and P, showed significant variability in response to FVs (Table 9). Despite the dominance of the year factor, confirmed by the variability of K availability classes, the effect of FVs was noticeable (Figure S6). The content of K in the NP plot, in each growing season, especially in the second and third, showed a clear decrease in relation to AC. The largest K gap was noted in 2017 (−20%) and the gap was much lower in the remaining two seasons. The most important characteristic obtained was the stable content of K in FV NPK+KK (CV = 13.1%). Firstly, in each season, it was in the medium content availability class. Secondly, in this FV, the CV value was in the low class. The greatest variability of K content was noted for FV NPK+KK (47%). The soil K balance (K_b) during the growing season showed a completely opposite trend to the K content at harvest (Table 9). Its greatest balance was recorded in the first season, on average, amounting to almost 58 mg kg^{-1} soil. In the third season, it was more than 10 times smaller (Figure 5). The highest inter-seasonal variability of the K balance K was noted for NPK+MPK (CV = 182%, Table S1). This resulted from the fact that only for this FV was a positive balance K noted (+29 mg kg^{-1} soil). On the other hand, the most stable was NPK-KK (CV = 43%) (Table S1).

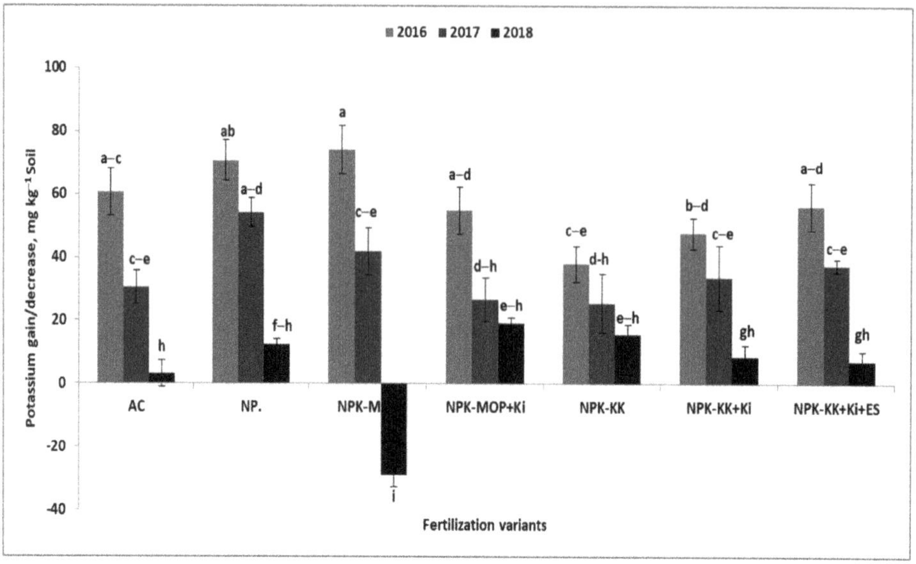

Figure 5. The effect of fertilization systems (FVs) on the balance of available K in subsequent years of the study. Means followed by the same letter within a column indicate the lack of a significant difference between the treatments. The vertical bar in the column is the standard error of the mean.

The content of available Mg was in the high class and did not show any significant response to the year factor and FVs. The variability in response to the Y × FV interaction was dominated by the year factor. The average value of CV was extremely low (<2%), being two times higher in two FVs, i.e., NPK-KK and NPK-KK+Ki+ES (5%; Table 9). The Mg (Mg_b) balance showed a completely different trend. The Mg_b in 2016 was more than four times and 90 times higher than that recorded in 2017 and 2018, respectively. The CV, averaged over FVs, was 60%, ranging from 90% for AC to 159% for NPK+KK (Table S1). The net increase in the available Mg content was noted for three FVs, including AC, NP, and NPK+KK.

The content of N and K at harvest was negatively associated with its balance. The strongest correlation was found for K (r = −0.99 ***) and a slightly weaker one for N (r = −0.56 **) (Table A4). However, for P and Mg, these relationships were nonsignificant. The amount of residual N at harvest was negatively correlated with the content of available P and K, but not with Mg. The content of P was positively correlated with the content of K, but not with Mg. At the same time, it was negatively correlated with K_b and Mg_b. The most important are, however, the relationship between K soil characteristics and Mg_b balance, as shown below:

K content:

$$Mg_b = -0.34K + 51.7 \; for \; n = 21, \; R^2 = 0.66, \; p \leq 0.001$$

K balance:

$$Mg_b = 0.38K_b - 1.45 \; for \; n = 21, \; R^2 = 0.63, \; p \leq 0.001$$

The first equation clearly shows that the higher the content of available K in the soil at harvest, the lower the value of the Mg balance. The second equation clearly indicates the synergy in the decrease in the content of both nutrients in the soil during the winter wheat vegetation period.

Among the soil nutrient characteristics, the greatest influence on the basic characteristics of winter wheat was exerted by the content and balance of K (Table A5). The latter trait, although in opposite ways, affected GY, TGW, and SY. P content showed a negative effect on NE, TGW, and SY and P balance on HI. Mg balance had a positive effect on NE and TGW. N management indices showed a variable response to both the soil nutrient content at harvest and to their balance during the growing season (Table A5). The highest sensitivity was shown for both NUP indices (five characteristics). They were positively correlated with N mass at harvest and K_b and Mg_b and, at the same time, were negatively correlated with P and K content. The same set of soil characteristics was noted for NUA-TN, but the correlation was reversed. PFP-N_f showed only a relationship with N_b (positive). However, for NHI, no significant relationship with the studied soil characteristics was noted. Nitrogen efficiency indices showed the greatest and a generally negative response to both P content and its balance (Table A6). The strongest relationship with soil properties was shown by both PhNE indices. Positive correlations occurred for the residual N mass and were especially strong for K and Mg balance. Negative correlations were found for the content of both of these nutrients in the soil.

4. Discussion

A sustainable crop production system is based on several pillars, assuming efficient use of production inputs [2,41]. This basic goal is achieved but only provided that first, N_f and secondly, inorganic N present in the soil or released from soil resources during the growing season are effectively used by the currently grown crop. Achieving these two main goals is only possible if N_f is used efficiently and its utilization rate does not lead to yield reduction [9]. The assumption of effective N management cannot ignore the sustainability of soil fertility [13,53].

4.1. Balanced Fertilization—Grain Yield–Yield Components—Stability

In food production systems, the most important indicator is the stability of staple crop production, which is decisive for the continuous global food supply [54]. The grain yield of winter wheat harvested in the NPK-KK variant was the most stable among the examined FVs, reaching, on average, 7.2 t ha^{-1} (CV = 9.4%). It was 13% (+0.84 t ha^{-1}) higher compared to NP and almost two-fold higher (+95%) compared to AC. Such high stability for all these three FVs, as well as the difference in yield, indicates, first of all, high natural soil fertility and good fore-crop, which was winter oilseed rape [55] From the farmer's production strategy view, the FV with K applied in the form of Korn–Kali seems to be the most stable production option. The moderate but, at the same, stable yields for NPK-KK were found with the fertilizer formula of Korn–Kali, which contains a certain amount of Mg and S [37]. However, the amount of N in grain was below the threshold value of 21.7 kg N t^{-1} grain (11.5% protein content) (Table S2). A too-low value of this winter bread wheat trait in this particular FV clearly indicates N_f as a critical factor in bread wheat production [56]. The response of winter wheat to fresh K clearly indicates that the soil K content at the border between the low and medium availability class for soil developed from sandy loam is high enough for this crop [57]. Moreover, the obtained relationships are consistent with observations made in other geographical areas [58,59].

A much higher yield increase was obtained in FVs with Kieserite, containing Mg and S, applied together with K. However, these FVs were moderately instable for GY. The effect of Kieserite was revealed only in years with favorable weather in the spring part of the winter wheat growing season. In these years (2016 and 2017), soil with Kieserite applied resulted in a GY (on average) of 0.72 t ha^{-1} (10%) and, supported by foliar spray with Epsom salt, of 1.11 t ha^{-1} (15%). The increase in GY did not limit nitrogen accumulation under optimal weather conditions, which, in the NPK-KK variants with Kiserit, exceeded the threshold value of 11.5%. However, in the dry year 2018, no increase in GY or in protein content in these FVs was observed. The obtained results confirm the complex response of winter wheat to the use of fertilizers containing Mg and S [37,60]. In regions where winter wheat is grown, with variable weather conditions, it is rational to use these nutrients together with K during the growing season, an example of which is Korn–Kali. Additional application of these nutrients, soil-applied (Kieserite) or foliar-applied (Epsom salt), is justified only in optimal weather conditions during the spring growing season of winter wheat [61–63].

Among the main yield components, the strongest impact on GY was found for the number of grains per ear (GE). This wheat trait is formed in the period extending from the booting phase to the watery stage of the grain growth [6,28]. The GE value emphasizes the key role of N in forming grain yield: first, because N supply defines the size of GE [64]; and secondly, GE feedback generates the sink for both the grain mass and simultaneously for the N mass [65]. In the studied case, the N use resulted in GE increase by 73% compared to the absolute control (19 vs. 33 grains per ear). The relationship between both traits was not linear but curvilinear and was best described by a quadratic regression model (Figure S1). The effect of applied K and Mg fertilizers on GE cannot be explained without taking into account the compensation mechanism between the number of ears per m^2 (NE) and GE. This is the yield component that mainly determines wheat grain yield is NE [6,28]. It was observed that an increase in NE in the N fertilized plots resulted in a decrease in GE (Figure S7).

The in-season interaction of these two yield components (NE and GE) is responsible for the number of grains per m^2 (grain density, GD). This yield component is considered as the crucial GY predicator, which was fully proved in the conducted study [66,67]. The significant increase of GD compared to NP, and even more to NPK-MOP, was recorded in treatments with Kieserite (NPK-MOP+Ki and NPK-KK+Ki+ES). The discussion cannot ignore the role of nutrients in the formation of TGW [6]. The size of this yield reflects the conditions of the kernel growth after flowering [63]. These conditions were the best and, therefore, the most stable, on NPK-KK+Ki, which is emphasized by the extremely low CV value (0.9%). The obtained effect of Kieserite application to winter wheat fully corroborates the conclusions by Potarzycki et al. [68]. These authors clearly stated that Kieserite application within the main period of yield formation extended from the beginning of shooting to the beginning of flowering results in GD increase, regardless of the status of primary yield components.

4.2. Balanced Fertilization—Control of Soil Fertility Sustainability

The second goal of sustainable N management in crop production is the efficient use of available N resources in the soil. Furthermore, the N_{min} content after harvest of the currently cultivated crop should be at a level that does not pose a threat to the environment [2,8,12]. The content of residual N_{min} in the soil at wheat harvest was significantly lower than the initial period (before the application of fertilizer) in spring. The efficiency of N_f in relation to N accumulation in grain was 83.4%, and it was 114.6% for the total N mass in wheat biomass. This difference indirectly indicates the significant contribution of inorganic soil N to wheat biomass and grain.

The first operational goal of sustainable N management is to increase the productivity of the applied N_f, which was achieved. It was even met applying N together with P fertilizer (114%). The further GY increase was due to the action of K and Mg fertilizers. The application of K in the form of MOP increased N utilization by winter wheat by 30 kg ha^{-1}, which doubled when Korn–Kali was applied together with Kieserite and supported with Epsom Salt. The obtained N use efficiency clearly indicates the important role of inorganic nitrogen (N_{min}) present in the soil at the beginning of the spring vegetation of winter wheat. The amount of N_{in} released from soil resources during the growing season of winter wheat cannot be ignored [44]. The N_{min} efficiency in the soil/wheat system, averaged over FVs, was 101%. This system was most effective in the AC plot, but yields were low. Based on the net nitrogen balance, the tested FVs can be divided into two main groups. The first, the low-effect group, included two FVs, i.e., NP and NPK-MOP. In these two variants, wheat used the available N resources in 91%. In the second, the high-effect system, N_{min} use efficiency reached 106%. This group included KK variants with Kieserite. The NPK-KK variant, despite its high stability for GY, showed only a moderate N_{min} utilization, which was 98%.

The amount of depleted N_{min} during the winter wheat growing season was significantly associated with the amounts of depleted K and Mg, but not with P. This relationship clearly indicates that the uptake of N by plants was conditioned by the available resources of K and Mg. This is consistent with the general mechanism of the uptake of nitrate N from the soil by plants [69,70]. Therefore, low values of residual N_{min} at harvest indicate an optimal state and availability of K and Mg for its utilization by winter wheat. It is worth emphasizing one highly important fact, namely, that the uptake of both of these nutrients from the soil showed mutual synergism. Therefore, it can stated that both of these soil nutrients met the assumptions of sustainable N management in winter wheat, and the conditioning factor was the weather.

The role of K and Mg was evident in each growing season. In the wet season (2016/2017), plants absorbed almost twice as much N from the soil as in the dry season (2017/2018). The main reason for the increase in GY in response to K was the conditions of its uptake from soil resources. The highest K depletion was obtained in 2016 under optimal weather conditions, despite the lowest content of available K in the soil. The lowest K depletion was revealed in the dry 2018, despite the optimal content of K in the soil. K ions are taken up in significant amounts by plant roots from the soil solution by diffusion, the efficiency of which depends on soil moisture [71]. Hence, in drought conditions, this process was significantly inhibited, as shown by the obtained results. Despite strong K depletion, the K balance was not potentially disturbed. Wheat crop residues left in the field would have fully compensated for the negative its balance in the soil. The average value of K replacement ratio was 1.37, exceeding the threshold value of 1.0 in each growing season.

The GY increase compared to the NPK-MOP variant was significantly greater in the KK variants with Kieserite. The stimulating effect of Mg fertilizer was revealed regardless of the level of available Mg, but only under optimal weather conditions. Therefore, the key to effective N management, even in conditions of high soil Mg content, turned out to be Mg management. The significant factor determining N uptake in this particular season was not its soil resources, but the uptake conditions. Under water stress conditions, a series of processes occur that limit the efficient uptake of Mg ions from the soil solution. The first is a decrease in both transpiration and the soil diffusion coefficient. The second is a decrease in the rate of Mg ion uptake by roots from the soil, caused by the high water requirements of these ions and competition with Ca ions [14,37,72]. As a result, Mg uptake from soil was significantly inhibited in the dry 2017/2018 season. This is confirmed not only by the low balance of available Mg in the soil but also by the net increase in this nutrient in the variants with intensive Mg fertilizer application. In the non-Mg variants, maintaining a balanced level of soil fertility may be disturbed because crop residues are too poor a source of this nutrient.

The yield-forming effect of nutrients in the tested FVs results from their influence on the processes responsible for yield components formation [63]. The critical period for the formation of this wheat trait extends from the beginning of stem elongation to full heading [28]. The nutritional determinant of ear and grain formation in the ear is the availability of N, but supported by other nutrients, including in the lines K and Mg [11]. The role of Mg, or more precisely, its balance in the growing season, is presented by the equation

$$NE = 430 + 3.5Mg_b \; for \; n = 21, \; R^2 = 0.41, \; p \leq 0.01$$

This equation clearly shows that the positive Mg balance during the winter wheat growing season was important for the number of formed ears. This process was limited in 2018 due to drought. The discussion cannot ignore the role of nutrients in the formation of TGW [6]. The size of this yield component reflects the conditions of the kernel growth after flowering [63,66]. These conditions were the best, and, therefore, the most stable, on NPK-KK+Ki, which is emphasized by the extremely low CV value (0.9%). The role of nutrients, or more precisely, the balance, is presented by the equation

$$TGW = 48.6 - 0.15P_b + 0.21Mg_b \; for \; n = 21, \; R^2 = 0.64, \; p \leq 0.001$$

This equation clearly confirms the limiting effect of Mg balance on TGW, which was noticeable in the dry 2018. It is obvious that Mg plays an important role due to its role in the current photosynthesis and in the transport of assimilates to the growing grain [35,60,61].

4.3. Balanced Fertilization—Evaluation of NUE Indicators

The impact of balanced N fertilization on grain yield was assessed based on direct and aggregated N traits in winter wheat, management N indicators, and nitrogen use efficiency (NUE) indicators. Direct traits concern the mass of N in grain and straw. Aggregated traits include the total N amount in the wheat canopy at harvest and its division between grain and straw.

The first two groups of N traits at winter wheat maturity responded significantly to the studied FVs, but they were variable in subsequent growing seasons. However, no interaction between them was found. Th total N amount (TN) was significantly correlated with GN (r = 0.99 ***) and SN (r = 0.95 ***). As a consequence, the obtained value of N redistribution between grain and straw, known as the nitrogen harvest index, was 72.6 ± 3.1%. It is not so high, because in optimal conditions, it can even approach 85%, and it is an important GY predictor [66]. Moreover, the CV was only 4.3%, which means high NHI stability in response to the factors studied. The low variability of both HI and NHI confirms the conservative status of this wheat trait [67].

The most stable GN over years was recorded for AC and NPK-KK (6.2%). The latter FV was, therefore, the most stable FV considered for both, GY and GN, which confirmed the close relationship between both traits of winter wheat. However, the achieved GY stability did not guarantee protein content at the formal level of 11.5% as is required for flour production [73,74]. This condition was met, provided that Kieserite was included in fertilization system for winter bread wheat. On average, for the years studied, it concerned two FVs, namely, NPK-KK+Ki and NPK-KK+Ki+ES. It can be concluded that the N dose of 150 kg ha^{-1} was too low to obtain high quality grain. Therefore, a higher N dose is required to reach two goals of bread wheat production: yield and protein content [74].

The efficiency of N accumulated in winter wheat during the growing season was evaluated using 10 indicators. Three questions were formulated to assess the practical and methodological usefulness of the studied indicators. The key question is as follows: Does balanced N_f fertilization, supported by the use of K and Mg fertilizers, result in increased N productivity? The second question concerns the sensitivity of the N indicators to the current state and change of the resources of basic soil nutrients. The methodological question is as follows: Which NUE indicator best reflects the effect of both tested nutrients on N productivity?

Based on the strength of associations (R^2) between the evaluated N indicators, they were divided into two groups. The first one includes four indicators, basically evaluating productivity of N_f: (i) the partial factor productivity of N_f (PFP-N_f); (ii) the agronomic nitrogen efficiency (ANE), and (iii) the nitrogen recovery (R) for GN and TN. ANE was found to be equivalent to PFP-N_f in assessing the effect of K and Mg on N productivity. These indicators clearly confirmed an increase in N_f productivity in response to the application of K and Mg. A greater increase was recorded for Mg. At the same time, they were positively and strongly correlated with GN and SN. For these three indices, the best FV was NPK-KK+Ki+ES.

The second group of N indicators characterizes the utilization efficiency of N accumulated in the plant during the growing season [44]. It comprises three double indices, calculated based on GN or TN. They were (i) the unit N productivity (NUP), (ii) the unit nitrogen accumulation (NUA), and (iii) the physiological N efficiency (PhNE). All discussed indices were strongly correlated each other (Table S4). The relationships between the obtained indexes and GY were, in fact, not linear, but best fitted a quadratic regression model (Figures S8 and S9). These relationships clearly indicate the sensitivity of these indexes to changes in the directions of physiological processes in wheat. The value of both PhNE indices showed a significant and very strong but negative relationship with the content of available P. The same relationship was obtained for both NUP indices. This trend clearly indicates that the P content at wheat harvest was not a factor limiting the physiological efficiency of N accumulated in the plant. However, the values of NUA increased progressively with the increase in the content of available P in the soil. However, this is only a superficial assessment of the observed phenomenon, which may lead to an erroneous evaluation. The criterion correcting the assessment of the reaction of the indicators turned out to be the soil nutrient balance. The stepwise analysis for PhNE-GN clearly showed that the increase in the P balance (actual loss from the soil during the growing season due plant uptake) led to a decrease in the index value, i.e., the physiological productivity of N accumulated in the grain. On the other hand, the component determining the increase was Mg. The developed dependency is as follows:

$$PhNE = 46.5 - 0.42P_b + 0.7Mg_b \; for \; n = 18, \; R^2 = 0.80, \; p \leq 0.01$$

Two indices, NUP and NUA, which are mutually mathematically reversible, deserve special attention (Figure S10). Both have practical significance, as they serve basic data to determine the productivity and nutritional N requirement for N of the cultivated plant [2]. The highest NUA increase compared to the NPK-MOP plot was recorded for NPK-KK+Ki, which was 10.2% for NUA-GN and 6.9% for NUA-TN. This difference clearly indicates the yield-forming role of Mg. All three indices showed a significant relationship with both the N mass in the grain and the Mg and K balance, which were strongly correlated with each other. This relationship fully confirms the yield-forming role of Mg during the crop formation period and grain filling [63]. In the assessment of NUP and NUA, it is necessary to indicate their specific feature resulting from the calculation methodology, i.e., reversibility in accordance with the exponential function (Figure S6). The relationship obtained by analyzing the effect of Mg on both indicators can be defined as *the scissors effect*. The unit productivity of nitrogen accumulated in grain increased linearly, while the value of the unit nitrogen accumulation decreased (Figure 6).

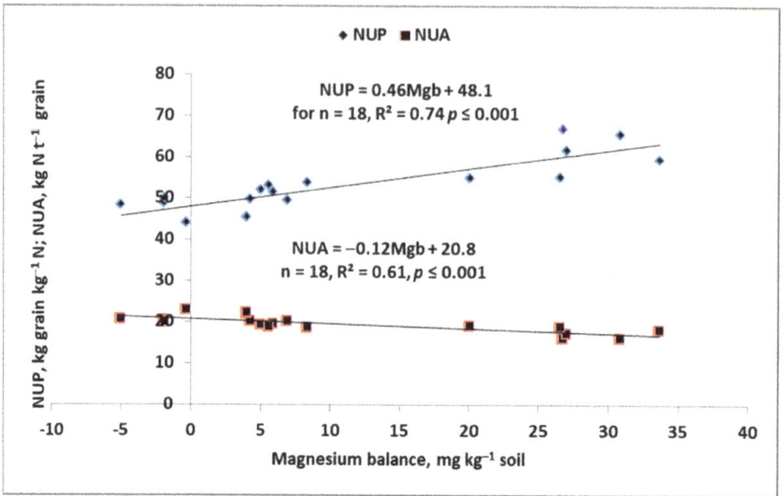

Figure 6. *Scissors effect* of nitrogen productivity on the background of available Mg balance during the growing season of winter wheat.

5. Conclusions

A sustainable model of crop production comes down to controlling the efficiency of nitrogen in the soil/plant system, but under the condition of maintaining soil fertility. This general assumption was fully realized for the tested project. The N_f efficiency in relation to N accumulation in grain was 83.4%, and for the total N in wheat biomass, it was 114.6%. This difference indirectly indicates the significant contribution of inorganic N to wheat biomass and grain. Meeting the second condition depends on the farmer's actions, as there has been expected a decrease in the content of available forms of P, K, and Mg. The mass of nutrients in wheat crop residues is able to replenish K and Mg resources in the soil after winter wheat harvest.

The key nutritional factors for effective nitrogen management in the soil/winter wheat production system were the resources of available K and Mg in the soil. The mobilization of these resources determined the effective management of N, both introduced in the fertilizer and present in the soil during the wheat vegetation period. The grain yield obtained in the NPK variant with Korn–Kali as a K carrier guaranteed the most stable grain yield, regardless of the weather conditions during the growing season. However, this fertilization variant did not provide the required grain quality. The effect of K and Mg generally oriented on GY was less effective in increasing the accumulation of N in grain. Maximizing the amount of nutrients in applied fertilizers to winter wheat, focused on grain yield increase, may have a return effect, resulting in the dilution N effect, which has been termed *the scissor effect*. The amount of N in the grain, and, therefore, its potential milling value, can be increased, as studies have shown, by using a potash fertilizer manufactured with Mg (Korn–Kali) or by additionally using a Mg fertilizer, for example, Kieserite. The positive effect of K and Mg on grain yield was fully confirmed by all tested nitrogen use efficiency indicators. The conducted studies showed the presence of two groups of indices, which were separated on the basis of a high degree of mutual correlation. All the tested indicators meet the requirements for evaluating the results of field experiments. The nitrogen unit accumulation index has proven its usefulness in science and practice. The analysis of this indicator shows that the mass of N in grain exceeding 22 kg 1 t^{-1} grain also meets the minimum quality requirement. Two fertilizer variants met this requirement, i.e., NPK-

KK+Ki and NPK-KK+Ki+ES. The study clearly showed that three of the six FVs fully met the three basic conditions for sustainable crop production: (i) stabilization and even an increase in grain yield; (ii) a decrease in the mass of inorganic N in the soil at harvest, potentially susceptible to leaching; and (iii) stabilization of soil fertility of P, K, and Mg.

Supplementary Materials: The following supporting information can be downloaded at: https://www.mdpi.com/article/10.3390/su17156705/s1, Table S1. Values of the coefficient of variation for selected characteristics for fertilization variants for the years of the study. Table S2. Descriptive statistics of grain nitrogen, GN, kg N ha^{-1}; Table S3. Descriptive statistics of straw nitrogen, GN, kg N ha^{-1}; Table S4. Correlation matrix for nitrogen indicators, n = 18; Figure S1. Grain yield as a function of the number of grains per ear; Figure S2. The effect of fertilization systems (FVs) on the crop residue nitrogen replacement ratio in subsequent years of the study; Figure S3. The effect of fertilization systems (FVs) on the phosphorus balance in winter wheat in subsequent years of the study; Figure S4. The effect of fertilization systems (FVs) on the crop residue phosphorus replacement ratio in subsequent years of the study; Figure S5. The effect of fertilization systems (FVs) on the crop residue magnesium replacement ratio in subsequent years of the study; Figure S6. The effect of fertilization systems (FVs) on the content of available K at harvest in subsequent years of the study; Figure S7. Compensation phenomenon: the relationship between the number of ears per m^2 and the number of grains per ear; Figure S8. Grain yield as function of nitrogen unit productivity based on grain nitrogen (GN); Figure S9. Grain yield as function of nitrogen unit productivity based on total nitrogen (TN); Figure S10. Mathematical relationship between nitrogen unit productivity and nitrogen unit accumulation.

Author Contributions: Conceptualization, A.A. and W.G.; methodology, A.A. and W.G.; software, K.P.-C.; validation, A.A. and K.P.-C.; formal analysis, W.G.; investigation, A.A. and K.P.-C.; resources, K.P.-C.; data curation, A.A.; writing—original draft preparation, A.A.; writing—review and editing, W.G.; visualization, K.P.-C.; supervision, A.A.; project administration, K.P.-C. All authors have read and agreed to the published version of the manuscript.

Funding: The publication was financed by the Polish Minister of Science and Higher Education as part of the Strategy of the Poznan University of Life Sciences for 2024–2026 in the field of improving scientific research and development work in priority research areas.

Institutional Review Board Statement: Not applicable.

Informed Consent Statement: Not applicable.

Data Availability Statement: The original contributions presented in the study are included in the article/Supplementary Materials, and further inquiries can be directed to the corresponding author.

Conflicts of Interest: The authors declare no conflicts of interest.

Abbreviations

The following abbreviations are used in this manuscript:

Acronyms	Full name
ANE	Agronomic nitrogen efficiency
CRRR	Crop residue replacement ratio
CNB	Crop nutrient balance
GD	Grain density, number of grains per m^2
GE	Number of grains per ear
GY	Grain yield
GN	Grain nitrogen, nitrogen accumulated in grain
HI	Harvest index
KK	Korn–Kali
MOP	Muriate of potash
NE	Number of ears per m^2

NUA-GN	Nitrogen unit accumulation in grain
NUA-TN	Nitrogen unit accumulation in total wheat biomass
PFP-N_f	Partial factor productivity of fertilizer nitrogen
PhNE-GN	Physiological nitrogen efficiency of N in grain
PhNE-TN	Physiological nitrogen efficiency of N in winter wheat biomass
NNb	Net nitrogen balance
NUP-GN	Nitrogen unit productivity in grain nitrogen
NUP-TN	Nitrogen unit productivity in total winter wheat biomass
NHI	Nitrogen harvest index
R-GN	Nitrogen recovery in grain nitrogen
R-TN	Nitrogen recovery in total N in winter wheat biomass
SN	Straw nitrogen—nitrogen accumulated in straw
SY	Straw yield
TN	Total N mass in winter wheat at harvest

Appendix A

Table A1. Correlation matrix for the grain yield and yield components, n = 21.

Traits	NE	GE	GD	TGW	SY	TB	HI
GY	0.47 *	0.64 **	0.96 ***	0.15	0.84 ***	0.96 ***	0.44 *
NE	1.00	0.30	0.36	0.41	0.74 ***	0.63 **	−0.32
GE		1.00	0.77 ***	−0.46 *	0.32	0.49 *	0.62 **
GD			1.00	−0.13	0.78 ***	0.90 ***	0.47 *
TGW				1.00	0.25	0.21	−0.16
SY					1.00	0.96 ***	−0.11
TB						1.00	0.17

***, **, * indicate significant differences between yield traits at $p < 0.001$, $p < 0.01$, and $p < 0.05$, respectively. Legend: GY—grain yield; NE—number of ears per m^2; GE—number of grains per ear; GD—grain density, number of grains per m^2;.TGW—thousand grain weight; SY—straw yield; TB—total biomass of winter wheat at harvest; HI—harvest index.

Table A2. Correlation matrix of the nitrogen indicators with grain yield and yield components, n = 21.

Traits	SN	TN	NHI	PFP-N_f	NUP-GN	NUP-TN	NUA-GN	NUA-TN	GY	NE	GE	TGW	SY	HI
GN	0.91 ***	0.99 ***	0.22	0.93 ***	−0.73 ***	−0.70 ***	0.71 ***	0.64 **	0.93 ***	0.22	0.82 ***	−0.11	0.69 **	0.55 *
SN	1.00	0.95 ***	−0.16	0.82 ***	−0.70 ***	−0.80 ***	0.69 **	0.73 ***	0.82 ***	0.26	0.75 ***	−0.20	0.74 ***	0.29
TN		1.00	0.12	0.92 ***	−0.74 ***	−0.74 ***	0.71 ***	0.68 **	0.92 ***	0.23	0.81 ***	−0.13	0.72 ***	0.49 *
NHI			1.00	0.17	−0.29	0.08	0.21	−0.07	0.17	−0.34	0.29	0.05	−0.29	0.82 ***
PFP-N_f				1.00	−0.44 *	−0.42	0.40	0.33	1.00	0.47 *	0.64 **	0.15	0.84 ***	0.44 *
NUP-GN					1.00	0.93 ***	−0.98 ***	−0.92 ***	−0.44 *	0.35	−0.85 ***	0.56 **	−0.12	−0.61 **
NUP-TN						1.00	−0.95 ***	−0.99 ***	−0.42	0.21	−0.79 ***	0.59 **	−0.27	−0.32
NUA-GN							1.00	0.96 ***	0.40	−0.34	0.83 ***	−0.58 **	0.12	0.53 *
NUA-TN								1.00	0.34	−0.28	0.76 ***	−0.63 **	0.17	0.32

***, **, * indicate significant differences between nitrogen indicators and yield traits at $p < 0.001$, $p < 0.01$, and $p < 0.05$, respectively. Legend: GN—N mass in grain; SN—N mass in straw; TN—total N mass in winter wheat at harvest; NHI—nitrogen harvest index; PFP-N_f—partial factor productivity of fertilizer nitrogen; NUP-GN—nitrogen unit productivity in grain N; NUP-TN—nitrogen unit productivity in total N in winter wheat biomass; NUA-GN—nitrogen unit accumulation in grain N; NUA-TN—nitrogen unit accumulation in total N in winter wheat biomass at harvest; GY—grain yield; NE—number of ears per m^2; GE—number of grains per ear; GD—grain density, number of grains per m^2; TGW—thousand grain weight; SY—straw yield; HI—harvest index.

Table A3. Correlation matrix of the nitrogen use efficiency indicators with grain yield and yield components, n = 21.

Traits	R-GN	R-TN	PhNE-GN	PhNE-TN	GY	NE	GE	TGW	SY	HI
ANE	0.86 ***	0.77 ***	0.63 **	0.65 **	0.99 ***	0.51 *	0.06	0.65 **	0.80 **	0.35
R-GN	1.00	0.94 ***	0.16	0.24	0.86 ***	0.17	0.38	0.35	0.49 *	0.62 **
R-TN		1.00	0.02	0.02	0.73 **	0.04	0.47	0.20	0.40	0.54 *
PhNE-GN			1.00	0.96 ***	0.63 ***	0.75 ***	−0.52 *	0.79 ***	0.81 ***	−0.23
PhNE-TN				1.00	0.69 **	0.72 **	−0.46	0.82 ***	0.76 ***	−0.06
GY					1.00	0.52 *	0.06	0.66 **	0.79 ***	0.38

Table A3. *Cont.*

Traits	R-GN	R-TN	PhNE-GN	PhNE-TN	GY	NE	GE	TGW	SY	HI
NE						1.00	−0.76 ***	0.52 *	0.77 ***	−0.32
GE							1.00	−0.41	−0.33	0.56 *
TGW								1.00	0.63 **	0.09
SY									1.00	−0.26

***, **, * indicate significant differences between nitrogen use efficiency indicators and yield traits at $p < 0.001$, $p < 0.01$, and $p < 0.05$, respectively. Legend: ANE—agronomic nitrogen efficiency; R-GN—nitrogen recovery in grain; R-TN—nitrogen recovery for total N in winter wheat biomass at harvest; PhNE—physiological nitrogen efficiency of N in grain; PhTN—physiological nitrogen efficiency of N in TN in winter wheat biomass; GY—grain yield; NE—number of ears per m^2; GE—number of grains per ear; GD—grain density, number of grains per m^2; TGW—thousand grain weight; SY—straw yield; HI—harvest index.

Table A4. Correlation matrix of relationships between soil nutrient characteristics and basic winter wheat traits, n = 21.

Traits	N	N_b	NNb	P	P_b	K	K_b	Mg	Mg_b
GY	0.22	0.14	0.09	−0.28	−0.38	−0.25	0.25	0.25	0.11
NE	0.44 *	0.21	0.14	−0.73 ***	0.08	−0.58 **	0.55 *	0.28	0.64
GE	−0.15	−0.14	0.11	0.40	−0.29	0.31	−0.28	0.14	−0.49 *
TGW	0.31	0.43	−0.15	−0.69 **	−0.31	−0.66 **	0.65 **	0.02	0.60 **
SY	0.34	0.18	0.18	−0.55 *	−0.11	−0.55 *	0.54 *	0.38	0.36
HI	−0.14	−0.05	−0.03	0.37	−0.54 *	0.41	−0.40	−0.20	−0.38
N_{minh}	1.00	−0.56 **	0.04	−0.64 **	0.22	−0.62 **	0.60	0.02	0.71 ***
N_b		1.00	−0.10	−0.24	−0.44 *	−0.13	0.15	0.14	−0.08
NNb			1.00	0.03	0.19	−0.14	0.12	0.01	0.26
P				1.00	−0.01	0.89 ***	−0.87 ***	−0.26	−0.81 ***
P_b					1.00	−0.17	0.13	0.31	0.36
K						1.00	−0.99 ***	−0.26	−0.81 ***
K_b							1.00	0.25	0.77 ***
Mg								1.00	0.05

***, **, and * indicate significant differences at $p < 0.001$, $p < 0.01$, and $p < 0.05$, respectively; ns—non-significant. Legend: GY—grain yield; NE—number of ears per m^2; GE—number of grains per ear; GD—grain density, number of grains per m^2; TGW—thousand grain weight; SY—straw yield; TB—total biomass at harvest; HI—harvest index; N, P, K, Mg—nutrients; b—soil nutrient balance; NNb—net nitrogen balance.

Table A5. Correlation matrix of relationships between soil nutrient characteristics and indicators of nitrogen management in winter wheat, n = 21.

Traits	N	N_b	NNb	P	P_b	K	K_b	Mg	Mg_b
GN	−0.11	0.40	−0.94 ***	−0.13	−0.58 *	0.06	−0.04	0.17	−0.28
SN	−0.40	0.52 *	−0.68 **	0.11	−0.69 **	0.19	−0.14	−0.08	−0.54 *
TN	−0.20	0.46	−0.94 ***	−0.08	−0.65 **	0.10	−0.07	0.12	−0.374
NHI	0.11	−0.03	−0.58 *	−0.07	−0.13	0.09	−0.11	0.27	−0.03
PFP-N_f	0.18	0.55 *	−0.62 **	−0.65 **	−0.50 *	−0.44	0.44	0.25	0.24
NUP-GN	0.50 *	0.14	0.53 *	−0.78 ***	0.23	−0.80 ***	0.76 ***	0.09	0.86 ***
NUP-TN	0.54 *	0.18	0.33	−0.84	0.14	−0.79 ***	0.75 ***	0.20	0.86 ***
NUA-GN	−0.44	−0.26	−0.44	0.81 ***	−0.09	0.79 ***	−0.78 ***	−0.13	−0.77 ***
NUA-TN	−0.52 *	−0.25	−0.29	0.87 ***	−0.07	0.80 **	−0.76 ***	−0.20	−0.83 ***

***, **, * indicate significant differences between nitrogen use efficiency indicators and yield traits at $p < 0.001$, $p < 0.01$, and $p < 0.05$, respectively. Legend: GN—amount of nitrogen in grain; SN—amount of nitrogen in straw; TN—total amount of nitrogen in winter wheat biomass; NHI—nitrogen harvest index; PFP-N_f—partial factor productivity of fertilizer nitrogen; NUP-GN—nitrogen unit productivity in grain N; NUP-TN—nitrogen unit productivity in total N in winter wheat biomass; NUA-GN—nitrogen unit accumulation for grain N; NUA-TN—nitrogen unit accumulation in total N in winter wheat biomass; TB—total winter wheat biomass; N, P, K, Mg—nutrients; b—soil nutrient balance; NNb—net nitrogen balance.

Table A6. Correlation matrix of relationships between soil nutrient characteristics and indicators of nitrogen use efficiency in winter wheat, n = 21.

Traits	N	Nb	NNb	P	Pb	K	Kb	Mg	Mgb
ANE	0.17	0.57 *	−0.64 **	−0.64 **	−0.58 *	−0.430	0.43	0.19	0.19
R-GN	−0.02	0.43	−0.91 ***	−0.27	−0.58 *	−0.07	0.08	0.18	−0.16
R-TN	−0.21	0.48 *	−0.92 ***	−0.07	−0.71 **	0.11	−0.08	0.07	−0.41
PhNE-GN	0.49 *	0.35	0.13	−0.89 ***	−0.16	−0.82 ***	0.80 ***	0.09	0.74 ***
PhNE-TN	0.54 *	0.28	0.07	−0.90 ***	−0.08	−0.81 ***	0.77 ***	0.19	0.79 ***

***, **, and * indicate significant differences at $p < 0.001$, $p < 0.01$, and $p < 0.05$, respectively; ns—non-significant. Legend: ANE—agronomic nitrogen efficiency; R-GN—nitrogen recovery in grain; R-TN—nitrogen recovery in total N in winter wheat biomass; PhNE—physiological efficiency of N in grain; PhTN—nitrogen efficiency of N in winter wheat biomass; N, P, K, Mg—nutrients; b—soil nutrient balance; NNb—net nitrogen balance.

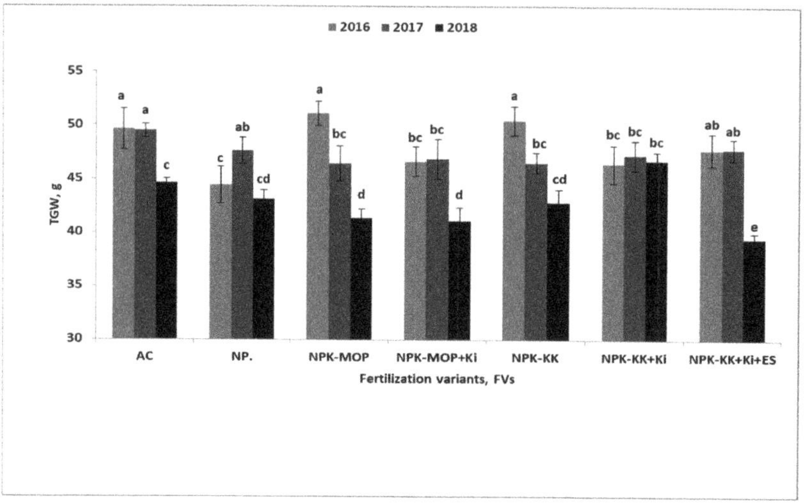

Figure A1. Effect of fertilization variant in consecutive years of the study on thousand grain weight. [a] Similar letters in the column mean no significant differences in the Tukey test.

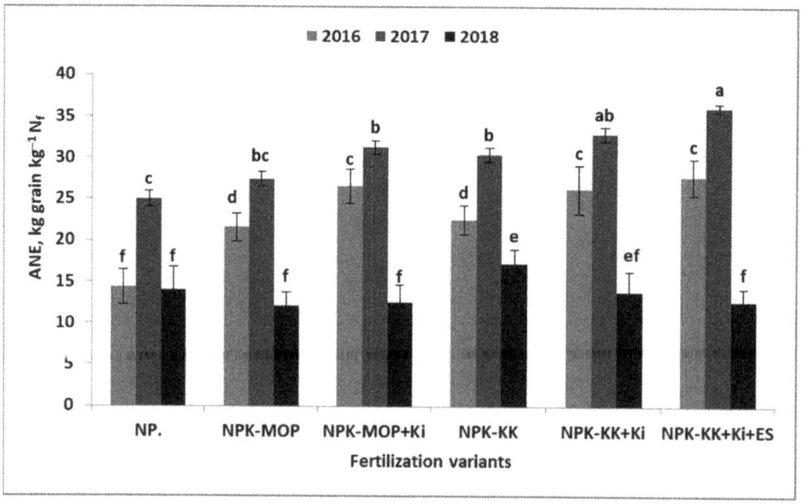

Figure A2. Effect of fertilization variant in consecutive years of the study on the agronomic nitrogen efficiency. [a] Similar letters in the column mean no significant differences in the Tukey test.

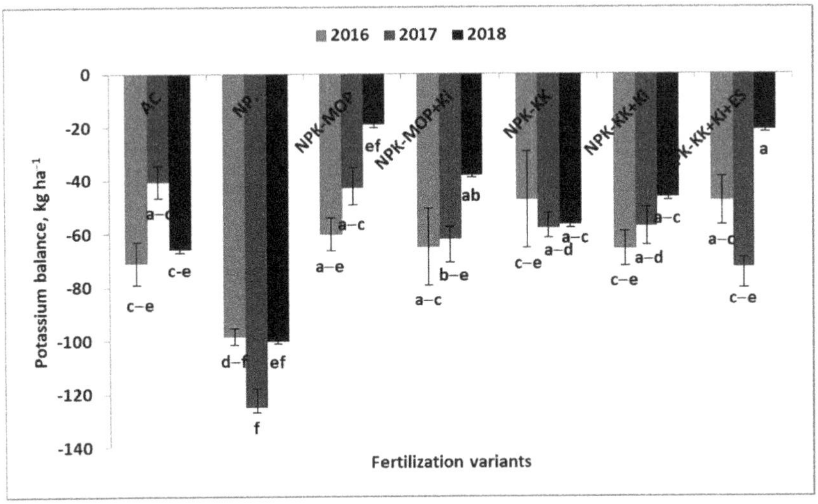

Figure A3. The effect of fertilization variants in consecutive years of the study on the potassium winter wheat balance. Means followed by the same letter within a column indicate the lack of a significant difference between the treatments. The vertical bar in the column is the standard error of the mean.

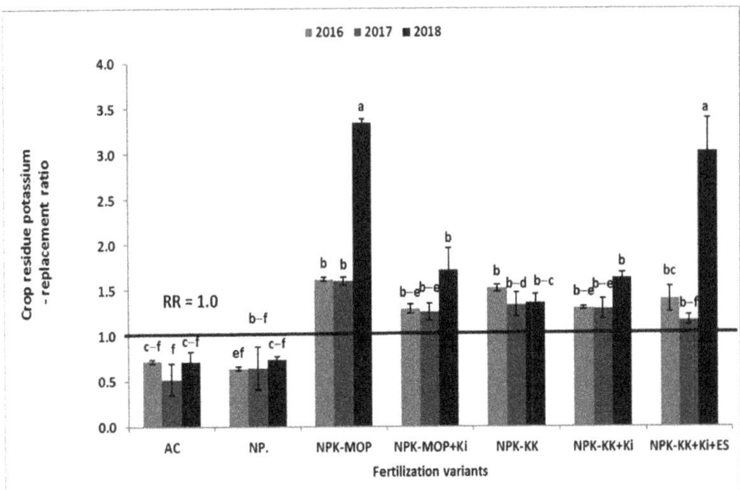

Figure A4. The effect of fertilization variants in consecutive years of the study on the potassium replacement ratio. Means followed by the same letter within a column indicate the lack of a significant difference between the treatments. The vertical bar in the column is the standard error of the mean.

References

1. Sharma, L.K.; Bali, S.K. A review of methods to improve nitrogen use efficiency in agriculture. *Sustainability* **2018**, *10*, 51. [CrossRef]
2. Barłóg, P.; Grzebisz, W.; Łukowiak, R. Fertilizers and fertilization strategies mitigating soil factors constraining efficiency of nitrogen in plant production. *Plants* **2022**, *11*, 1855. [CrossRef] [PubMed]
3. Luo, L.; Zhang, Y.; Xu, G. How does nitrogen shape plant architecture. *J. Exp. Bot.* **2020**, *71*, 4415–4427. [CrossRef] [PubMed]
4. Ötvös, K.; Marconi, M.; Vega, A.; O'brian, J.; Johnston, A.; Abualla, R.; Antonielli, L.; Montesinos, J.C.; Zhang, Y.; Yan, S.; et al. Modulation of plant root growth by nitrogen source-defined regulation of polar auxin transport. *EMBO J.* **2021**, *40*, e106862. [CrossRef] [PubMed]

5. Lynch, J.L. Root phenotypes for improved capture: An underexploited opportunity for global agriculture. *New Phytol.* **2019**, *223*, 548–564. [CrossRef] [PubMed]
6. Klepper, B.; Rickman, R.W.; Waldman, S.; Chevalier, P. The physiological life cycle of wheat: Its use in breeding and crop management. *Euphytica* **1998**, *100*, 341–347. [CrossRef]
7. Anas, M.; Liao, F.; Verma, K.K.; Sarwar, M.A.; Mahmood, A.; Chen, Z.-L.; Li, Q.; Zeng, X.-P.; Liu, Y.; Li, Y.-R. Fate of nitrogen in agriculture and environment: Agronomic, eco-physiological and molecular approaches to improve nitrogen use efficiency. *Biol. Res.* **2020**, *53*, 47. [CrossRef] [PubMed]
8. Rockström, J.; Williams, J.; Daily, G.; Noble, A.; Matthews, N.; Gordon, L.; Wetterstrand, H.; DeClerck, F.; Shah, M.; Steduto, P.; et al. Sustainable intensification of agriculture for human prosperity and global sustainability. *Ambio* **2017**, *46*, 4–17. [CrossRef] [PubMed]
9. Grzebisz, W.; Łukowiak, R. Nitrogen gap amelioration is a core for sustainable intensification of agriculture—A concept. *Agronomy* **2021**, *11*, 419. [CrossRef]
10. Asseng, S.; Guarin, J.R.; Raman, M.; Monje, O.; Kiss, G.; Despomnier, D.D.; Meggers, F.M.; Gaither, P.P.G. Wheat yield potential in controlled—Environment vertical farms. *Proc. Natl. Acad. Sci. USA* **2020**, *117*, 19131–19135. [CrossRef] [PubMed]
11. Grzebisz, W.; Szczepaniak, W.; Przygocka-Cyna, K.; Biber, M.; Spiżewski, T. Sources of nutrients for the growing ear of winter wheat in the Critical Cereal Window. *Agronomy* **2024**, *14*, 3018. [CrossRef]
12. Johnston, A.M.; Bruulsema, T.W. 4R nutrient stewarship for improved nutrient use efficiency. *Procedia Engin.* **2014**, *83*, 365–370. [CrossRef]
13. Grzebisz, W.; Diatta, J.; Barłóg, P.; Biber, M.; Potarzycki, J.; Łukowiak, R.; Przygocka-Cyna, K.; Szczepaniak, W. Soil fertility clock–crop rotation as a paradigm in nitrogen fertilizer productivity control. *Plants* **2022**, *11*, 2841. [CrossRef] [PubMed]
14. Marschner, H. *Mineral Nutrition of Higher Plants*, 2nd ed.; Academic Press: London, UK, 1995; 899p.
15. Krouk, G.; Kiba, T. Nitrogen and phosphorus interactions in plants: From agronomic to physiological and molecular insights. *Curr. Opin. Plant Biol.* **2020**, *57*, 104–109. [CrossRef] [PubMed]
16. MacDonald, G.K.; Bennett, E.M.; Potter, P.A.; Ramankutty, N. Agronomic phosphorus imbalances across the world's croplands. *Proc. Natl. Acad. Sci. USA* **2011**, *108*, 3086–3091. [CrossRef] [PubMed]
17. Holmes, M.R.J. *Nutrition of the Oilseed Rape Crop*; Applied Science Publishers Ltd.: London, UK, 1980; 158p.
18. Grzebisz, W.; Szczepaniak, W.; Barłóg, P.; Przygocka-Cyna, K.; Potarzycki, J. Phosphorus sources for winter oilseed rape (*Brassica napus* L.) during reproductive growth—Magnesium sulfate management impact on P use efficiency. *Arch. Agron. Soil Sci.* **2018**, *64*, 1646–1662. [CrossRef]
19. FAOSTAT. Food and Agriculture Organization of the United Nations. Available online: http://faostat.fao.org/site/567/default.aspx#ancor (accessed on 5 April 2025).
20. Grzebisz, W.; Čermák, P.; Rroco, E.; Szczepaniak, W.; Potarzycki, J.; Füleky, G. Potassium impact on nitrogen use efficiency in potato—A case study from the Central-East Europe. *Plant Soil Environ.* **2017**, *63*, 422–427. [CrossRef]
21. Grzebisz, W.; Potarzycki, J. The in-season nitrogen concentration in the potato tuber as the yield driver. *Agron. J.* **2020**, *112*, 1287–1308. [CrossRef]
22. Barraclough, P.B. The growth and activity of winter wheat roots in the field: Nutrient uptakes of high-yielding crops. *J. Agric. Sci. Camb.* **1986**, *106*, 45–52. [CrossRef]
23. Ciampitti, I.A.; Camberato, J.J.; Murrel, S.T.; Vyn, T.J. Maize nutrient accumulation and partitioning in response to plant density and nitrogen rate: I. Macronutrients. *Agron. J.* **2013**, *10593*, 7830. [CrossRef]
24. Conijn, J.G.; Bindraban, P.S.; Schröder, J.J.; Jongschaap, R.E.E. Can our global food system meet food demand within planetary boundries? *Agric. Ecosys. Environ.* **2018**, *251*, 244–256. [CrossRef]
25. Jones, J.B., Jr. *Agronomic Handbook*; CRC Press LLC: Boca Raton, FL, USA, 2003; 450p.
26. Sardans, J.; Peñuelas, J. Potassium control of plant functions: Ecological and agricultural implications. *Plants* **2021**, *10*, 419. [CrossRef] [PubMed]
27. Rawat, J.; Pandey, N.; Saxena, J. Role of Potassium in Plant Photosynthesis, Transport, Growth and Yield. In *Role of Potassium in Abiotic Stress*; Iqbal, N., Umar, S., Eds.; Springer: Singapore, 2022; pp. 1–14.
28. Guo, Z.; Chen, D.; Schnurbusch, T. Plant and floret growth at distinct developmental stages during the stem elongation phase in wheat. *Front. Plant Sci.* **2018**, *9*, 330. [CrossRef] [PubMed]
29. Hubbard, K.E.; Webb, A.A.R. Circadian Rhythms in Stomata: Physiological and Molecular Aspects. In *Rhythms in Plants*; Mancuso, S., Shabala, S., Eds.; Springer: Cham, Switzerland, 2015; pp. 231–255.
30. Ragel, P.; Raddatz, N.; Leidi, E.O.; Quintero, F.J.; Pardo, J.M. Regulation of K+ nutrition in plants. *Front. Plant Sci.* **2019**, *10*, 281. [CrossRef] [PubMed]
31. Wang, M.; Zheng, Q.S.; Shen, Q.R.; Guo, S.W. The critical role of potassium in plant stress response. *Int. J. Mol. Sci.* **2013**, *14*, 7370–7390. [CrossRef] [PubMed]

32. Zörb, C.; Ludewig, U.; Hawkesford, M.J. Perspective on wheat yield and quality with reduced nitrogen supply. *Trends Plant Sci.* **2018**, *23*, 1019–1037. [CrossRef] [PubMed]
33. Bindraban, P.S.; Dimkpa, C.; Nagarajan, L.; Roy, A.; Rabbinge, R. Revisiting fertilisers and fertilisation strategies for improved nutrient uptake by plants. *Biol. Fertil. Soils* **2015**, *51*, 897–911. [CrossRef]
34. Grzebisz, W.; Zielewicz, W.; Przygocka-Cyna, K. Deficiencies of secondary nutrients in crop plants—A real challenge to improve nitrogen management. *Agronomy* **2023**, *13*, 66. [CrossRef]
35. Gerendás, J.; Führs, H. The significance of magnesium for crop quality. *Plant Soil* **2013**, *368*, 101–128. [CrossRef]
36. Klikocka, H.; Marks, M. Sulphur and nitrogen fertilization as a potential means of agronomic biofortification to improve the content and uptake of microelements in spring wheat grain DM. *J. Chem.* **2018**, *2018*, 9326820. [CrossRef]
37. Härdter, R.; Rex, M.; Orlovius, K. Effects of different Mg fertilizer sources on the magnesium availability in soils. *Nutr. Cycl. Agroecos.* **2004**, *70*, 249–259. [CrossRef]
38. Shaul, O. Magnesium transport and function in plants: The tip of the iceberg. *Biometals* **2002**, *15*, 309–323. [CrossRef] [PubMed]
39. Jamal, A.; Moon, Y.-S.; Abdin, M.Z. Sulphur—A general overview and interaction with nitrogen. *Aust. J. Crop Sci.* **2010**, *4*, 523–529.
40. Grant, C.; Hawkesford, M.J. Sulfur. In *Handbook od Plant Nutrition*, 2nd ed.; Barker, A.V., Pilbeam, D.J., Eds.; CRC Press, Taylor and Francis Group: Boca Raton, FL, USA, 2015; pp. 261–304.
41. The Royal Society. *Reaping the Benefits: Science and the Sustainable Intensification of Global Agriculture*; RS Policy Document; The Royal Society: London, UK, 2009; p. 86.
42. Wallace, A.; Wallace, G.A. *Closing the Crop—Yield Gap through Better Soil and Better Management*; Wallace Laboratories: Los Angeles, CA, USA, 2003; p. 162.
43. Congreves, K.A.; Otchere, O.; Ferland, D.; Farzadfar, S.; Williams, S.; Arcand, M.M. Nitrogen use efficiency definitions of today and tomorrow. *Front. Plant Sci.* **2021**, *12*, 637108. [CrossRef] [PubMed]
44. Grzebisz, W.; Potarzycki, J. A realistic approach to calculating the nitrogen use efficiency index in cereals with winter wheat (*Triticum aestivum* L.) as an example. *Agronomy* **2025**, *15*, 161. [CrossRef]
45. Mehlich, A. Mehlich 3 soil test extractant: A modification of Mehlich 2 extractant. *Com. Soil Sci. Plant Anal.* **1984**, *15*, 1409–1416. [CrossRef]
46. Kęsik, K. Application of the Mehlich 3 method in the fertilizer advisory system. *Stud. I Rap. IUNG-PIB* **2016**, *48*, 95–104. (In Polish)
47. Trávník, K.; Zbíral, J.; Nemec, P. *Agrochemical Soil Testing—Mehlich III*; Central Institute for Supervising and Testing in Agriculture: Brno, Czech Republic, 1999.
48. Zbíral, J. Determination of plant-available micronutrients by the Mehlich 3 soil extractant—A proposal of critical values. *Plant Soil Environ.* **2016**, *62*, 527–531. [CrossRef]
49. *PN-EN ISO 20483:2014-02*; Cereal Grains and Pulses—Determination of Nitrogen Content and Conversion to Crude Protein—Kjeldahl Method. PKN: Warsaw, Poland, 2015; p. 24.
50. Szczepaniak, W.; Nowicki, B.; Bełka, D.; Kazimierowicz, A.; Kulwicki, M.; Grzebisz, W. Effect of foliar application of micronutrients and fungicides on the nitrogen use efficiency in winter wheat. *Agronomy* **2022**, *12*, 257. [CrossRef]
51. Wilding, L.P.; Dress, L.R. Spatial variability and pedology. In *Pedogenesis and Soil Taxonomy*; Wilding, L.P., Smeck, N., Hall, G.F., Eds.; Elsevier: Wageningen, The Netherlands, 1983; pp. 83–116.
52. Smith, G. Step away from stepwise. *J. Big Data* **2018**, *5*, 32. [CrossRef]
53. Schröder, J.J.; Schulte, R.P.O.; Cremer, R.E.; Delgado, A.; van Leeuwen, J.; Lehtinen, T.; Rutgers, M.; Spiegel, H.; Staes, J.; Tóth, G.; et al. The elusive role of soil quality in nutrient cycling: A review. *Soil Use Manag.* **2016**, *32*, 476–486. [CrossRef]
54. Ahrends, H.E.; Siebert, S.; Rezaei, E.E.; Seidel, S.J.; Hüging, H.; Ewert, F.; Döring, T.; Rueda-Ayala, V.; Eugster, W.; Gaiser, T. Nutrient supply affects the yield stability of major European crops–A 50 year study. *Environ. Res. Lett.* **2020**, *16*, 014003. [CrossRef]
55. Wójcik-Gront, E.; Iwańska, M.; Wnuk, A.; Oleksiak, T. The analysis of wheat yield variability based on experimental data from 2008–2018 to understand the yield gap. *Agriculture* **2022**, *12*, 32. [CrossRef]
56. Shewry, P.R.; Hey, S.J. The contribution of wheat to human diet and health. *Food Energy Sec.* **2015**, *493*, 178–202. [CrossRef] [PubMed]
57. Szczepaniak, W.; Barłóg, P.; Łukowiak, R.; Przygocka-Cyna, K. Effect of balanced nitrogen fertilization in four-year rotation on plant productivity. *J. Cent. Eur. Agric.* **2013**, *14*, 64–77. [CrossRef]
58. Hlisnikovsky, L.; Menšik, L.; Kunzová, E. The development of winter wheat yield and quality under different fertilizer regimes and soil-climatic conditions in the Czech Republic. *Agronomy* **2020**, *10*, 1160. [CrossRef]
59. Messaoudi, A.; Labdelli, F.; Rebouh, N.Y.; Djerbaoui, M.; Kucher, D.E.; Hadjout, S.; Ouaret, W.; Zakharova, O.A.; Latati, M. Investigating the potassium fertilization effect on morphological and agrophysical indicators of durum wheat under Mediterranean rain-fed conditions. *Agriculture* **2023**, *13*, 1142. [CrossRef]
60. Wang, Z.; Hassan, M.U.; Nadeem, F.; Wu, L.; Zhang, F.; Li, X. Magnesium fertilization improves crop yield in most production systems: A meta–analysis. *Front. Plant Sci.* **2020**, *10*, 1727. [CrossRef] [PubMed]

61. Potarzycki, J. Influence of nitrogen and magnesium fertilization at the flag leaf stage of winter wheat development on the yield and grain quality. *Fertil. Fertil.* **2008**, *32*, 100–121.
62. EL-Metwally, A.E.; Abdalla, F.E.; Saady, A.M.; Safina, S.A.; EI-Sawy, S.S. Response of wheat to magnesium and copper foliar feeding under sandy soil conditions. *J. Am. Sci.* **2010**, *6*, 818–823.
63. Grzebisz, W.; Potarzycki, J. Effect of magnesium fertilization systems on grain yield formation by winter wheat (*Triticum aestivum* L.) during the grain filling period. *Agronomy* **2022**, *12*, 12. [CrossRef]
64. González, F.G.; Miralles, D.J.; Slafer, G.A. Wheat floret survival as related to pre-anthesis spike growth. *J. Exp. Bot.* **2011**, *62*, 4889–4901. [CrossRef] [PubMed]
65. Duan, J.; Wu, Y.; Zhou, Y.; Ren, X.; Shao, Y.; Feng, W.; Zhu, Y.; Wang, Y.; Guo, T. Grain number response to pre-anthesis dry matter and nitrogen in improving wheat yield in the Huang-Huai Plain. *Sci. Rep.* **2018**, *8*, 7126. [CrossRef]
66. Serrago, R.A.; Alzueta, I.; Savin, R.; Slafer, G.A. Understanding grain yield responses to source–sink ratios during grain filling in wheat and barley under constraining environments. *Field Crops Res.* **2013**, *150*, 42–51. [CrossRef]
67. Slafer, G.A.; Garcia, G.A.; Serrago, R.A.; Miralles, D.J. Physiological drivers of responses of grains per m^2 to environmental and genetic factors in wheat. *Field Crops Res.* **2022**, *285*, 108593.
68. Potarzycki, J.; Grzebisz, W.; Szczepaniak, W. Magnesium fertilization increases nitrogen use efficiency in winter wheat (*Triticum aestivum* L.). *Plants* **2022**, *11*, 2600. [CrossRef] [PubMed]
69. Grzebisz, W.; Szczepaniak, W.; Potarzycki, J.; Łukowiak, R. Sustainable management of soil potassium—A crop rotation oriented concept. In *Soil Fertility*; Issaka, R.N., Ed.; InTech: London, UK, 2012; 38p.
70. Yawson, D.O.; Amoach, K.K.; Asare, P.A.; Techie-Menson, J.; Afutu, E.; Atiah, K.; Sanieri, A.K.; Sare-Larbi, S.; Adu, M.O. A meta-analysis of modifications of root system traits of crop plants to potassium (K) deprivation. In *Plant Roots*; Yildirim, E., Turan, M., Ekinci, M., Eds.; InTech: London, UK, 2021; 27p.
71. Kuchenbuch, R.; Claassesn, N.; Jungk, A. Potassium availability in relation to soil moisture. II. Calculation by means of a mathemantical model simulation model. *Plant Soil* **1986**, *95*, 233–243. [CrossRef]
72. Gransee, A.; Führs, H. Magnesium mobility in soils as a challenge for soil and plant analysis, magnesium fertilization and root uptake under adverse growth conditions. *Plant Soil* **2013**, *368*, 5–21. [CrossRef]
73. Klikocka, H.; Szczepaniak, W. Energy crises—Alternative use of winter bread wheat grain depending on protein content. *Agronomy* **2023**, *13*, 861. [CrossRef]
74. Burton, A.; Levy Haner, L.; Schaad, N.; Strebel, S.; Vuille-dit-Bille, N.; de Figueiredo Bongiovani, P.; Holzkamper, A.; Pellet, D.; Herrera, J.M. Evaluating nitrogen fertilization strategies to optimize yield and grain nitrogen in top winter wheat varieties across Switzerland. *Field Crops Res.* **2024**, *307*, 109251. [CrossRef]

sustainability

Article

Effect of Farming System and Irrigation on Physicochemical and Biological Properties of Soil Under Spring Wheat Crops

Elżbieta Harasim and Cezary A. Kwiatkowski *

Department of Herbology and Plant Cultivation Techniques, University of Life Sciences in Lublin, Akademicka 13, 20-950 Lublin, Poland; elzbieta.harasim@up.lublin.pl
* Correspondence: cezary.kwiatkowski@up.lublin.pl

Abstract

A field experiment in growing spring wheat (*Triticum aestivum* L.—cv. 'Monsun') under organic, integrated and conventional farming systems was conducted over the period of 2020–2022 at the Czesławice Experimental Farm (Lubelskie Voivodeship, Poland). The first experimental factor analyzed was the farming system: A. organic system (control)—without the use of chemical plant protection products and NPK mineral fertilization; B. conventional system—the use of plant protection products and NPK fertilization in the range and doses recommended for spring wheat; C. integrated system—use of plant protection products and NPK fertilization in an "economical" way—doses reduced by 50%. The second experimental factor was irrigation strategy: 1. no irrigation—control; 2. double irrigation; 3. multiple irrigation The aim of the research was to determine the physical, chemical, and enzymatic properties of loess soil under spring wheat crops as influenced by the factors listed above. The highest organic C content of the soil (1.11%) was determined in the integrated system with multiple irrigation of spring wheat, whereas the lowest one (0.77%)—in the conventional system without irrigation. In the conventional system, the highest contents of total N (0.15%), P (131.4 mg kg^{-1}), and K (269.6 mg kg^{-1}) in the soil were determined under conditions of multiple irrigation. In turn, the organic system facilitated the highest contents of Mg, B, Cu, Mn, and Zn in the soil, especially upon multiple irrigation of crops. It also had the most beneficial effect on the evaluated physical parameters of the soil. In each farming system, the multiple irrigation of spring wheat significantly increased moisture content, density, and compaction of the soil and also improved its total sorption capacity (particularly in the integrated system). The highest count of beneficial fungi, the lowest population number of pathogenic fungi, and the highest count of actinobacteria were recorded in the soil from the organic system. Activity of soil enzymes was the highest in the integrated system, followed by the organic system—particularly upon multiple irrigation of crops. Summing up, the present study results demonstrate varied effects of the farming systems on the quality and health of loess soil. From a scientific point of view, the integrated farming system ensures the most stable and balanced physicochemical and biological parameters of the soil due to the sufficient amount of nutrients supplied to the soil and the minimized impact of chemical plant protection products on the soil. The multiple irrigation of crops resulting from indications of soil moisture sensors mounted on plots (indicating the real need for irrigation) contributed to the improvement of almost all analyzed soil quality indices. Multiple irrigation generated high costs, but in combination with fertilization and chemical crop protection (conventional and integrated system), it influenced the high productivity of spring wheat and compensated for the incurred costs (the greatest profit).

 https://doi.org/10.3390/su17146473

Keywords: loess soil; organic system; conventional system; integrated system; irrigation; soil quality properties; soil enzymatic activity

1. Introduction

Wheat (*Triticum aestivum* L.), as one of the most important food crops globally, serves as a primary source of daily protein and provides 20% of the calories consumed by man [1]. The yield of both winter and spring wheat varieties is influenced by numerous factors, including climatic conditions, farming systems (conventional, integrated, organic), as well as the physical, chemical, and biological properties of the soil. Proper soil pH and structure are fundamental factors that ensure optimal growth and yield of crops. Therefore, investigating the relationship between soil quality and crop yields is crucial, particularly in the context of changing farming systems [2–5].

Soil quality is determined by its physical, chemical, and biological properties and is closely related to its fertility and health. Various indicators provide early insights into the processes occurring in the soil and the availability of nutrients to plants. They also reflect the impacts of changes in agricultural land use or farming practices [6,7]. Improvements in soil quality are often equated with its enhanced physicochemical properties, such as the size and stability of aggregates, organic matter content, reduced bulk density, and soil resistance. In contrast, biological indicators of soil quality, such as overall diversity and biomass of organisms, genetic diversity of beneficial microorganisms, and activity of enzymes involved in nutrient cycling, are dynamic indicators closely linked to farming systems and agricultural practices [2,8]. They play a significant role in the proper development of plants, ultimately affecting not only the quantity but also the quality of the harvested crops [9,10]. Soil organisms perform numerous essential functions, like ensuring nutrient cycling and supplying nutrients to plants, modifying soil physical structure and water conditions, and influencing the elimination of undesirable organisms in cultivated lands. Therefore, a good quality of the soil promotes its productivity and, ultimately, impacts plant health [11,12].

Organic and integrated systems respond to the intensification of agricultural production (excessive use of mineral fertilizers and pesticides), which is considered an underlying cause of soil degradation and the resulting environmental pollution [13]. Evaluating the effectiveness of crop cultivation in various farming systems under changing climatic conditions has shown that organic and integrated agricultural systems are generally more resilient to climate change than the conventional system. This resilience is primarily due to farmers' careful management of soil, cultivated crop biodiversity, and care over the status of water resources. Pro-ecological and sustainable farming practices entail regular soil nourishment with natural fertilizers and reduced use of pesticides and mineral fertilizers, as well as employing various practices to maintain soil fertility. They also aid the preservation of soil biodiversity [14].

Contemporary development trends demand innovative solutions not only in agricultural technology but also in the precise dosage of water. On a global scale, the primary criterion for applying irrigation is the one associated with the climate conditions, particularly the quantity and distribution of precipitation. Pursuant to the global warming theory, manifested primarily by rising air temperatures, the frequency of droughts across various geographical regions is expected to increase, while research findings indicate that these changes are already in progress [15].

Thus, irrigation aims to maintain optimal conditions for plant growth, whereas its effectiveness is driven by the intensity and quantity of practices employed in different farming systems, as well as the location and amount of water in the soil [16]. Calow et al. [17] also emphasized the importance of integrated field irrigation management in ensuring optimal quality parameters of both the soil and agricultural crops. Soil water content directly and indirectly influences wheat growth. Its direct influence is related to the water available around the roots through which it is absorbed, while its indirect impact is associated with soil properties [18]. Water acts as a medium for transporting mineral particles, dissolved matter, and organic matter within the soil profile. The migration of water through soil pores affects the distribution of nutrients, minerals, and contaminants, thereby influencing soil fertility and the quality of the soil environment. Furthermore, water impacts physical soil properties such as texture, structure, and porosity, which in turn affect infiltration, water storage, and water availability for uptake by plants [19]. Understanding the long-term effects of irrigation on the fundamental characteristics and quality of soil is essential for sustainable land management and agricultural production, particularly in arid regions where water availability is limited [20].

The purpose of the study was to determine the effect of the conventional and integrated (economical) system versus the control (organic system), and irrigation levels on chemical, physical, and biological properties of the soil and soil enzyme activity under spring wheat crops. The hypothesis posited that the most favorable soil quality parameters would be provided by the integrated system. It was also assumed that multiple irrigation of wheat would provide the most favorable indicators of soil quality and health.

2. Materials and Methods

2.1. Experiment Design and Field Management

In the years 2020–2022, a strict field experiment was carried out with the cultivation of spring wheat (*Triticum aestivum* L.—cv. 'Monsun') at the Czesławice Experimental Farm (51°30′ N; 22°26′ E Lubelskie Voivodeship, Poland—Figure 1).

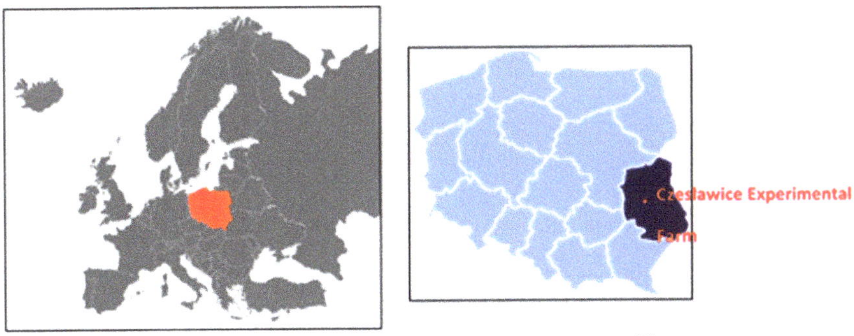

(A) (B)

Figure 1. Location of the research area: (**A**) Location of Poland against the background of Europe; (**B**) Location of the Czesławice Experimental Farm in the Lublin region in Poland.

The experiment was conducted on an area of 1350 m^2 (27 plots), with an area of a single plot of 50 m^2 (5 m × 10 m), in 3 replications, in a split-block design. The soil in the experiment belonged to quality class II (good wheat soil complex), defined as a loess-derived Luvisol (PWsp) with the grain size distribution of silt loam. Spring wheat served as a test plant in the experiment. The composition of available forms of macronutrients in the soil before the experiment was at an average level in each farming system; its organic C content fitted within the range of 0.94–0.97%, and its pH was 6.5. Hence, in each year of the study, the fields intended for the experiment were homogeneous in terms of the chemical composition of the soil, which ensured objective conditions for determining the effects of individual farming systems on changes in soil quality (Table 1).

Table 1. Soil characteristics prior to the experiment establishment in 2019–2021.

Specification	Organic System (Control)	Integrated System	Conventional System
Organic C (%)	0.95–0.97	0.94–0.96	0.95–0.96
Total N (%)	0.08–0.09	0.08–0.09	0.08–0.09
P (mg kg^{-1})	128.2–129.3	127.4–129.1	127.5–128.8
K (mg kg^{-1})	215.4–217.6	217.5–218.1	216.7–218.5
Mg (mg kg^{-1})	68.8–69.2	68.7–69.1	68.6–68.9
Soil pH (1M KCl)	6.5	6.5	6.5

The following factors were considered in the study: 1. Farming system: organic (control)—without pesticides or NPK fertilization; conventional—with the use of the 100% recommended doses of pesticides and NPK mineral fertilizers; and integrated (economical)—with the use of reduced (by 50%) recommended doses of pesticides and mineral fertilizers. 2. Irrigation of crops: control treatment—no irrigation, double irrigation (at the 2–3 leaf stage and at the shooting stage); and multiple irrigation, resulting from monitoring the drought status in the arable field. The water content of the soil was monitored throughout the growing season of wheat based on readings from moisture sensors mounted at two depth levels in the soil (0–15 and 15–20 cm), which enabled establishing the actual water content of the soil and replenishing water deficits by means of irrigation. A tubular irrigation system (drip irrigation) was used to irrigate the wheat crops, ensuring even distribution of water across the experimental plots. The readings of soil moisture sensors did not differ significantly due to similar amounts of precipitation in the individual growing seasons (434 mm—2020; 419 mm—2021; 402 mm—2022; multi-year average of 1985–2015—589 mm). Hence, similar volumes of water were used for irrigation in 2020–2022, i.e.,

a. double irrigation:

1. 250,000 L water/ha (2020); 240,000 L water/ha (2021); 260,000 L water/ha (2022)
2. 200,000 L water/ha (2020); 190,000 L water/ha (2021); 210,000 L water/ha (2022)

The total volume of water used was 450,000 L water/ha (2020); 430,000 L water/ha (2021); 470,000 L water/ha (2022).

b. multiple irrigation:

1. 250,000 L water/ha (2020); 230,000 L water/ha (2021); 250,000 L water/ha (2022)
2. 100,000 L water/ha (2020); 110,000 L water/ha (2021); 110,000 L water/ha (2022)
3. 200,000 L water/ha (2020); 190,000 L water/ha (2021); 210,000 L water/ha (2022)
4. 250,000 L water/ha (2020); 240,000 L water/ha (2021); 260,000 L water/ha (2022)
5. 150,000 L water/ha (2020); 140,000 L water/ha (2021); 160,000 L water/ha (2022)

The total volume of water used was: 950,000 L water/ha (2020); 910,000 L water/ha (2021); 990,000 L water/ha (2022).

Mineral fertilization applied in the conventional system (in kg ha^{-1}) included the following NPK doses: N = 80, P = 60, and K = 100, whereas in the integrated system, it included the following NPK doses: N = 40, P = 20, K = 30, and Mg = 50. Mineral fertilization with N was applied in the form of 34% ammonium nitrate, 17% nitrogen (N) in its nitrate form (NO_3), and 17% nitrogen (N) in its ammonium form (NH_4), whereas P was applied in the form of 46% granulated triple superphosphate (in the P_2O_5 form), while K was applied in the form of 50% potassium salt (in K_2O form). In the organic system, no fertilization allowed in this system was applied, which enabled precise capture of the effects of full (100%) and reduced (50%) doses of NPK mineral fertilizers (in the conventional and integrated system) on the soil quality of the test plant (spring wheat).

Wheat was cultivated in the conventional (ploughing) soil tillage system. The following plant protection treatments were applied in the conventional system (100% of recommended doses): Omnix 025 FS seed dressing (a.s. fludioxonil)—200 mL 100 kg^{-1} of grain; Chwastox Trio 390 SL herbicide (a.s. MCPA + mecoprop-P + dicamba) at 3 L ha^{-1}, Puma Universal 69 EW herbicide (a.s. fenoxaprop-P-ethyl) at 1 L ha^{-1}, Glora 633 EC fungicide (a.s. fenpropidin + prochloraz) at 1 L ha^{-1}, and Decis Mega 50 EW insecticide (a.s. deltamethrin) at 1 L ha^{-1}.

Treatments applied in the integrated (economical) system entailed 50% recommended doses of plant protection agents administered in the same terms as in the conventional system., i.e., mechanical weed eradication was performed twice in all farming systems—before the emergence and at the onset of tillering of wheat. During the study period, the sowing and harvesting dates of spring wheat were the same in the conventional, integrated, and ecological systems (ranging from 19 to 22 April—sowing, 20 to 24 August—harvesting). The sowing rate of spring wheat was the same in all systems and amounted to 200 kg ha^{-1}. Straw remaining after the spring wheat harvest was removed from the field in each year of the study (it was not ploughed under).

2.2. Cost Estimate for Irrigation, Fertilization and Chemical Protection of Spring Wheat

I. Spring wheat irrigation costs (average from the study period in each farming system):
Cost of 1 m^3 water = 1000 L water = USD 0.62

a. double irrigation:

1. 250,000 L water/ha
2. 200,000 L water/ha

Total: 450,000 L water/ha = USD 281.2

b. multiple irrigation:

1. 243,000 L water/ha
2. 107,000 L water/ha
3. 200,000 L water/ha
4. 250,000 L water/ha
5. 150,000 L water/ha

Total: 950,000 L water/ha = USD 593.7

II. Spring wheat fertilization costs (mean for the study period):
Conventional system (100% NPK doses):
ammonium nitrate – 80 kg ha^{-1} – cost = USD 154.0/ha
granulated triple superphosphate – 60 kg ha^{-1} – cost = USD 95.0/ha
potassium salt – 100 kg ha^{-1} – cost = USD 169.0/ha
Total fertilization cost in conventional system = USD 418.0/ha
Integrated (economical) system (50% NPK doses):
ammonium nitrate – 40 kg ha^{-1} – cost = USD 77.0/ha
granulated triple superphosphate – 30 kg ha^{-1} – cost = USD 47.5/ha
potassium salt – 50 kg ha^{-1} – cost = USD 84.5/ha
Total fertilization cost in integrated (economical) system = USD 209.0/ha
Organic system (control) (without NPK fertilization):
Total fertilization cost in organic system (control) = USD 0.0/ha

III. Costs of chemical protection of spring wheat (mean for the study period):
Conventional system (100% of pesticide doses)
Omnix 025 FS seed dressing 200 mL 100 kg^{-1} of grain − cost = USD 22.5/ha
Chwastox Trio 390 SL herbicide 3 L ha^{-1} − cost = USD 28.5/ha
Puma Universal 69 EW herbicide 1 L ha^{-1} − cost = USD 32.0/ha
Glora 633 EC fungicide 1 L ha^{-1} − cost = USD 25.5/ha
Decis Mega 50 EW insecticide 1 L ha^{-1} − cost = USD 41.0/ha
Total cost of chemical protection in conventional system = USD 149.5/ha
Integrated (economical) system (50% of pesticide doses)
Omnix 025 FS seed dressing 100 mL 100 kg^{-1} of grain − cost = USD 11.2/ha
Chwastox Trio 390 SL herbicide 1.5 L ha^{-1} − cost = USD 14.2/ha
Puma Universal 69 EW herbicide 0.5 L ha^{-1} − cost = USD 16.0/ha
Glora 633 EC fungicide 0.5 L ha^{-1} − cost = USD 12.8/ha
Decis Mega 50 EW insecticide 0.5 L ha^{-1} − cost = USD 20.5/ha
Total cost of chemical protection in an integrated (economical) system = USD 74.8/ha
Organic system (control) (without using pesticides)
Total cost of chemical protection in organic system (control) = USD 0.0/ha

Table 2 presents the economic profitability of spring wheat cultivation subjected to irrigation treatments (double irrigation and multiple irrigation) in various farming systems.

Table 2. Spring wheat grain yield, yield value, and cultivation cost estimates in various farming and irrigation systems—mean values of the three-year study period.

Irrigation	Organic System (Control)				Integrated (Economical) System				Conventional System			
	Grain Yield (t ha^{-1})	Grain Yield Value (USD/ha) *	Total Costs Incurred (USD/ha)	Profit (USD/ha)	Grain Yield (t ha^{-1})	Grain Yield Value (USD/ha) *	Total Costs Incurred (USD/ha)	Profit (USD/ha)	Grain Yield (t ha^{-1})	Grain Yield Value (USD/ha) *	Total Costs Incurred (USD/ha)	Profit (USD/ha)
NI	2.1	630.0	0.0	+630.0	4.5	1352.0	283.8	+1066.0	6.0	1800.0	567.5	+1232.0
2I	3.0	900.0	281.2	+618.8	6.3	1890.0	565.0	+1325.0	7.6	2280.0	848.8	+1431.0
MI	4.1	1230.0	593.7	+636.2	7.5	2250.0	877.5	+1372.0	8.8	2640.0	1161.0	+1478.0

NI—no irrigation, 2I—double irrigation, MI—multiple irrigation; * the calculations assumed an average price of wheat grain in Poland in 2020–2022 of USD 300/ton.

The data in Table 2 show that the efficiency of water used in spring wheat irrigation depended on the farming system (NPK mineral fertilization and pesticide doses). The highest profit measured by net yield value was found in the case of repeated irrigation of wheat in the conventional system (100% NPK fertilizer and pesticide doses). However, a slightly smaller profit was brought by irrigation of wheat in the integrated (economical) system, in which 50% fertilizer and pesticide doses were used. In the control objects (organic system), the smallest profit resulted from the lack of fertilizer and pesticide use, and consequently, from low spring wheat yields. The use of irrigation in the control plots, despite the increase in wheat yields, was economically unjustified (the profit was lower than in the absence of irrigation). On the other hand, in the integrated system, and especially in the conventional system, high costs of irrigation and chemical protection compensated for high grain yields of spring wheat.

2.3. *Analyses of Soil*

In order to determine the coupled effect of farming systems and irrigation strategies on the physicochemical and biological properties and enzyme activity of the soil under wheat crops, soil samples were collected from a layer of 0–25 cm and analyzed for the selected soil condition parameters. The soil samples were collected using a soil sampling tube from an area of 0.20 m^2 of each plot, in three replicates for each experimental treatment, 1 day after spring wheat harvest. Three soil samples were collected from each plot during the research season and combined into a collective sample for laboratory analysis. In total, 81 soil samples were collected from 27 plots in one season (243 soil samples were collected during the 3-year research period).

2.4. Chemical Properties of Soil

The following parameters were determined:

C organic content was determined using a carbon analyzer (SDCHN435), total nitrogen content was analyzed by the Kjeldahl method, the content of available forms of phosphorus and potassium was assayed with the Egner–Riehm method, magnesium content was analyzed by means of atomic absorption spectrometry (AAS), and micronutrient content (B, Cu, Mn, Zn) by flame photometry; finally, the total sorption capacity (cmol (+) kg^{-1}) of the soil was determined with Kappen's method.

2.5. Physical Properties of Soil

Soil moisture content and bulk density as well as total and capillary porosity in the layers of 5–10 and 15–20 cm were determined in two replicates per plot using a 100 mL cylinder. Soil total porosity was determined by the pycnometric method. Soil capillary porosity was established by capillary infiltration method. Soil compaction was examined using an electronic probe (penetrometer) in the 0–30 cm layer, every 5 cm in five replicates per plot.

2.6. Enzymatic Activity of Soil

The enzymatic activity of the soil was established based on determinations of activities of five enzymes, i.e., dehydrogenase with the Thalmann method [21], acid phosphatase and alkaline phosphatase with the Tabatabai and Bremner method [22], urease with the Zantua and Bremner method [23], and protease with the Ladd and Butler method [24].

Dehydrogenase activity (in mg TPF kg^{-1} d.m. of soil d^{-1}) was determined in 5 g soil samples using 2,3,5-triphenyltetrazolium chloride as a substrate, incubated for 48 h at 30 °C in 0.2 M trishydroxymethylaminomethane-HCl buffer (pH 7.4).

The activity of acid phosphatase and alkaline phosphatase, expressed in mg PNP kg^{-1} d.m. of soil h^{-1}, was determined in 1 g of soil samples using disodium p-nitrophenylpho sphate as a substrate, incubated for 1 h at 37 °C in universal buffer at pH 6.5 for acid phosphatase and pH 11 for alkaline phosphatase.

Urease activity, expressed as mg N-NH4 kg^{-1} dm of soil 18 h^{-1}, was determined in 10 g of soil samples using a urea solution substrate and incubated at 37 °C for 18 h.

Protease activity, expressed in mg tyrosine kg^{-1} d.m. soil h^{-1}, was determined in 2 g of soil samples using casein substrate, incubated at 50 °C in 0.2 M Tris-HCl buffer (pH 8.0) for 1 h.

2.7. Biological Properties of Soil

The counts of useful fungi *Trichoderma* ssp. and parasitic fungi *Fusarium* ssp. were determined with the plate method according to the procedure described by Foght and Aislabie [25]. The total count of fungi was determined after 3 days of incubation at a temperature of 270 °C, on Martin's culture medium [26], with antibiotics added to reduce bacterial contamination [27]. To this end, 50 mg of streptomycin and 4 mL of a 1% Bengal rose solution were added to the culture medium before it had been spread onto plates. Once the total count of fungi had been determined, they were isolated from the plates on the same culture media, and the grown colonies were purified to pure cultures. The isolated fungi were identified based on their morphological traits according to the key posited by Domsch et al. [28].

The total number of actinobacteria was determined with the plate method using culture media with the addition of nystatin (50 µg·mL) according to the method by Wallace and Lochhead [29]. The material was incubated for 14 days at a temperature of 28 °C, and the growing colonies of bacteria cultured on the particular media were counted after 3–5 days using a colony counter [30].

2.8. Statistical Analysis

Statistical analysis of the study results was conducted with Statistica PL 13.3 software (TIBCO Software Inc., Palo Alto, CA, USA). The Tukey's Honestly Significant Difference (HSD) test was deployed to establish the significance of differences between the values, at an adopted significance level of $p \geq 0.05$. Due to a lack of significant differences between subsequent growing seasons (2020–2022), the results presented in the tables (Tables 2–7) are mean values of the three-year study period. In addition, standard deviation (±SD) was provided for all mean values presented in tables.

Table 3. Contents of total nitrogen, phosphorus, potassium, magnesium, and organic C in thesoil—mean values for the three-year study period.

Farming System	Wheat Irrigation	Total N (%)	P (mg kg^{-1})	K (mg kg^{-1})	Mg (mg kg^{-1})	C-Organic (%)
Organic	NI	0.06 ±0.004	121.5 ±0.9	201.8 ±1.0	63.5 ±0.4	0.80 ±0.013
	2I	0.07 ±0.002	121.7 ±1.0	203.2 ±1.1	70.1 ±0.4	0.87 ±0.023
	MI	0.09 ±0.002	122.2 ±1.1	209.3 ±1.3	71.3 ±0.6	0.95 ±0.033
Mean		0.07	121.8	204.8	68.3	0.87
Integrated	NI	0.07 ±0.001	124.4 ±1.2	212.2 ±1.4	56.8 ±0.3	0.83 ±0.031
	2I	0.09 ±0.002	125.6 ±0.8	216.3 ±1.4	65.7 ±0.4	0.98 ±0.02
	MI	0.10 ±0.002	126.0 ±0.9	223.1 ±1.5	71.0 ±0.5	1.11 ±0.036
Mean		0.09	125.3	217.2	64.5	0.97
Conventional	NI	0.08 ±0.003	127.1 ±1.2	244.3 ±1.6	52.2 ±0.5	0.77 ±0.014
	2I	0.11 ±0.005	128.3 ±1.4	258.9 ±1.7	60.3 ±0.6	0.80 ±0.022
	MI	0.15 ±0.006	131.4 ±1.3	269.6 ±1.9	70.4 ±0.7	0.91 ±0.037
Mean		0.11	128.9	257.6	61.0	0.83
HSD ($p \geq 0.05$) for farming system (A)		0.016	7.06	12.11	3.67	0.095
HSD ($p \geq 0.05$) for wheat irrigation (B)		0.017	n.s.	14.38	3.79	0.098
HSD ($p \geq 0.05$) for (A × B) interaction		0.023	n.s.	19.37	4.53	0.124

NI—no irrigation, 2I—double irrigation, MI—multiple irrigation, ±SD—standard deviation, n.s.—not significant.

Table 4. Contents of boron, copper, manganese, and zinc, and total sorption capacity of the soil under spring wheat crops—mean values for the three-year study period.

Farming System	Wheat Irrigation	B (mg kg^{-1})	Cu (mg kg^{-1})	Mn (mg kg^{-1})	Zn (mg kg^{-1})	Total Sorption Capacity of Soil (cmol (+) kg^{-1})
Organic	NI	2.36 ±0.04	6.86 ±0.08	201 ±2.7	8.44 ±0.08	33.5 ±0.6
	2I	2.40 ±0.03	6.95 ±0.10	211 ±3.5	8.62 ±0.09	35.7 ±0.8
	MI	2.56 ±0.05	7.27 ±0.11	218 ±4.2	8.91 ±0.08	36.9 ±1.0
Mean		2.44	7.03	210	8.66	35.3
Integrated	NI	2.30 ±0.04	6.64 ±0.08	184 ±2.5	8.14 ±0.04	33.4 ±0.9
	2I	2.33 ±0.04	6.72 ±0.09	188 ±2.2	8.26 ±0.05	38.0 ±1.1
	MI	2.38 ±0.05	6.79 ±0.07	193 ±3.0	8.40 ±0.06	42.0 ±1.1
Mean		2.34	6.72	188	8.27	37.8
Conventional	NI	2.09 ±0.02	6.41 ±0.06	168 ±2.1	8.01 ±0.05	34.3 ±0.8
	2I	2.11 ±0.03	6.47 ±0.07	171 ±2.2	8.09 ±0.06	35.6 ±0.7
	MI	2.16 ±0.03	6.54 ±0.08	177 ±2.4	8.11 ±0.07	39.3 ±1.2
Mean		2.12	6.47	172	8.07	36.4
HSD ($p \geq 0.05$) for farming system (A)		0.199	0.515	14.4	0.563	2.46
HSD ($p \geq 0.05$) for wheat irrigation (B)		n.s.	n.s.	8.9	0.461	2.17
HSD ($p \geq 0.05$) for interaction (A × B)		0.154	n.s.	n.s.	n.s.	2.59

NI—no irrigation, 2I—double irrigation, MI—multiple irrigation, ± SD—standard deviation, n.s.—not significant.

Table 5. Moisture content and total and capillary porosity of the soil—mean values for the three-year study period.

Farming System	Wheat Irrigation	Soil Moisture Content (%)		Total Soil Porosity (%) in the 0–25 cm Layer	Capillary Soil Porosity (%) in the 0–25 cm Layer
		0–20 cm	20–35 cm		
Organic	NI	5.62 ±0.016	5.32 ±0.014	42.2 ±1.06	34.0 ±0.62
	2I	14.87 ±0.028	16.21 ±0.030	41.8 ±1.04	33.8 ±0.58
	MI	15.56 ±0.034	19.32 ±0.039	41.1 ±1.01	33.0 ±0.54
Mean		12.01	13.61	41.7	33.6

Table 5. *Cont.*

Farming System	Wheat Irrigation	Soil Moisture Content (%)		Total Soil Porosity (%) in the 0–25 cm Layer	Capillary Soil Porosity (%) in the 0–25 cm Layer
		0–20 cm	20–35 cm		
Integrated	NI	5.43 ±0.018	5.15 ±0.015	40.6 ±0.94	31.3 ±0.49
	2I	14.65 ±0.037	16.05 ±0.039	39.7 ±0.97	30.7 ±0.47
	MI	15.17 ±0.041	18.81 ±0.055	38.7 ±1.02	30.4 ±0.38
Mean		11.75	13.33	39.7	30.8
Conventional	NI	5.24 ±0.028	5.08 ±0.019	38.2 ±1.09	30.9 ±0.40
	2I	14.10 ±0.033	15.90 ±0.038	37.5 ±1.11	30.6 ±0.35
	MI	14.98 ±0.041	17.79 ±0.052	36.5 ±1.10	30.0 ±0.29
Mean		11.44	12.92	37.4	30.5
HSD ($p \geq 0.05$) for farming system (A)		n.s.	0.685	2.28	1.86
HSD ($p \geq 0.05$) for wheat irrigation (B)		0.865	0.987	n.s.	n.s.
HSD ($p \geq 0.05$) for interaction (A × B)		n.s.	n.s.	n.s.	n.s.

NI—no irrigation, 2I—double irrigation, MI—multiple irrigation, ± SD—standard deviation, n.s.—not significant.

Table 6. Density and compaction of the soil under spring wheat crops—mean values for the three-year study period.

Farming System	Wheat Irrigation	Soil Density (g cm^{-3}) in the 0–25 cm Layer	Soil Compaction (MPa) in the 0–25 cm Layer
Organic	NI	1.41 ± 0.014	1.51 ± 0.017
	2I	1.58 ± 0.018	1.67 ± 0.022
	MI	1.64 ± 0.026	1.97 ± 0.032
Mean		1.54	1.72
Integrated	NI	1.46 ± 0.013	1.54 ± 0.019
	2I	1.54 ± 0.022	1.75 ± 0.025
	MI	1.65 ± 0.024	1.95 ± 0.027
Mean		1.55	1.75
Conventional	NI	1.51 ± 0.016	1.59 ± 0.019
	2I	1.61 ± 0.019	1.88 ± 0.032
	MI	1.65 ± 0.021	2.09 ± 0.034
Mean		1.59	1.85
HSD ($p \geq 0.05$) for farming system (A)		n.s.	0.096
HSD ($p \geq 0.05$) for wheat irrigation (B)		0.095	0.142
HSD ($p \geq 0.05$) for interaction (A × B)		n.s.	0.104

NI—no irrigation, 2I—double irrigation, MI—multiple irrigation, ± SD—standard deviation, n.s.—not significant.

Table 7. Enzymatic activity of the soil under spring wheat crops—mean values for the three-year study period.

Farming System	Wheat Irrigation	Dehydrogenase (mg TPF kg^{-1} d.m.)	Acid Phosphatase(mg PNP kg^{-1} d.m.)	Alkaline Phosphatase(mg PNP kg^{-1} d.m.)	Urease (mg $N\text{-}NH_4$ kg^{-1} d.m.)	Protease (mg Tyrosine kg^{-1} d.m.)
Organic	NI	2.07 ± 0.029	61.07 ± 1.63	73.64 ± 2.19	35.32 ± 1.17	12.14 ± 0.08
	2I	2.49 ± 0.039	69.03 ± 1.66	75.85 ± 2.24	41.03 ± 1.45	14.80 ± 0.07
	MI	2.60 ± 0.052	69.72 ± 1.72	78.05 ± 2.34	44.00 ± 1.52	13.40 ± 0.06
Mean		2.42	66.61	75.85	40.12	13.44
Integrated	NI	2.62 ± 0.038	64.63 ± 1.83	82.88 ± 2.19	56.00 ± 1.73	16.47 ± 0.09
	2I	2.87 ± 0.027	71.44 ± 1.88	84.10 ± 2.27	63.93 ± 1.81	19.64 ± 0.16
	MI	3.25 ± 0.045	75.52 ± 1.91	85.10 ± 2.32	66.69 ± 1.87	17.42 ± 0.12
Mean		2.91	70.53	84.03	62.20	17.84
Conventional	NI	1.71 ± 0.026	55.18 ± 1.71	66.30 ± 1.99	24.74 ± 1.08	8.63 ± 0.06
	2I	2.28 ± 0.041	59.63 ± 1.75	67.77 ± 2.08	36.73 ± 1.12	10.74 ± 0.08
	MI	2.63 ± 0.047	59.72 ± 1.78	68.69 ± 2.11	31.72 ± 1.09	9.23 ± 0.07
Mean		2.21	58.18	67.59	31.06	9.53
HSD ($p \geq 0.05$) for farming system (A)		0.207	3.901	4.453	6.021	2.146
HSD ($p \geq 0.05$) for wheat irrigation (B)		0.241	4.024	4.391	5.442	0.944
HSD ($p \geq 0.05$) for interaction (A × B)		0.371	4.045	n.s.	5.981	2.189

NI—no irrigation, 2I—double irrigation, MI—multiple irrigation, ± SD—standard deviation, n.s.—not significant.

3. Results

3.1. Chemical Properties of Soil

The total nitrogen content of the soil under spring wheat crops was significantly affected both by experimental factors and their interaction. Regardless of irrigation level, the highest total N content of the soil was determined under conditions of the conventional farming system—0.11% on average, i.e., being 0.04 percentage points (p.p.) higher than in the organic system and 0.02 p.p. higher than in the integrated system. In each farming system, the multiple irrigation of wheat crops contributed to the highest content of total nitrogen in the soil, compared to the control treatment without irrigation, whereas in the case of the conventional and organic systems—also compared to the double irrigation. Significantly the highest total N content (0.15%) was determined under conditions of the conventional system × multiple irrigation interaction (Table 3).

The content of phosphorus in the soil depended on the farming system and reached 128.9 mg kg^{-1} in the conventional system, thus being ca. 6% higher compared to that determined in the organic system. In the integrated system, the phosphorus content of the soil showed only an ascending trend compared to the organic system. Irrigation level had no statistically significant effect on P content of the soil; only a tendency for the highest phosphorus content was noted in each farming system upon multiple irrigation (Table 3).

Potassium content of the soil differed significantly between the farming systems. Under conventional system conditions, it reached 257.6 mg kg^{-1} on average, thus being ca. 16% higher than in the integrated system and as much as ca. 21% higher than in the organic system. Also worth noting was the content of K in the soil from the integrated system, which was significantly higher than in the soil from the organic system—by ca. 6% on average. Wheat crop irrigation (double or multiple) elicited a statistically significant effect on potassium content of the soil only in the conventional system—with ca. 6%

higher potassium content determined upon double irrigation and ca. 10% higher content noted under conditions of multiple irrigation, compared to the control without irrigation. Among all experimental treatments, significantly the highest content of potassium in the soil (reaching 269.6 mg kg^{-1}), was recorded under conditions of the conventional system × multiple irrigation interaction (Table 3).

Different correlations were observed between the farming systems and magnesium (Mg) content of the soil. The most beneficial in this respect turned out to be the organic system, where the mean Mg content of the soil was at 68.3 mg kg^{-1}, being nearly 6% higher than in the integrated system and ca. 11% higher than in the conventional system. Irrespective of the farming system, no irrigation of wheat crops resulted in a significantly lower content of Mg in the soil compared to both multiple and double irrigation. Significantly the lowest Mg content of the soil (barely 52.2 mg kg^{-1}) was determined under conditions of the conventional system × no irrigation interaction (Table 3).

The integrated system had the most beneficial effect on organic C content of the soil under spring wheat, which reached 0.97% on average and was significantly higher compared to the organic and conventional systems (by ca. 11% and ca. 15%, respectively). Irrespective of the farming system, the multiple irrigation had a significant effect on organic C content of the soil compared to the control treatment, by 16% (organic system), 15% (integrated system), and 15% (conventional system). Significantly the highest content of organic carbon in the soil (1.11%) was noted under conditions of the integrated system × multiple irrigation interaction (Table 3).

The content of boron in the soil under spring wheat crops was significantly the lowest in the conventional system, and lower by ca. 10% compared to the integrated system and by ca. 13% compared to the organic system. Crop irrigation treatments had no statistically significant effect on B content of the soil. Only a tendency for its highest content was noted along with irrigation intensification. In turn, a significant effect was demonstrated for the interaction between the organic system and multiple irrigation, where B content of the soil reached 2.56 mg kg^{-1} (Table 4).

The content of copper in the soil was significantly modified only by the farming systems, with its significantly highest content determined in the organic system, compared to the conventional system. The multiple irrigation of spring wheat crops caused only a tendency for a higher Cu content of the soil, whereas the effect of the double irrigation on its content was even less noticeable (Table 4).

Both experimental factors significantly modified the content of manganese in the soil, which reached 210 mg kg^{-1} in the soil from the organic system and was ca. 11% and ca. 18% higher than in the soil from the integrated and conventional systems, respectively. At the same time, the Mn content in the integrated system was higher than in the conventional system. The multiple crop irrigation applied in each farming system caused a significant increase in Mn content of the soil, compared to the strategy of no irrigation (control), and the same effect was observed upon double irrigation applied in the organic system (Table 4).

The content of zinc in the soil under spring wheat crops was the highest in the organic system (8.66 mg kg^{-1} on average). The conditions provided in this system resulted in a significantly higher content of this element compared to the conventional system (by approx. 7%). In turn, Zn content of the soil from the integrated system showed statistically insignificant (intermediate) values compared to the conventional and organic systems. In each farming system, the double irrigation and, particularly, the multiple irrigation of wheat crops contributed to a significant increase in zinc content of the soil, compared to no irrigation strategy (Table 4).

The integrated system promoted the total sorption capacity of the soil, which was ca. 7% higher, on average, compared to that determined in the organic system. In turn, the total sorption capacity of the soil noted under conditions of the conventional system did not differ significantly when compared to both the integrated and organic system. Considering changes in this parameter under the influence of crop irrigation frequency, its highest significant values were noted upon multiple irrigation, compared to no irrigation, but also compared to the double irrigation (in the conventional and integrated systems). The highest significant total sorption capacity of the soil (42.0 (cmol (+) kg^{-1}) was determined under conditions of the integrated system × multiple irrigation interaction (Table 4).

3.2. *Physical Properties of Soil*

Farming systems caused no significant differences in the moisture content of the soil determined in a soil layer of 0–20 cm. Only a tendency was noted for a higher moisture content under conditions of the organic system. The moisture content in this soil layer was highly significantly affected by wheat crop irrigation level. On plots with no irrigation (control), the moisture content of the soil fitted within the range of 5.24–5.62% in all farming systems. In turn, the double irrigation caused a significant increase in the moisture content in this soil layer, to 14.10–14.87%, whereas under multiple irrigation, reached 14.98% in the conventional system, 15.17% in the integrated system, and 15.56% in the organic system. Differences determined in the moisture content of this soil layer between the double and multiple irrigation strategies were statistically insignificant (Table 5).

The moisture content analyzed in the deeper soil layer (20–35 cm) was statistically significantly influenced by the farming systems. The highest moisture content, reaching 13.61% on average, was determined in the soil from the organic system. It was higher by ca. 0.28 p.p. compared to that noted in the soil from the integrated system and significantly higher—by ca. 0.69 p.p.—compared to the soil from the conventional system. Irrigation of wheat crops resulted in a significant increase in the moisture content of the soil from all farming systems compared to the soil from control plots (without irrigation). In the case of the double irrigation, the moisture content increased by ca. 10.82–10.90 p.p., whereas in the case of the multiple irrigation—by ca. 14.00 p.p. in the organic system, 13.66 p.p. in the integrated system, and 12.71 p.p., in the conventional system. Analyses conducted in the 20–35 cm soil layer also demonstrated significant differences in its moisture content between the double and multiple irrigation strategies, reaching, on average, 3.11 p.p. in the organic system, 2.76 p.p. in the integrated system, and 1.89 p.p. in the conventional system (Table 5).

The total porosity of the soil analyzed in its 0–25 cm layer was significantly influenced by the farming system. Its highest value (41.1%) was determined in the soil from the organic system and was slightly lower (39.7%) in the soil from the integrated system. The total porosity of the soil from the conventional system was significantly lower (by ca. 2.3 p.p.) compared to that noted in the soil from the integrated system and ca. 4.3 p.p. lower compared to the soil from the organic system. Wheat crop irrigation had no statistically significant effect on the total porosity of the soil. Only a tendency toward lower values was noted along with irrigation intensification (with the lowest total porosity determined under conditions of multiple irrigation) (Table 5).

The correlations observed for the capillary porosity of the soil analyzed in its 0–25 cm layer differed slightly from those noted for the total porosity. Significantly the lowest capillary porosity of the soil was determined under conditions of conventional and integrated systems compared to the organic system (lower by 3.1 p.p. and 2.8 p.p., respectively). The double and, particularly, the multiple irrigation of wheat crops contributed to diminished capillary porosity of the soil; however, differences noted between the irrigation strategies were statistically insignificant (Table 5).

The farming systems caused no significant differences in soil density. Only a tendency was noted for greater soil density under conditions of the conventional system. Different was the case with wheat crop irrigation strategies, where even double irrigation caused a significant increase in soil density (under conditions of organic and conventional systems), compared to the control (no irrigation), while the multiple irrigation strategy produced the same effect in all farming systems, where soil density was higher by ca. 14% (organic system), ca. 12% (integrated system), and ca. 9% (conventional system) than in the control (Table 6).

Regardless of wheat crop irrigation, soil compaction determined under conditions of the conventional system (reaching 1.85 MPa on average) was significantly higher compared to that found under conditions of integrated and organic systems (by ca. 5% and ca. 7%, respectively). Even double irrigation of wheat crops caused a significant increase in soil compaction in each farming system, whereas multiple irrigation resulted in significantly greater soil compaction (by ca. 23–24%) not only compared to the control soil (without irrigation) but also compared to the soil from double-irrigated plots (conventional system—soil compaction increase by ca. 10%, integrated system—by ca. 10%, and organic system—by ca. 15%). Significantly the highest soil compaction (2.09 MPa on average) was determined under conditions of the conventional system × multiple irrigation interaction (Table 6).

3.3. Enzymatic Activity and Biological Properties of Soil

Enzymatic activity of the soil under spring wheat crops was found to be significantly affected by the adopted experimental factors (Table 7). All enzymes analyzed in the present study exhibited significantly the highest activities under conditions of the integrated system, compared to the organic system, and particularly compared to the conventional system. The activity of dehydrogenase determined in the integrated system as compared to the organic and conventional systems was significantly higher, by 26% and 32%, respectively, acid phosphatase—by approx. 12% and 23%, alkaline phosphatase—by approx. 10% and 20%, urease—by approx. 36% and 100%, and protease—by approx. 23% and 100%. The irrigation of spring wheat crops had varied effects on the enzymatic activity of the soil. In the case of dehydrogenase, significantly the most beneficial effect was caused by multiple irrigation (activity increase by ca. 20–35%), but also by double irrigation (activity increase by ca. 17–25%), compared to the control (no irrigation). In the case of acid phosphatase, multiple irrigation increased its activity compared to the control by ca. 8–13% (conventional and organic system) and 15% (integrated system), whereas double irrigation—by ca. 7% (conventional system) and 12% (integrated and organic system). In turn, differences in the activity of alkaline phosphatase under the influence of multiple and double irrigation were statistically insignificant. In contrast, the effect of multiple irrigation on the activity of these enzymes was statistically significantly positive compared to the control (no irrigation) only in the organic system—enzyme activity increased by ca. 6%. Activities of urease were at similar (statistically insignificant) levels in the soil from plots with multiple and double irrigation, but higher (a significantly positive correlation) compared to the soil from control plots without irrigation, i.e., by ca. 18% (conventional system), 15% (integrated system), and 17% (organic system). Significantly the lowest urease activity was determined under conditions of the conventional system × no irrigation interaction. Protease activity was significantly the highest under conditions of double irrigation in each farming system compared to the soil from plots with multiple irrigation (by ca. 10–15%), but most of all compared to the soil from control plots (without irrigation—by ca.

17–20%). The integrated system × multiple irrigation interaction promoted significantly the highest activities of dehydrogenase and acid phosphatase, whereas the integrated system × double irrigation interaction caused statistically the highest activity of protease. In turn, the lack of irrigation in the interaction with the conventional system contributed to significantly the lowest activity of urease (Table 7).

The counts of beneficial fungi in the soil under spring wheat crops were similar under conditions of the organic and integrated systems (a statistically insignificant difference) but significantly higher than those determined in the conventional system, i.e., by ca. 19% and ca. 15%, respectively. The counts of the beneficial fungi were significantly modified by irrigation strategies. Generally, significantly the lowest count of beneficial fungi was determined in the soil from plots without irrigation, compared to the plots with double irrigation and, particularly, compared to the plots with multiple irrigation (where the difference reached ca. 15% in all farming treatments). Multiple irrigation applied in the conventional and integrated systems also contributed to a significantly higher number of beneficial fungi compared to double irrigation. Significantly the lowest count of positive fungi was determined under conditions of the conventional system × no irrigation interaction (Table 8).

Table 8. Counts of beneficial and pathogenic fungi and actinobacteria in the soil under spring wheat crops—mean values for the three-year study period.

Farming System	Wheat Irrigation	Count of Beneficial Fungi (*Trichoderma* ssp.) in 1 g of Soil from 0–25 cm Layer	Count of Pathogenic Fungi (*Fusarium* ssp.) in 1 g of Soil from 0–25 cm Layer	Count of Actinobacteria in 1 g of Soil from 0–25 cm Layer
Organic	NI	17,142 ± 89	10,017 ± 52	47,495 ± 121
	2I	19,234 ± 94	9162 ± 49	48,377 ± 129
	MI	20,144 ± 99	9058 ± 38	50,289 ± 135
Mean		18,840	9412	48,720
Integrated	NI	16,393 ± 84	11,237 ± 59	38,165 ± 104
	2I	18,344 ± 91	10,350 ± 51	38,790 ± 108
	MI	19,424 ± 78	10,152 ± 55	39,254 ± 112
Mean		18,054	10,580	38,736
Conventional	NI	14,205 ± 69	12,071 ± 62	29,615 ± 86
	2I	15,368 ± 75	11,710 ± 58	31,316 ± 92
	MI	16,337 ± 81	10,986 ± 54	33,452 ± 97
Mean		15,303	11,589	31,461
HSD ($p \geq 0.05$) for farming system (A)		975.4	710.6	1712.4
HSD ($p \geq 0.05$) for wheat irrigation (B)		961.5	709.4	1653.3
HSD ($p \geq 0.05$) for interaction (A × B)		1144.2	n.s.	1699.2

NI—no irrigation, 2I—double irrigation, MI—multiple irrigation, ±SD—standard deviation, n.s.—not significant.

Significantly the lowest count of pathogenic fungi was determined in the soil from the organic system. In the integrated system, their number was higher by ca. 11%, whereas in the conventional system, by ca. 19%. In addition, a statistically significant difference was noted in the count of these fungi (by ca. 8%) between the soil samples from integrated and conventional systems. Intensification of irrigation of spring wheat crops generally contributed to diminished counts of pathogenic fungi in the soil; however, the differences in their numbers determined upon multiple and double irrigation were statistically insignificant. To recapitulate, the no irrigation strategy applied to spring wheat crops increased the count of pathogenic fungi by ca. 9% (conventional system) and 10% (integrated and conventional system), compared to the multiple irrigation approach (Table 8).

The numbers of actinobacteria in the soil under spring wheat crops varied depending on the farming system. Significantly the highest number of actinobacteria was determined in the soil from the organic system, i.e., higher by ca. 21% than in the soil from the integrated system and by ca. 36% than in the soil from the conventional system. Multiple and double irrigation of wheat crops caused significant differences (reaching ca. 5%) in the count of these bacteria in all farming systems except for the integrated one. In turn, the no irrigation strategy (control plots) caused a noticeable decrease in the number of actinobacteria in the soil compared to multiple irrigation in the conventional (by ca. 12%) and organic (by ca. 6%) systems. Interestingly, no irrigation in the integrated system caused statistically insignificantly lower number of actinobacteria in the soil (by only 3%) compared to the multiple irrigation strategy. Significantly the highest number of actinobacteria in 1 g of soil under spring wheat crops (48.7×10^3) was under conditions of the organic system × multiple irrigation interaction, whereas significantly the lowest count (29.6×10^3) was determined under conditions of the conventional system × no irrigation interaction (Table 8).

4. Discussion

4.1. Chemical Properties of Soil

The results obtained from this field experiment demonstrate that the effects of farming systems on the chemical composition of loess soil under spring wheat crops varied. The conventional system proved to have the most favorable effect on total nitrogen, phosphorus, and potassium contents of the soil. In contrast, the organic and integrated systems produced the highest levels of magnesium and organic carbon. Similarly to the present study, Wang et al. [31] noted a reduction in organic carbon content of the soils under conventionally grown crops, because the intensification of fertilization and chemical protection leads to the degradation of organic matter. In turn, Lal [32] emphasized that the organic and integrated systems positively influenced soil organic matter compared to the conventional system. However, results of a study by Maucieri et al. [33] showed that organic farming was not superior to conventional farming regarding the accumulation of organic matter (organic carbon) in the soil, in a situation when we do not use organic fertilization in this system (similarly to our own research). Other studies also pointed to improved soil quality indicators along with progressive agro-ecological practices (integrated and organic systems) [34,35]. The findings from these studies also correspond to observations made by other authors [36,37], who reported a decrease in phosphorus and potassium contents in the soil cultivated in the organic system. Our previous investigations [38,39] demonstrated that the organic system contributed to increased levels of magnesium, boron, copper, manganese, zinc, organic carbon, and total nitrogen in the soil. Moreover, organic farming promoted more favorable soil pH and a higher humus content, significantly improving the total sorption capacity compared to the conventional system [40]. Conversely, the conventional system produced higher phosphorus and potassium levels in the soil. Al-Busaidi et al. [41] found that long-term intensive agronomy (conventional farming) negatively impacted phosphorus content of the soil, leading to its excess accumulation.

Wang et al. [31] emphasized that conventional farming practices, characterized by intensive tillage and a high input of synthetic chemicals, critically depleted soil carbon resources. In contrast, alternative practices, such as reducing doses of agrochemicals and adopting organic systems, were found to enhance soil carbon content. Another study [42] indicated no impact of organic farming on organic carbon levels of the soil.

In the present research, the highest levels of boron, copper, manganese, and zinc were determined in the soil from the organic system, while the lowest ones were assayed in the soil from the conventional system. This relationship was also supported by findings from other studies. Bhanuvally et al. [43] demonstrated that in the case of various crops, ecological farming improved soil chemical composition by increasing the availability of macro- and microelements and the content of organic carbon in the soil. The dynamics and transformations of microelements (Zn, Cu, Fe, Mn, B, and Mo) in the soil are regulated by various factors, such as pH, electrical conductivity, and organic matter content, which consequently modify various physicochemical reactions, thereby affecting the availability of microelements. Soil organic matter fosters a reduced environment (a lower redox potential) and increases the availability of cations of microelements in the soil. It fixes more Zn, Cu, B, and Mo compared to Fe and Mn, as the former are less sensitive to redox changes [44].

The results of a study by Wang et al. [45] indicated that appropriate water resource management in the soil, coupled with rational fertilization, significantly increased ammonium nitrogen, phosphorus, and potassium levels, and substantially enhanced soil moisture content. These changes, in turn, facilitated the earlier availability of nutrients in the soil for the cultivated plants. The above findings correspond with the results of the present study, as the most favorable soil chemical composition was noted under irrigation management (multiple irrigations based on soil moisture sensor readings) in the interaction with the integrated farming system. Also, D'Odorico et al. [46] and Slaboch and Malý [47] emphasized that optimizing soil moisture content had a significant impact on ensuring physicochemical parameters of the soil that are beneficial for cultivated crops, particularly under conditions of the integrated farming system [48]. In the study by Leogrande et al. [49], the application of crop irrigation significantly increased contents of organic carbon as well as available forms of P, Mg, and Na in the soil. In turn, the results of a study by Fadl et al. [20] demonstrated that irrigation led to an 8.00% increase in soil organic matter content and a 7.22% increase in nitrogen resources compared to the control treatment without irrigation.

4.2. Physical Properties of Soil

In the current study, the physical properties of the soil were more closely related to the irrigation strategies applied to the spring wheat crops than to the farming systems adopted. However, the conventional system resulted in a lower soil moisture content and reduced total and capillary porosity. Woldeyohannis et al. [50] pointed to a correlation between the intensification of cultivation practices typical of the conventional system and increasing soil density. In turn, Gajda et al. [51] observed that the increase in organic matter content of the soil under organic cultivation led to a decrease in its bulk density. In the present study, a tendency towards higher soil density was noted in the conventional system compared to the organic system, while the compaction of soil under conventionally cultivated wheat was significantly the greatest. Sainju et al. [52] have claimed that soil aggregation (overall and capillary porosity) is associated with agricultural practices that result in a high organic matter content (including organic carbon) and with optimal soil moisture content maintained. This thesis was also supported by the current study findings, as the most favorable values of these soil quality indicators were recorded under multiple irrigation of the wheat crops coupled with their cultivation in the organic system.

Works in the literature on the subject [53–55] indicate that organic farming improves certain physical properties of soil compared to conventional farming. In particular, it increases the stability of wet soil aggregates, saturated hydraulic conductivity, and water resources available to plants. Also, it generally increases soil porosity, thereby enhancing cumulative water infiltration. However, organic farming exerts either a varied or no effect on soil compaction indicators (i.e., bulk density and compaction), which partially corresponds with the results of the present study.

Previous studies have proved that soil density and compaction are fundamentally associated with the soil's moisture levels. In the absence of precipitation, rational irrigation of crops proves to be an effective method for managing these parameters [56]. An important aspect of the organic system is also the limited use of chemicals. In the conventional system, these chemicals can adversely affect soil structure and reduce its water-holding capacity. Methods deployed in organic farming, such as crop rotation, reduce soil water demand while increasing yields over the long term [57–59].

4.3. Biological Properties and Enzymatic Activity of Soil

Furtak and Gałązka [60] have claimed that organic agriculture positively influences soil quality compared to conventional farming systems. The organic system favorably modifies the structure of soil microbial communities and stimulates soil enzymatic activity. Agronomic practices associated with the organic and integrated systems positively impact the organic carbon content of the soil, which translates into microbial biomass (including the biomass of saprophytic fungi) and soil enzyme activity compared to the conventional system [35,51]. Also, Wang et al. [31] reported that organic farming significantly increased soil microbial biomass by 63–139% (depending on the year of study), and boosted soil microbial activity by 52–117% compared to conventional farming. Organic farming was also reported to contribute to reducing the occurrence of *Fusarium oxysporum* in the soil due to a higher organic carbon content [61]. Also, Nannipieri et al. [62] demonstrated that the composition of soil microbial communities was more desirable (dominance of beneficial microorganisms) under conditions of a higher organic carbon content in the soil. This is confirmed by the results of the current study, where the most favorable parameters of beneficial microbial communities were determined in the organic system, followed by the integrated system. Conversely, the soil enzyme activity in this study was highest in the integrated system, followed by the organic system. Thus, the conventional system contributed to a significantly lower number of beneficial fungi and actinobacteria, while promoting the highest count of pathogenic fungi in the soil. Furthermore, the intensive chemicalization of agriculture resulting from the conventional farming system had adversely affected enzyme activity in the soil under spring wheat crops.

A deficiency of organic and mineral nutrients in the soil can lead to nearly complete inactivity of soil enzymes [63]. In the study by Fließbach et al. [34], the intensification of agricultural practices generally contributed to diminished activities of soil enzymes. The microbial biomass of the soil determined in the integrated system was 25% lower compared to that noted in the organic system. Regarding enzyme activity, dehydrogenase levels were 39–42% lower in the soil from the integrated system compared to the organic system soils. Our previous study [38] demonstrated that analyses of the enzymatic activity of soil in a five-crop rotation showed significantly higher activities of all enzymes tested (particularly of dehydrogenase, protease, and urease) in the organic system compared to the conventional one, regardless of the crop species. It is noteworthy that the high enzymatic activity in the organic and integrated systems was partially due to the influence of plant biomass in crops where pesticides were not used or were applied in limited doses. This corroborates previous findings reported by other authors [40,64–66], and was also confirmed in the present study. The activity of soil enzymes, including, in particular, dehydrogenases, is dependent on the high organic matter content of the soil, its stratification, and extent of degradation. Additionally, the degradation of organic matter and enzymatic activity are more efficient under good soil moisture and temperature conditions [67,68]. These findings are confirmed by the results of the present study, where the highest enzymatic activity in loess soil was observed as a result of rational irrigation of spring wheat crops in the integrated system (with the highest organic carbon content recorded). Gajda et al. [51] noted that arable soil under organic cultivation exhibited 2–2.5 times greater microbial activity than the soil under conventional cultivation. In turn, Terashima and Mihara [12], who compared organic and conventional systems in terms of the overall soil quality, found the most significant differences in biological properties (microbial communities) and enzymatic activity of the soil, with more beneficial changes observed in the organic system. Lesser differences were noted in soil chemical properties, while the smallest differences pertained to its physical properties. The current study results confirm these relationships to some extent.

4.4. *Profitability of Irrigation in the Context of Farming Systems—Summary*

As can be seen from the above subsections, irrigation (especially multiple irrigation) had a positive effect on the quality parameters of the soil under spring wheat crops (physicochemical, biological, soil enzyme activity) compared to the lack of irrigation. Moreover, the use of irrigation in interaction with the integrated (economical) system influenced the greatest stability of the determined quality features of the soil in relation to the control object (organic system) and conventional system. In the organic system, despite the lack of fertilization and pesticides, many favorable quality features of the soil were found (including a high content of some macro- and microelements), especially in connection with the irrigation of wheat crops. This resulted from the low yield of spring wheat on the control objects (Table 2), and therefore a lower depletion of nutrients taken up by wheat plants from the soil. Similarly, the lowest content of microelements in the soil under conventional system conditions was probably a consequence of the very high productivity of spring wheat (8.8 t ha^{-1} as a result of multiple irrigation).

In the three analyzed research seasons, a much lower level of total precipitation was observed during the growing season (408–434 mm), compared to the multi-annual average precipitation in the research area (589 mm). This situation caused a high demand for water in the soil (which is clearly illustrated by the amount of water used in multiple irrigation) resulting from the readings of soil moisture sensors. Thus, on the one hand, irrigation had a positive effect on the quality and health of the soil, but generated high costs that were not always compensated by the obtained grain yield of spring wheat. It should be noted that in the case of higher total precipitation during the growing season (than observed in our studies), the profitability of irrigation would be much higher due to lower water consumption for irrigation. Tolimir et al. [69] found, using the example of corn irrigation, that the effectiveness of such treatments was closely related to high variability from year to year of the total amount and distribution of precipitation.

Irrigation efficiency is also closely related to the amounts of fertilizers and pesticides used, as emphasized by Özocak [70]. Water shortages can reduce yields due to reduced photosynthesis and other physiological limitations of the crop [71,72]. Our findings (Table 2) show that satisfactory profitability of spring wheat irrigation can be achieved by using half of the recommended doses of NPK mineral fertilization and pesticides, under the conditions of an integrated (economic) system. Zuber et al. [73] claim that climate change will have a significant impact on water demand and yields, which is why it is necessary to quantify the climate risk and develop climate-resistant field management practices (deficit irrigation). In the case of wheat, water demand may increase by as much as 13% in the coming years. Hence, it is recommended to apply deficit irrigation at the species-specific recommended fertilization rates in a given geographical area to obtain the best irrigation option to increase the resilience of cereal crops to climate change.

A separate issue concerning the profitability of the applied irrigation of crops is the price of water used for irrigation (different in many countries), water availability, infrastructure related to irrigation, climatic conditions of a given region, as well as the type of soil and the plant grown on it [74]. For example, the prices of wheat grain (a global crop on a global scale) are not as high as the prices of the harvested yield of some minor crops (e.g., vegetable plants, herbal plants), generating high profits from a small area of crops (and therefore areas requiring much smaller amounts of water for irrigation). Our research on carrot irrigation [75] shows that to a much greater extent than in the case of wheat, multiple irrigation affects the yield of carrot roots (the price of which is several times higher than that of cereal grains). Similarly, Ali et al. [76] demonstrated, using the example of tomato, the profitability of drip irrigation at the level of USD 1720/ha (profit). One way to reduce the cost of water for crop irrigation may be to use treated wastewater [77]. Wang et al. [78] noticed that with the increase in the amount of irrigation, the productivity parameters of winter wheat improved. The authors also state that in order to conserve the amount of water used for irrigation (and reduce its related costs), the use of activated water, e.g., water with magnetization and oxygenation, can be considered. Generally, in many countries there is poor irrigation infrastructure, and low doses of fertilization are used, which does not support the efficiency of irrigation and satisfactory grain yields [79]. Our own studies confirm that the highest irrigation efficiency is obtained in interaction with conventional farming (100% doses of fertilizers and pesticides), but is slightly lower in the integrated (economical) system, with the use of agrochemical doses reduced by half.

5. Conclusions

The results of this study on the quality of loess soil under spring wheat crops indicate that farming systems exerted varied effects on its chemical, physical, and biological properties, as well as its enzymatic activity.

The organic system produced the highest levels of magnesium, boron, copper, manganese, and zinc in the soil. It also contributed to the most favorable biological properties of the soil, such as the highest number of beneficial actinobacteria and fungi, alongside the lowest number of pathogenic fungi. Additionally, it positively affected certain physical properties of the soil, such as its moisture content, density, and porosity.

In turn, the integrated system yielded the highest organic carbon content, the highest sorption capacity, and the highest enzyme activity (including dehydrogenase, acid and alkaline phosphatase, protease, and urease) of the soil. At the same time, it ensured favorable chemical and physical parameters of the soil, as well as its microbiological composition.

In contrast, the conventional system produced the highest levels of total nitrogen, phosphorus, and potassium in the soil. However, its drawbacks included the lowest levels of organic carbon and microelements in the soil, unbeneficial biological parameters, and the weakest enzymatic activity, followed by a deterioration in most physical soil parameters.

Irrigation of spring wheat crops (particularly multiple irrigation) resulted in a noticeable improvement in all assessed soil quality indicators compared to the control plots (without irrigation), regardless of the farming system. This was due to the optimal water supply to the soil ensured based on the readings from soil moisture sensors. Particularly beneficial effects of multiple irrigation on soil quality and health were observed in the interaction with the organic or integrated systems.

The limitations of the study are related to the costs of irrigation (especially multiple irrigation) in the situation of progressive year-by-year soil drought and high water prices in many countries. Future studies should focus on long-term analysis, taking into account greater variability of weather conditions (amount of precipitation), different types of soil and test plant (e.g., small-area crops generating higher profits, with relatively lower water consumption for irrigation of crops).

Author Contributions: Conceptualization, E.H. and C.A.K.; methodology, C.A.K. and E.H.; software, C.A.K.; validation, E.H. and C.A.K.; formal analysis, E.H. and C.A.K.; investigation, E.H. and C.A.K.; resources, E.H. and C.A.K.; writing—original draft preparation, C.A.K. and E.H.; visualization, E.H. and C.A.K. All authors have read and agreed to the published version of the manuscript.

Funding: Researchers supported by National Center for Research and Development in Poland (Project number ZKB/U-389/RiO/2021).

Institutional Review Board Statement: Not applicable.

Informed Consent Statement: Not applicable.

Data Availability Statement: The data supporting the results of this study are included in the manuscript.

Conflicts of Interest: The authors declare no conflicts of interest.

References

1. Uhlmann, N.K.; Beckles, D.M. Storage products and transcriptional analysis of the endosperm of cultivated wheat and two wild wheat species. *J. Appl. Genet.* **2010**, *51*, 431–447. [CrossRef] [PubMed]
2. Bastida, F.; Moreno, J.L.; Hernandez, T.; Garcia, C. Microbiological degradation index of soils in a semiarid climate. *Soil Biol. Biochem.* **2006**, *38*, 3463–3473. [CrossRef]
3. Cooper, J. Soil Structure, Management and Effect on Nutrient Availability and Crop Production. In Proceedings of the Organic Producer Conference, Facing Current and Future Challenges, Newcastle University, Newcastle upon Tyne, UK, 17 January 2011; Available online: https://www.organicresearchcentre.com/manage/authincludes/article_uploads/Julia%20Cooper%20-%20Soil%20structure.pdf (accessed on 10 June 2025).
4. Biswas, J.C.; Naveen, K.; Maniruzzaman, M.; Naher, U.A.; Haque, M.M. Soil Health Assessment Methods and Relationship with Wheat Yield. *Open J. Soil Sci.* **2019**, *9*, 189–205. [CrossRef]
5. El-Hassanin, A.S.; Samak, M.R.; Atta, Y.; Labib, S.M. Controlled Drainage and Irrigation Regime for Improving Some Soil Properties and Wheat Production, Sahl El-Tina, Sinai Peninsula, Egypt. *Water Conserv. Sci. Eng.* **2022**, *7*, 375–388. [CrossRef]
6. Gruver, J.B.; Weil, R.R. Farmer Perceptions of Soil Quality and Their Relationship to Management-Sensitive Soil Parameters. *Renew. Agric. Food Syst.* **2007**, *22*, 271–281. [CrossRef]
7. Loganathan, M.; Narendiran, J. Pollution of Soil Due To Leather Factory Near Ranipet, Tamil Nadu (India). *Int. J. Emerg. Trends Sci. Technol.* **2014**, *01*, 86–90. Available online: https://journals.indexcopernicus.com/api/file/viewByFileId/149571 (accessed on 10 June 2025).
8. Franchini, J.C.; Crispino, C.C.; Souza, R.A.; Torres, E.; Hungria, M. Microbiological parameters as indicators of soil quality under various soil management and crop rotation systems in southern Brazil. *Soil Tillage Res.* **2007**, *92*, 18–29. [CrossRef]
9. Shukla, M.K.; Lal, R.; Ebinger, M. Determining soil quality indicators by factor analysis. *Soil Tillage Res.* **2006**, *87*, 194–204. [CrossRef]
10. Parr, J.F.; Papendick, R.I.; Hornick, S.B.; Meyer, R.E. Soil quality: Attributes and relationship to alternative and sustainable agriculture. *Am. J. Altern. Agric.* **1992**, *7*, 5–11. [CrossRef]
11. Laishram, J.; Saxena, K.G.; Maikhuri, R.K.; Rao, K.S. Soil quality and soil health: A review. *Int. J. Ecol. Environ. Sci.* **2012**, *38*, 19–37. Available online: https://www.researchgate.net/publication/232237296_Soil_quality_and_soil_health_A_review#fullTextFileContent (accessed on 10 June 2025).
12. Terashima, M.; Mihara, M. Interactions among soil physical, chemical and biological properties under different farming systems. *Int. J. Environ. Rural Develop.* **2022**, *13*, 1–6. Available online: https://iserd.net/ijerd131/13-1-1.pdf (accessed on 5 June 2025).
13. Medan, D.; Torretta, J.P.; Hodara, K.; de la Fuente, E.B.; Montaldo, N.H. Effects of agriculture expansion and intensification on the vertebrate and invertebrate diversity in the Pampas of Argentina. *Biodivers. Conserv.* **2011**, *20*, 3077–3100. [CrossRef]
14. Komatsuzaki, M.; Ohta, H. Soil Management Practices for Sustainable Agro-Ecosystems. *Sustain. Sci.* **2007**, *2*, 103–120. [CrossRef]
15. Ouda, S.; Noreldin, T.; Alarcón, J.J.; Ragab, R.; Caruso, G.; Sekara, A.; Abdelhamid, M.T. Response of Spring Wheat (*Triticum aestivum*) to Deficit Irrigation Management under the Semi-Arid Environment of Egypt: Field and Modeling Study. *Agriculture* **2021**, *11*, 90. [CrossRef]

16. Carvalho, M.; Carvalho, S.; Rodrigues, G.; Ferreira, L. Efeito do estresse hídrico na fisiologia e no crescimento inicial de plantas de batata. *Rev. Bras. Eng. Agríc. Ambient.* **2018**, *22*, 38–44. [CrossRef]
17. Calow, R.C.; Howarth, S.E.; Wang, J. Irrigation Development and Water Rights Reform in China. *Int. J. Water Resour. Dev.* **2009**, *25*, 227. [CrossRef]
18. Ranjbar, A.; Sepaskhah, A.R.; Emadi, S. Relationships between wheat yield, yield components and physico-chemical properties of soil under rain-fed conditions. *Int. J. Plant Prod.* **2015**, *9*, 433–466. Available online: https://www.researchgate.net/publication/282315376_Relationships_between_wheat_yield_yield_components_and_physico-chemical_properties_of_soil_under_rain-fed_conditions (accessed on 5 June 2025).
19. Gong, F. The Role of Water in Soil Formation and Development. *J. Soil Sci. Plant Nutr.* **2023**, *7*, 3.
20. Fadl, M.E.; Sayed, Y.A.; El-Desoky, A.I.; Shams, E.M.; Zekari, M.; Abdelsamie, E.A.; Drosos, M.; Scopa, A. Irrigation Practices and Their Effects on Soil Quality and Soil Characteristics in Arid Lands: A Comprehensive Geomatic Analysis. *Soil Syst.* **2024**, *8*, 52. [CrossRef]
21. Thalmann, A. Zur Methodik der Bestimmung der Dehydrogenase Aktivität in Boden mittels Triphenyl tetrazolium chlorid (TTC). *Landwirtsch. Forsch.* **1968**, *21*, 249–258.
22. Tabatabai, M.A.; Bremner, J.M. Use of p-nitrophenyl phosphate for assay of soil phosphatase activity. *Soil Biol. Biochem.* **1969**, *1*, 301–307. [CrossRef]
23. Zantua, M.I.; Bremner, J.M. Comparison of methods of assaying urease activity in soils. *Soil Biol. Biochem.* **1975**, *7*, 291–295. [CrossRef]
24. Ladd, N.; Butler, J.H.A. Short-term assays of soil proteolytic enzyme activities using proteins and dipeptide derivatives as substrates. *Soil Biol. Biochem.* **1972**, *4*, 19–30. [CrossRef]
25. Foght, J.; Aislabie, J. Enumeration of Soil Microorganisms. In *Monitoring and Assessing Soil Bioremediation. Soil Biology Part 5*; Margesin, R., Schinner, F., Eds.; Springer: Berlin/Heidelberg, Germany, 2005; pp. 261–280. [CrossRef]
26. Martin, J.P. Use of acid rose bengal and streptomycin in the plate method for estimating soil fungi. *Soil Sci.* **1950**, *69*, 215–232. [CrossRef]
27. Choi, Y.W.; Hyde, K.D.; Ho, W.W.H. Single spore isolation of fungi. *Fungal Divers.* **1999**, *3*, 29–38. Available online: https://www.researchgate.net/publication/228482364_Single_spore_isolation_of_fungi (accessed on 10 June 2025).
28. Seifert, K.A. Compendium of soil fungi—by K.H. Domsch, W. Gams & T.-H. Anderson. *Eur. J. Soil Sci.* **1980**, *59*, 1007. [CrossRef]
29. Wallace, R.; Lochhead, A. Qualitative studies of soil microorganisms IX. Amino acid requirements of rhizosphere bacteria. *Can. J. Res. Sect. C Bot. Sci.* **1950**, *28*, 1–6. [CrossRef]
30. Badura, L.; Smyłła, A. *Wybrane Metody Izolowania Promieniowców Z Gleby [Selected Methods Of Isolating Actinomycetes From Soil]*; Prace Komisji Naukowych: Warszawa, Poland, 1979; p. 15. (In Polish)
31. Wang, Y.; Tu, C.; Cheng, L.; Li, C.; Gentry, L.F.; Hoyt, G.D.; Zhang, X.; Hu, S. Long-term impact of farming practices on soil organic carbon and nitrogen pools and microbial biomass and activity. *Soil Till. Res.* **2011**, *117*, 8–16. [CrossRef]
32. Lal, R. Challenges and opportunities in soil organic matter research. *Eur. J. Soil Sci.* **2009**, *60*, 158–169. [CrossRef]
33. Maucieri, C.; Tolomio, M.; Raimondi, G.; Toffanin, A.; Morari, F.; Berti, A.; Borin, M. Organic versus conventional farming: Medium-term evaluation of soil chemical properties. *Ital. J. Agron.* **2022**, *17*, 2114. [CrossRef]
34. Fließbach, A.; Oberholzer, H.R.; Gunst, L.; Mäder, P. Soil organic matter and biological soil quality indicators after 21 years of organic and conventional farming. *Agric. Ecosyst. Environ.* **2007**, *118*, 273–284. [CrossRef]
35. Joergensen, R.G.; Mader, P.; Fließbach, A. Long-term effects of organic farming on fungal and bacterial residues in relation to microbial energy metabolism. *Biol. Fertil. Soils* **2010**, *46*, 303–307. [CrossRef]
36. Askegaard, M.; Eriksen, J.; Olesen, J.E. Exchangeable potassium and potassium balances in organic crop rotations on a coarse sand. *Soil Use Manag.* **2003**, *19*, 96–103. [CrossRef]
37. Gosling, P.; Shepherd, M. Long-term changes in soil fertility in organic arable farming systems in England. with particular reference to phosphorus and potassium. *Agric. Ecosyst. Environ.* **2005**, *105*, 425–432. [CrossRef]
38. Kwiatkowski, C.A.; Harasim, E.; Feledyn-Szewczyk, B.; Antonkiewicz, J. Enzymatic Activity of Loess Soil in Organic and Conventional Farming Systems. *Agriculture* **2020**, *10*, 135. [CrossRef]
39. Kwiatkowski, C.A.; Harasim, E. Chemical Properties of Soil in Four-Field Crop Rotations under Organic and Conventional Farming Systems. *Agronomy* **2020**, *10*, 1045. [CrossRef]
40. Wesołowska, S.; Futa, B.; Myszura, M.; Kobyłka, A. Residual Effects of Different Cropping Systems on Physicochemical Properties and the Activity of Phosphatases of Soil. *Agriculture* **2022**, *12*, 693. [CrossRef]
41. Al-Busaidi, W.; Janke, R.; Menezes-Blackburn, D.; Khan, M.M. Impact of long-term agricultural farming on soil and water chemical properties: A case study from Batinah regions (Oman). *J. Saudi Soc. Agric. Sci.* **2022**, *21*, 397–403. [CrossRef]
42. Marinari, S.; Mancinelli, R.; Campiglia, E.; Grego, S. Chemical and biological indicators of soil quality in organic and conventional farming systems in Central Italy. *Ecol. Indic.* **2006**, *6*, 701–711. [CrossRef]

43. Bhanuvally, M.; Sunitha, N.H.; Sharanabasava; Ravi, S.; Mahadevaswamy. Effect of Organic Farming Practices on Soil Chemical Properties. *Int. J. Environ. Clim. Change* **2024**, *14*, 470–481. [CrossRef]
44. Dhaliwal, S.S.; Naresh, R.K.; Mandal, A.; Singh, R.; Dhaliwal, M.K. Dynamics and transformations of micronutrients in agricultural soils as influenced by organic matter build-up: A review. *Environ. Sustain. Indic.* **2019**, *1–2*, 100007. [CrossRef]
45. Wang, H.; Huo, Q.; Wang, T.; Lv, G.; Li, W.; Ren, J.; Zhang, S.; Li, J. Effects of Combination of Water-Retaining Agent and Nitrogen Fertilizer on Soil Characteristics and Growth of Winter Wheat under Subsoiling Tillage in South Loess Plateau of China. *Agronomy* **2024**, *14*, 1287. [CrossRef]
46. D'Odorico, P.; Chiarelli, D.D.; Rosa, L.; Bini, A.; Zilberman, D.; Rulli, M.C. The Global Value of Water in Agriculture. *Proc. Natl. Acad. Sci. USA* **2020**, *117*, 21985–21993. [CrossRef] [PubMed]
47. Slaboch, J.; Malý, M. Land Valuation Systems in Relation to Water Retention. *Agronomy* **2023**, *13*, 2978. [CrossRef]
48. Bondarovich, A.; Illiger, P.; Schmidt, G.; Ponkina, E.; Nugumanova, A.; Maulit, A.; Sutula, M. Effects of Agricultural Cropping Systems on Soil Water Capacity: The Case in Cross-Border Altai. *Span. J. Soil Sci.* **2023**, *13*, 11493. [CrossRef]
49. Leogrande, R.; Pedrero, F.; Nicolas, E.; Vitti, C.; Lacolla, G.; Stellacci, A.M. Reclaimed Water Use in Agriculture: Effects on Soil Chemical and Biological Properties in a Long-Term Irrigated Citrus Farm. *Agronomy* **2022**, *12*, 1317. [CrossRef]
50. Woldeyohannis, Y.S.; Hiremath, S.S.; Tola, S.; Wako, A. Influence of soil physical and chemical characteristics on soil compaction in farm field. *Heliyon* **2024**, *10*, e25140. [CrossRef] [PubMed]
51. Gajda, A.M.; Czyż, E.A.; Dexter, A.R. Effects of long-term use of different farming systems on some physical, chemical and microbiological parameters of soil quality. *Int. Agrophys.* **2016**, *30*, 165–172. [CrossRef]
52. Sainju, U.M.; Liptzin, D.; Jabro, J.D. Relating soil physical properties to other soil properties and crop yields. *Sci. Rep.* **2022**, *12*, 22025. [CrossRef] [PubMed]
53. de Cima, D.S.; Luik, A.; Reintam, E. Organic farming andcover crops as an alternative to mineral fertilizers to improvesoil physical properties. *Int. Agroph.* **2015**, *29*, 405–412. [CrossRef]
54. Williams, D.M.; Blanco-Canqui, H.; Francis, C.A.; Galusha, T.D. Organic farming and soil physical properties: An assessment after 40 years. *Agron. J.* **2017**, *109*, 600–609. [CrossRef]
55. Blanco-Canqui, H.; Ruis, S.J.; Francis, C.A. Do organic farming practices improve soil physical properties? *Soil Use Manag.* **2024**, *40*, e12999. [CrossRef]
56. Hu, W.; Shao, M.A.; Si, B.C. Seasonal changes in surface bulk density and saturated hydraulic conductivity of natural landscapes. *Eur. J. Soil Sci.* **2012**, *63*, 820–830. [CrossRef]
57. Papadopoulos, A.; Bird, N.R.A.; Whitmore, A.P.; Mooney, S.J. Does organic management lead to enhanced soil physical quality? *Geoderma* **2014**, *213*, 435–443. [CrossRef]
58. Minasny, B.; McBratney, A.B. Limited effect of organicmatter on soil available water capacity. *Eur. J. Soil Sci.* **2018**, *69*, 39–47. [CrossRef]
59. Bagnall, D.K.; Morgan, C.L.S.; Cope, M.; Bean, G.M.; Cappellazzi, S.; Greub, K.; Liptzin, D.; Norris, C.L.; Rieke, E.; Tracy, P.; et al. Carbon-sensitive pedotransfer functions for plant available water. *Soil Sci. Soc. Am. J.* **2022**, *86*, 612–629. [CrossRef]
60. Furtak, K.; Gałązka, A. Effect of organic farming on soil microbiological parameters. *Pol. J. Soil Sci.* **2019**, *52*, 259. [CrossRef]
61. Liu, L.; Kong, J.; Cui, H.; Zhang, J.; Wang, F.; Cai, Z.; Huang, X. Relationships of decomposability and C/N ratio in different types of organic matter with suppression of Fusarium oxysporum and microbial communities during reductive soil disinfestation. *Biol. Control* **2016**, *101*, 103–113. [CrossRef]
62. Nannipieri, P.; Ascher, J.; Ceccherini, M.T.; Landi, L.; Pietramellara, G.; Renella, G. Microbial diversity and soil functions. *Eur. J. Soil Sci.* **2003**, *54*, 655–670. [CrossRef]
63. Gianfreda, L.; Rao, A.M.; Piotrowska, A.; Palumbo, G.; Colombo, C. Soil enzyme activities as affected by anthropogenic alterations: Intensive agricultural practices and organic pollution. *Sci. Total Environ.* **2005**, *341*, 265–279. [CrossRef] [PubMed]
64. Bielińska, E.J.; Pranagal, J. Enzymatic activity of soil contaminated with triazine herbicides. *Pol. J. Environ. Stud.* **2007**, *16*, 295–300. Available online: https://www.pjoes.com/pdf-87989-21848?filename=21848.pdf (accessed on 5 June 2025).
65. Tian, Y.; Zhang, X.; Liu, J.; Chen, Q.; Gao, L. Microbial properties of rhizosphere soils as affected by rotation, grafting, and soil sterilization in intensive vegetable production systems. *Sci. Hortic.* **2009**, *123*, 139–147. [CrossRef]
66. Niewiadomska, A.; Sulewska, H.; Wolna-Maruwka, A.; Klama, J. Effect of organic fertilization on development of proteolytic bacteria and activity of proteases in the soil for cultivation of maize (*Zea mays* L.). *Arch. Environ. Prot.* **2010**, *36*, 47–56. Available online: https://www.researchgate.net/publication/289068590_Effect_of_organic_fertilization_on_development_of_proteolytic_bacteria_and_activity_of_proteases_in_the_soil_for_cultivation_of_maize_ZEA_MAYS_L (accessed on 10 June 2025).
67. Wallenstein, M.D.; Haddix, M.L.; Lee, D.D.; Conant, R.T.; Paul, E.A. A litter-slurry technique elucidates the key role of enzyme production and microbial dynamics in temperature sensitivity of organic matter decomposition. *Soil Biol. Biochem.* **2012**, *47*, 18–26. [CrossRef]
68. Brzostek, E.R.; Finzi, A.C. Seasonal variation in the temperature sensitivity of proteolytic enzyme activity in temperate forest soils. *J. Geophys. Res.* **2012**, *117*, G01018. [CrossRef]

69. Tolimir, M.; Kresović, B.; Gajić, K.; Anđelković, V.; Brankov, M.; Dugalić, M.; Gajić, B. Integrated effect of irrigation rate and plant density on yield, yield components and water use efficiency of maize. *Plant Soil Environ.* **2024**, *70*, 475–482. [CrossRef]
70. Özocak, M. The Effect of Irrigation on Crop Yield Change in Some Cereals in Drought Conditions Determined Using SPI and PNI: Ankara Province Example. *Turk. J. Agric.-Food Sci. Technol.* **2024**, *12*, 2579–2589. [CrossRef]
71. Gavrilescu, M. Water, Soil, and Plants Interactions in a Threatened Environment. *Water* **2021**, *13*, 2746. [CrossRef]
72. Liu, W.Z.; Hunsaker, D.J.; Li, Y.S.; Xie, X.Q.; Wall, G.W. Interrelations of yield, evapotranspiration, and water use efficiency formmarginal analysis of water production functions. *Agric. Water Manag.* **2002**, *56*, 142–151. [CrossRef]
73. Zuber, M.; Kalauni, N.; Shrestha, N.; Pandey, V.P.; Pokharel, B. Cereal yield and water requirements in response to irrigation and soil fertility management in a changing climate: A case of Tulsipur, Western Nepal. *J. Water Clim. Change* **2025**, *16*, 837–859. [CrossRef]
74. Mu, L.; Su, K.; Zhou, T.; Yang, H. Yield performance, land and water use, economic profit of irrigated spring wheat/alfalfa intercropping in the inland arid area of northwestern China. *Field Crops Res.* **2023**, *303*, 109116. [CrossRef]
75. Harasim, E.; Kwiatkowski, C.A.; Buczek, J. Effect of Farming System and Irrigation on Nutrient Content and Health-Promoting Properties of Carrot Roots. *Agronomy* **2025**, *15*, 1289. [CrossRef]
76. Ali, A.; Xia, C.; Jia, C.; Faisal, M. Investment profitability and economic efficiency of the drip irrigation system: Evidence from Egypt. *Irrig. Drain.* **2020**, *69*, 1033–1050. [CrossRef]
77. Sdiri, W.; AlSalem, H.S.; Al-Goul, S.T.; Binkadem, M.S.; Ben Mansour, H. Assessing the Effects of Treated Wastewater Irrigation on Soil Physico-Chemical Properties. *Sustainability* **2023**, *15*, 5793. [CrossRef]
78. Wang, H.; Fan, J.; Fu, W. Effect of Activated Water Irrigation on the Yield and Water Use Efficiency of Winter Wheat under Irrigation Deficit. *Agronomy* **2022**, *12*, 1315. [CrossRef]
79. Basukala, A.K.; Rasche, L. Model-based yield gap assessment in Nepal's diverse agricultural landscape. *Land* **2022**, *11*, 1355. [CrossRef]

sustainability

MDPI

Article

Soil Health Practices and Decision Drivers on Diversified Vegetable Farms in Minnesota

Natalie Hoidal [1,*], Shane M. Bugeja [1], Emily Lindenfelser [1] and Paulo H. Pagliari [2]

[1] Department of Agricultural and Natural Resource Systems, University of Minnesota Extension, St. Paul, MN 55108, USA; sbugeja515@gmail.com (S.M.B.); hans6005@umn.edu (E.L.)
[2] Department of Soil Water and Climate, Southwest Research and Outreach Center, Lamberton, MN 56152, USA; pagli005@umn.edu
* Correspondence: hoida016@umn.edu

Abstract: Soil health is at the root of agricultural sustainability, and small-scale vegetable farmers are becoming an increasingly important part of the US food system. These farmers face unique challenges when it comes to managing soil on their farms. These challenges include reliance on intensive production practices, the use of primarily organic inputs with difficult to calculate nutrient concentrations, and lack of access to formal education tailored to their needs. We surveyed farmers at 100 small-scale vegetable farms in Minnesota to (1) develop a better baseline understanding of how small-scale vegetable farmers utilize key soil health practices including nutrient management, cover crops, and tillage; (2) explore how farm demographics influence the adoption of soil health practices; and (3) determine educational priorities to better support these growers. Here, we report a lack of understanding about the nutrient contributions of compost, which is often applied at very large volumes without guidance from soil test results, with implications for nutrient loading in the environment. Farmers in our study had high rates of cover crop adoption relative to other farmers in the region despite several barriers to using cover crops. More experienced farmers were more likely to utilize more tillage, with more use of deep tillage implements on larger farms. Overall, organic certification was correlated with higher adoption of soil health practices including utilization of soil tests and cover crop use, but it was not correlated with tillage. Other demographic variables including land access arrangement and race did not meaningfully influence soil health practices. Our findings suggest a need for more research, outreach, and education targeted to vegetable farmers about how to interpret laboratory soil test results, and how to responsibly utilize organic inputs including vegetative compost and composted manure at rates appropriate for crop production in a diversified farm setting. We also report a need to compensate farmers for their labor to incentive cover crop use on small farms, and a need for more research and support for farmers in the 3–50-acre range to utilize reduced tillage methods.

Keywords: compost; cover crops; emerging farmer; tillage; nutrient management; small-scale farming; fruit and vegetable

1. Introduction

Sustainable agriculture is defined as the ability of a crop production system to continuously produce food without environmental degradation [1]. A sustainable agricultural system integrates biological, physical, chemical, and ecological principles to utilize practices that are not harmful to the environment [2]. Soil is at the center of sustainable farming systems: sustaining soil health and fertility is key to maintaining food production as well as agricultural ecosystems [3].

Small-scale vegetable production is a growing industry in the Upper Midwest, but the soils and soil management strategies on these farms are understudied. With each recent census cycle, Midwest states such as Minnesota and Wisconsin lost thousands of farmers overall but gained hundreds of small-scale (<50 acres) fruit and vegetable farmers [4–6]. Due to lower barriers to entry than other types of farm enterprises, beginning farmers who do not come from farming backgrounds and who lack formal training often grow vegetables [7–10]. Increasingly, small-scale vegetable farms grow in both open fields and high tunnels. High tunnels are defined as plastic-covered enclosed growing structures without supplemental heating, and in-ground crop production [11]. To date, the United States Natural Resource Conservation Service has funded 26,216 high tunnels nationally, including 694 in Minnesota [12].

Studies of beginning small-scale vegetable farmers with limited agricultural backgrounds have characterized them as people with a desire to positively contribute to the natural environment while providing nutrient-dense fresh fruits and vegetables to their communities [13]. These farmers often lack both formal and informal training in production, stewardship, and business practices [9]. In Minnesota, these farmers are often referred to as "emerging farmers", defined as "those who traditionally face barriers to the education and resources necessary to build profitable agricultural businesses, including immigrant farmers and farmers of color" [7]. These farms tend to be highly diverse, with farmers growing many, sometimes dozens of crops at a time [7,9,13].

The United States Natural Resource Conservation Service has determined key soil health principles to guide their work, which include minimizing disturbance, maximizing soil cover, maximizing biodiversity, and maximizing the presence of living roots [7]. The agency posits that healthy soils contribute to agricultural sustainability by producing high yields, reducing production costs, and therefore improving profits, protecting natural resources on and off the farm, reducing nutrient loading and sediment runoff, and sustaining habitat for wildlife and microorganisms [14]. Two key practices for achieving these goals are the use of reduced tillage and cover crops. The benefits of these practices are well documented: conservation tillage practices can increase soil microbial activity, soil moisture, organic matter, aggregate stability, cation exchange capacity, and crop yield [1,15–18].

While the benefits of reduced tillage and cover crops are well documented, small-scale vegetable farmers face unique barriers to implementing these practices. Many small-scale farmers do not inherit land from family and, therefore, must farm in a way that allows them to be profitable with limited land, infrastructure, and equipment [9,13]. Therefore, small-scale direct market vegetable production is typically characterized by intensive cultivation methods to produce high yields on a limited scale with tools that allow for season extension (e.g., high tunnels), succession planting to achieve multiple crop cycles per season, and reliance on hand tools or light machinery [19]. Due to the intensive nature of production, small-scale vegetable farms often have few opportunities for fallow periods (including cover crops), particularly in the context of high tunnels, which are most profitable when farmers take advantage of off-season production windows [20]. Despite the intensive nature of production, these farmers often prioritize soil health and reduced chemical inputs, and conceptualize their own practices as being directly in opposition to a more harmful paradigm of industrial agriculture [10,13]. For all of the above reasons, small-scale vegetable farmers represent a unique demographic of farmers with unmet educational needs. Understanding their practices and motivations can inform educational outreach programs, as well as research priorities and the development of best practices.

University Extension programs across the country have begun to invest in local foods programs with dedicated local foods educators and researchers, urban agriculture programming, and small farms content [21–23]. However, small-scale direct market vegetable farmers have historically been underrepresented in university outreach programs. In Minnesota, three needs assessments shed light on the lack of contact between university and extension programs and this growing sector of the farming population. A 2019 needs assessment of mostly white fruit and vegetable farmers in Minnesota found that only 19% sought information from Extension or a university entity about farming practices [24]. Another survey of Hmong and Hispanic farmers in Minnesota found that only 5% would turn to Extension or a university entity about farming practices [25]. Finally, the 2020 emerging farmers report to the Minnesota state legislature cited that while Extension is a key resource, small-scale farmers felt that Extension needs more educators who could better support nontraditional farms [7]. Rather than seeking support from traditional "experts" or support people, farmers are likely to turn to each other as sources of information [24]. They also often turn to "master classes", YouTube, and podcasts led by "celebrity" farmers who have an outsized influence on small-scale vegetable production, as evidenced by subscriptions to these channels. A study of beginning farmers in the Southeastern United States found that YouTube videos, inspirational speakers, and books were some of the most common sources of inspiration for first-generation farmers, both to start farming and to influence their practices [13]. Another study of farmers in Hungary and the UK found that farmers increasingly trust one another over "experts", and that farmer influencers are becoming a more important source of information and influence [26]. As of March 2024, The No-Till Growers YouTube channel had 315K subscribers, Neversink Farms channel had 74.1k subscribers, and JM Fortier's Market Gardener Institute's channel had 59.8k subscribers. A common theme tying these channels and celebrity farmers together is vegetable production systems based on reduced or no tillage, intensive production on a small scale, and heavy inputs of compost.

One practice in particular, deep compost mulch, has emerged as a unique soil management strategy among many influential farmers. Deep compost mulch involves using large volumes of compost to quickly increase soil organic matter, suppress weeds, and improve soil moisture retention [27]. This system was first popularized by books like *Regenerative Agriculture: A Practical Whole Systems Guide to Making Small Farms Work* [28] and *The Living Soil Handbook: The No-Till Grower's Guide to Ecological Market Gardening* [29]. While this system is aesthetically pleasing and can result in many soil health benefits (e.g., increased organic matter, increased soil available water, improved nutrient retention, and increased biodiversity such as more earthworms), researchers have begun to identify concerns about the accumulation of nutrients, particularly nitrates and risks to surface and groundwater from this practice [27].

A 2019 needs assessment of 315 fruit and vegetable farmers in Minnesota identified soil health and fertility as the top educational priority [24]. Despite relatively low rates of organic certification, the use of organic practices is common among emerging farmers in Minnesota [7] and among fruit vegetable farmers in Minnesota [24]. Soil fertility can be particularly challenging for farmers relying primarily on organic practices due to nutrient imbalances between organic inputs and crop needs [30] and because macronutrient concentrations are highly variable across compost and manure products, and, thus, more expensive and time-consuming to calculate [31]. A recent survey and focus groups with emerging farmers in Minnesota identified significant confusion about inputs. The term "compost" was used interchangeably among participants to describe composted yard waste and food scraps, composted manure, and commercial fertilizer products containing composted animal products [32]. Additionally, Extension educators at the University of

Minnesota reported working with farmers who, between 2021 and 2022 experienced challenges after applying large volumes of compost on their farms. Examples included two urban farms that applied multiple cubic meters of compost to plots less than ¼ acre in size to mitigate compaction, resulting in high soluble salts that killed their plants. In another situation, a farmer added >6 tons of composted poultry manure to a single high tunnel attempting to use a deep compost mulch system, resulting in excessively high nutrient concentrations and high soluble salts.

Small-scale vegetable farmers, therefore, face a series of unique challenges when it comes to managing the soil on their farms: the influx of new farmers means this group may have more educational needs than a typical farmer. Reliance on intensive production due to space constraints limits opportunities for cover crops and fallow periods. The tendency to use organic inputs reduces opportunities to employ reduced tillage strategies and makes nutrient management more complicated. Finally, the lack of formal outreach and education for these farmers is often replaced by previously mentioned "celebrity" or otherwise influential farmers and "master classes", which may or may not be based on best management practices and research-based information.

Based on these experiences and identified needs, a project was developed to study the soils at 100 diversified vegetable farms growing their crops in open fields or under high tunnels. The objectives of this project were the following:

1. To develop a better baseline understanding of how small-scale vegetable farmers utilize key soil health practices including cover crops, reduced tillage, and nutrient management (including how the use of organic inputs like vegetative compost and composted manure factor into nutrient management decision-making).
2. To explore how farm demographics such as size, organic status, experience, land ownership, race and ethnicity, and production system (open fields vs. high tunnels) influence soil health practices.
3. To determine educational priorities for Extension and other farmer-focused education programs.

2. Materials and Methods

2.1. Recruitment

A team of 16 University of Minnesota Extension educators visited 100 small-scale (<50 acres) diversified vegetable farms between April and June 2023 to conduct soil tests and a soil management survey. In each region of the state, testing was completed as soon as fields were dry enough to sample following spring snowmelt. The county-based educators on the team first reached out to farmers in their local areas through local newsletters and personal contact. Following this local outreach, statewide staff posted a recruitment flier in the "University of Minnesota Fruit and Vegetable News" (1757 subscribers). Farmers were accepted into the trial until a total of 100 participants joined. Recruitment was carried out in this way versus a more random approach to ensure the geographic distribution of farm sites and to match educator capacity for conducting soil tests. A total of 200 vegetable sites (100 open fields, and 100 high tunnels) were identified on the participating farms. Participants with both high tunnel and field-grown vegetables were prioritized and made up the majority of trial sites (83 farms). On these 83 farms, one open field and at least one adjacent high tunnel were sampled, reducing variability by sampling one of each treatment group on the same soil type and location. The additional 17 farms did not have high tunnels, so a second high tunnel from one of the nearby 83 farms was included. Figure 1 illustrates the geographic location of the sites used in this study.

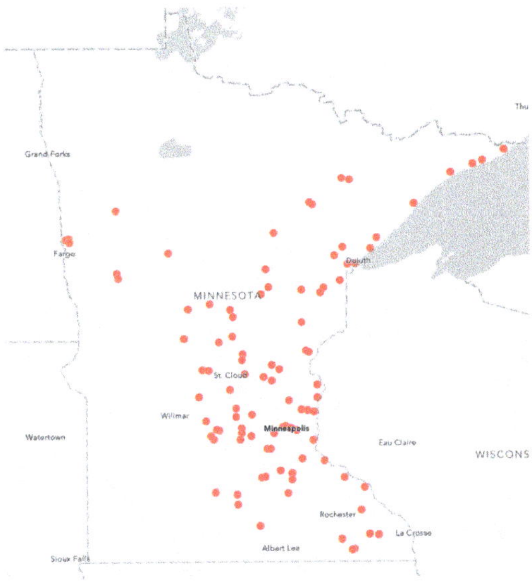

Figure 1. Location of 100 participating farm sites. Each farm site is indicated with a red dot. Names of major cities are included in the map for reference, and dark gray areas on the map show major bodies of water.

Extension educators collected soil and water for testing at each site, and during each visit, participants completed a survey collecting information on agricultural practices and land use history, administered via Qualtrics. Participants were not directly compensated for completing the project. Rather, they received free soil testing and consultation.

2.2. Survey

A 20-question survey questionnaire was developed using Qualtrics to learn about soil health and nutrient management practices on small-scale vegetable farms in Minnesota. The first draft of the survey was developed by the PI, and then it was reviewed for technical content and functionality by three University of Minnesota soil scientist colleagues, as well as four local Extension educators. The questionnaire was approved for distribution by the University of Minnesota Institutional Review Board (study #00018630).

The survey was divided into five sections: demographics and site history, fertility decisions and practices, soil health practices, compost use, and land tenure. It included a mix of multiple-choice and fillable text questions.

Participants completed the survey online via smartphone while the educator visiting their farm completed soil testing. In areas with limited internet access, they were given the link via email and encouraged to complete the survey when they had access to the internet. Participants completed a survey for each site sampled at their farm (e.g., one survey for their field and another for their high tunnel). The survey had an 88% response rate. Survey data were cleaned by the PI, which included de-identifying participants with a code for each farm and formatting the data for analysis in R. Unanswered questions were left blank in the dataset, and later filtered during statistical analysis.

2.3. Soil Tests

University of Minnesota Extension educators collected soil tests at each site between 7 April and 5 June 2023. Soil samples were collected with a standard soil probe from 0 to 15 cm, moving away any mulch present. Between 15 and 20 cores were collected from each site and aggregated into a single composite sample. The composite samples from each location were sent to the University of Minnesota Research Analytical Laboratory (RAL) to quantify Bray-P1 phosphorus, exchangeable potassium, and nitrate [33,34]. Additional soil testing was completed, which will be reported in a future publication.

2.4. Tillage Score

The tillage score was calculated based on survey results and an equation developed by the project team based loosely on the STIR rating [35], which attempts to assign a quantitative value to tillage intensity rather than classifying farms into vague qualitative groups like "no till", "low till", or "conventional till". Following the approach of Büchi and colleagues [36], a weight was assigned to different types of tillage, and participants were asked to report the average frequency of each type of tillage. They reported on the number of deep (rototiller or moldboard plow), shallow (e.g., harrow or tilther), and manual (broadfork) tillage passes per year. The equation used was as follows:

$$\text{Tillage score} = (\#\text{ deep tillage passes per year} \times 2) + (\#\text{ shallow tillage passes per year} \times 1) + (\#\text{ manual tillage passes per year} \times 0.5) \quad (1)$$

To further assess tillage differences, we created a farm size variable. Sites were grouped by size and production type: high tunnels, ≤1-acre fields, 1.1–3-acre fields, 3.1–5-acre fields, and >5-acre fields. These categories were chosen based on equipment suitability at each scale (based on author experience and observations): one can feasibly manage a 1-acre farm using only hand tools, whereas two-wheel tractors and other motorized equipment such as rototillers become more common at the 1–3-acre scale and larger equipment such as tractors become more common around the 5-acre scale.

2.5. Statistics

Survey and soil data were analyzed using R v4.2.2 [37]. Graphs were generated with the ggplot2 package, and summary statistics were generated using basic commands within the dplyr package. A basic chi-square test was used to assess differences between non-ordered categorical variables. For ordered categorical variables, we also used the Spearman correlation coefficient. Variable levels were assigned a rank in R (e.g., for organic certification, conventional = 1, using mostly organic practices = 2, using exclusively organic practices = 3, and certified organic = 4). Finally, Kruskal–Wallis was used to compare categorical values with continuous variables.

3. Results

3.1. Farm Demographics

Participating farms had an average of five acres in production, with fields in vegetable production for an average of seven years. A total of 23% of participants had less than 5 years of experience farming, 25% had 5–10 years of experience, and 49% had more than 10 years of experience. Three percent did not share their level of experience. Participants self-identified as belonging to the following racial and ethnic groups: white (75%), American Indian or Alaska Native (6%), Asian, Native Hawaiian, or other Pacific Islander (4%), Black or African American (5%), and Hispanic or Latino (1%). Nine percent chose not to answer. While only 14% of participants were certified organic, 44% claimed to use exclusively organic practices, and an additional 18% used mostly organic practices. In total, 11% of participants used conventional practices, while 13% chose not to specify.

Land use history (how the site was managed prior to growing vegetables, as reported by study participants) varied substantially. The largest category of previous land use was row crop farming (32%), followed by "other" (16%), fallow (13%), grazing animals (9%), specialty crop production under a different manager or owner (6%), woodland (5%), prairie (5%), and former industrial site (1%). The remaining respondents did not specify (13%). Write-in responses included hay or alfalfa, lawn, mixed use, and fruit production (7, 4, 3, and 2% of the total responses, respectively).

3.2. Demographics and Soil Testing

Farmers in our study were equally likely to test their soil in high tunnels versus fields, and the decision to collect soil for testing was not impacted by the farmer's race or ethnicity (Table 1). Across all testing sites (including fields and high tunnels), organic certification, farmer experience, and land access situation were all significantly correlated with the frequency of soil testing (Table 1). In general, certified organic farmers were the most likely to test their soil at regular intervals. Farmers who claimed to use organic practices without certification were not more likely than those who self-identified as conventional farmers to do regular soil tests. Less experienced farmers were also less likely to have tested their soil than those with at least five years of experience. While land access was significantly correlated with soil testing frequency, it did not follow a clear pattern, and having more stable land access did not necessarily make someone more or less likely to test their soil (Table 1).

Table 1. Impacts of various demographic factors of survey participants on how often they test their soil. For each variable, the chi-square test is reported. For ordered variables (e.g., experience), the Spearman correlation coefficient is also reported. Responses (% of total response) for each category are listed for variables that were significantly correlated with soil test frequency. Variables with significant correlations are shaded in gray.

Soil Test Frequency	Never	Once	4+ Years	2–3 Years	Every Year
Tunnel vs. field: $\chi^2 = 0.609$, $p = 0.9620$					
Race or ethnicity: $\chi^2 = 31.51$, $p = 0.1396$					
Organic certification $\chi^2 = 32.063$, $p = 0.0014$; $r_s = 0.26$, $p = <0.001$					
Conventional	12%	24%	24%	29%	12%
Mostly organic	34%	14%	6%	40%	6%
Exclusively organic, not certified	20%	16%	22%	31%	11%
Certified organic	7%	0%	7%	52%	34%
Experience: $\chi^2 = 27.823$, $p = 0.0005$; $r_s = 0.087$, $p = 0.265$					
<5 years	41%	16%	0%	38%	6%
5–10 years	16%	11%	11%	36%	27%
>10 years	11%	15%	24%	38%	11%
Land access: $\chi^2 = 47.843$, $p = 0.03556$					
Informal lease	27%	18%	0%	18%	36%
Formal lease	30%	0%	0%	50%	20%
Own	21%	14%	17%	36%	12%
Land trust or community farm	0%	33%	0%	33%	33%
Tribal land	0%	33%	0%	67%	0%

3.3. Nutrient Management and Input Decision Drivers

Survey participants reported using soil tests as one of their top three decision-making factors more than any other source of information when deciding which fertility products to use (and how much) (61% in high tunnels, 55% in fields). Soil tests were followed by observations of previous crop performance (55% in high tunnels, 52% in fields). Farmers were more likely to seek advice from other farmers about fertility (40% in high tunnels, 41% in fields) than they were to ask private consultants (20% in high tunnels, 22% in fields), Extension (14% in high tunnels, 16% in fields), co-ops or input sellers (10% in high tunnels, 9% in fields), or nonprofit support organizations (5% in high tunnels, 18% in fields). A total of 14% of participants selected "other" in high tunnels and 18% selected it in fields as one of their top decision-making factors. Write-in responses for "other" included applying based on observation of plant performance (n = 3), performing the same practices they have used before (n = 2), not applying any fertilizer (n = 4), based on university or agronomist recommendations (n = 4), personal or family knowledge (n = 5), "plant needs" (n = 3), personal research based on books or online articles (n = 4), and applying as much as is available n = 1) or as much as the participant could afford (n = 1).

3.3.1. Soil Testing

Since participants ranked their soil tests as the primary driver of nutrient management decisions, we created a new binary variable based on whether a survey respondent indicated soil tests as a factor in their decisions, then did basic *t*-tests to determine whether using a soil test for fertility decisions impacted the amount of each macronutrient in the soil. There were no significant differences in soil phosphorus (p = 0.1378), potassium (p = 0.1284), or nitrate (p = 0.1285) between farmers who claimed to use soil tests as a basis for fertility decisions and those who did not. This is consistent with the findings above, indicating that the use of soil tests is not significantly correlated with soil macronutrient values, except for people who had never completed a soil test in their high tunnels.

We also looked at whether soil test frequency influenced soil nutrient concentrations. If someone had never taken a soil test in their high tunnel, their soil nitrate concentrations were likely to be higher than someone who had taken a soil test, but farmers who tested more frequently had similar nitrate concentrations to those who tested less frequently (Figure 2). This same dynamic was true for phosphorus, but only in high tunnels (Figure 3). Using the Bray-PI test for phosphorus, 41 ppm is considered "very high" for most vegetable crops, and no additional phosphorus is recommended after 50 [38]. Of the 100 tunnels and 100 fields we sampled, 87% of tunnels and 84% of fields exceeded the "very high" threshold for soil phosphorus. Even on farms that tested their soil every year, soil phosphorus levels were on average well above the "very high" threshold at which no additional added phosphorus is recommended (Figure 3). Potassium concentrations were similar across all groups, including those who had never tested their soil (Figure 4).

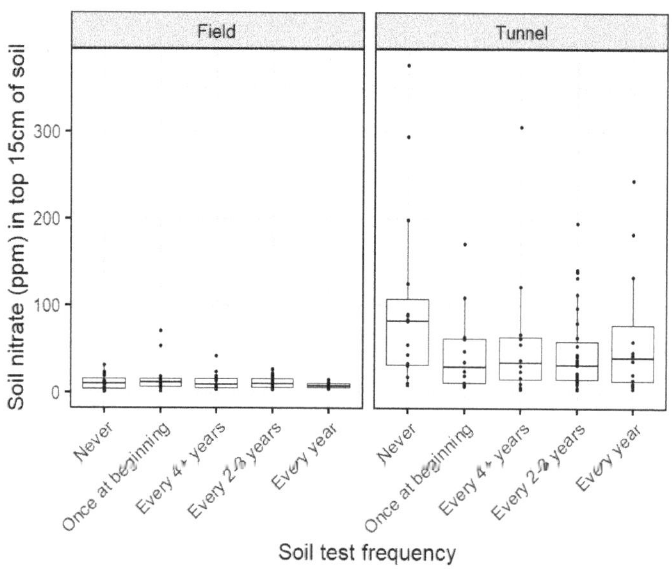

Figure 2. Frequency of soil testing vs. nitrate concentrations in the top 15 cm of soil (ppm) in 100 vegetable fields and 100 high tunnels in Minnesota. Box plots indicate the median, 1st, and 3rd quartile for each group.

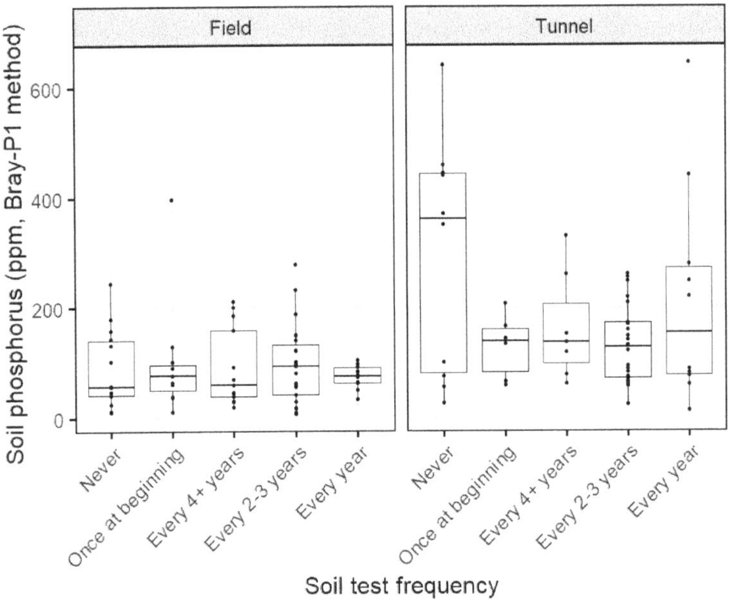

Figure 3. Frequency of soil testing vs. soil phosphorus concentrations in the top 15 cm of soil (ppm) using the Bray-P1 extraction method in 100 vegetable fields and 100 high tunnels in Minnesota. Box plots indicate the median, 1st, and 3rd quartile for each group.

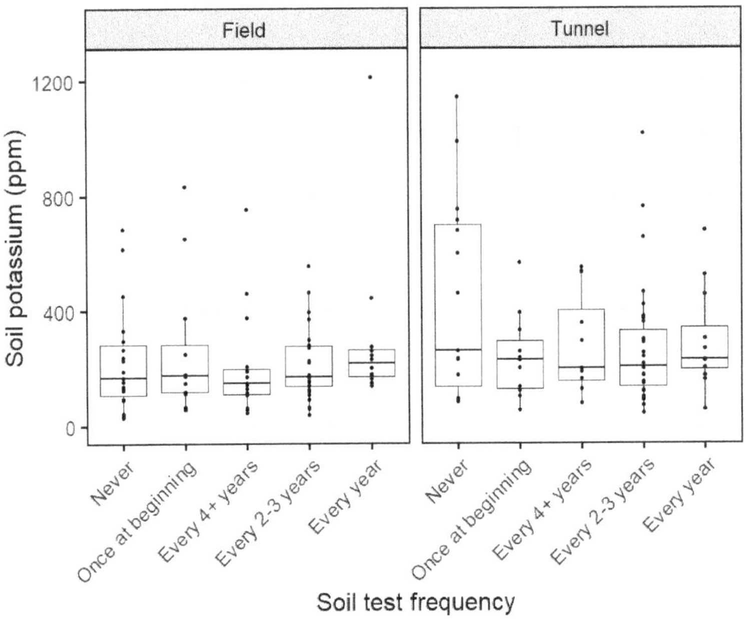

Figure 4. Frequency of soil testing vs. potassium concentrations in the top 15 cm of soil (ppm) in 100 vegetable fields and 100 high tunnels in Minnesota. The y-axis was limited to 1250 ppm, occluding one outlier (high tunnel, once at the beginning, 2464 ppm) to improve readability of graph. Box plots indicate the median, 1st, and 3rd quartile for each group.

3.3.2. Inputs

Inputs were similar across high tunnels and fields. Composted manure was the most frequently used input in both systems, with over half of respondents using it every year or more than once per year. This was closely followed by "organic supplemental fertilizers like bone meal, fish meal, blood meal, feather meal, etc." (Figures 5 and 6). Because educators have noticed confusion about the use of the term "compost", and the fact that it is often used interchangeably to describe composted manure and vegetable-based compost, we asked participants to share where they source their compost, and what type of compost they most often use. A total of 29% of respondents said they use animal-based compost, 16% said they use plant-based compost, and 54% said they use a mix of the two. Most compost was generated on the farm. Off-farm compost primarily came from commercial facilities. Farmer participants were more likely to purchase compost from a cooperative or input store for use in their high tunnel, whereas, for use in fields, they were more likely to source it from neighbors (Table 2).

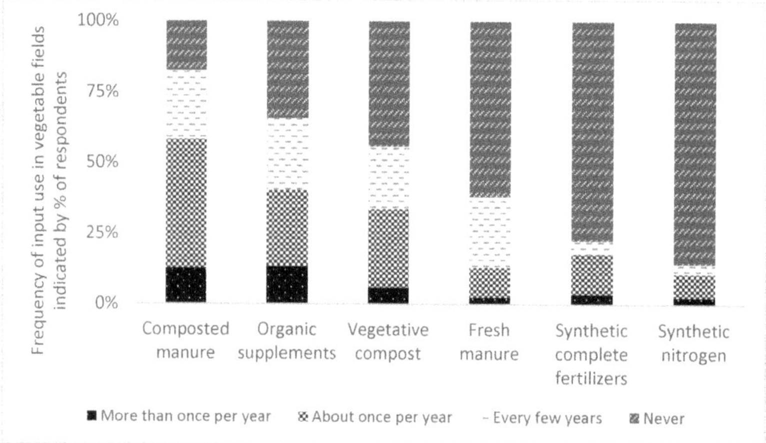

Figure 5. Input use frequency in fields at 100 Minnesota vegetable farms as reported by farmer participants.

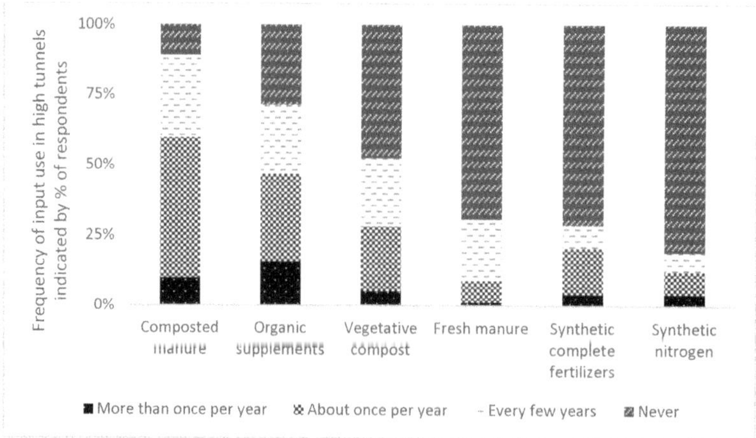

Figure 6. Input use frequency in high tunnels at 100 Minnesota vegetable farms as reported by farmer participants.

Table 2. Survey responses to questions about compost sourcing in both high tunnel and field environments at 100 Minnesota vegetable farms. Percentages reflect the percentage of participants who selected each option on a multiple-choice survey.

	High Tunnels	Fields
Source of compost (multiple responses allowed)		
Generated on farm	62%	62%
Commercial compost facility	35%	41%
Neighbors	10%	15%
County or city delivery program	7%	8%
Agricultural cooperative or input store	9%	4%
Yard waste site	4%	3%
Reasons for adding compost (multiple responses allowed)		
To add fertility (nutrients)	83%	81%
To improve soil structure	70%	79%
To add organic matter	72%	76%
To lessen soil compaction	32%	32%
To bury weeds	7%	8%
To fill beds	10%	6%
To remediate contamination	0%	0%
Other	10%	9%
How much compost do you apply each year?		
A set amount (e.g., 1 inch per bed)	41%	34%
As much as I can get	24%	34%
Based on soil tests	16%	16%
Enough to bury weeds	3%	4%
Other	16%	12%

The highest-ranked motivation for using compost among survey participants was to add fertility to the soil, indicating that many farmers do consider compost to be a source of nutrients. This was closely followed by the desire to improve soil structure and increase organic matter, with fewer participants using compost for compaction remediation, to bury weeds, or to fill beds. No one reported using compost to remediate contamination (Table 2). All of the "Other" responses were related to increasing crop yields, except for one write-in response from an open field indicating that the producer adds compost to improve soil biodiversity.

While participants ranked soil tests as their primary motivating factor in making fertility decisions, soil tests were less important than other factors when it came to deciding how much compost to apply. According to survey responses, participants were most likely to apply a set amount of compost each year (41% in tunnels, 34% in fields), or to apply as much as they could source (24% in tunnels, 34% in fields) (Table 2). In both tunnels and fields, only 16% of respondents applied compost based on soil test results. Write-in responses for "Other" included comments about the limited experience and still figuring out compost applications, amounts based on equipment (e.g., "one spreader full" or one truckload), purchasing as much as people could afford, and as much compost as is generated in the respondent's chicken coop.

3.4. Cover Crop Use

Overall, farmers were significantly more likely to use cover crops in open fields than in high tunnels (χ^2 = 17.208 p = 0.0006). While 72% of the farmer participants had planted a cover crop in their fields, only 41.5% had done so in their high tunnel (Table 3).

Table 3. Survey responses to questions about cover crop use in high tunnels and fields at 100 Minnesota vegetable farms, percentages reflect the percentage of participants who selected each option on a multiple-choice survey.

	High Tunnels	Fields
	Cover crop use frequency	
Never	58.50%	28%
Occasionally	19.50%	26%
Every 2–3 years	10%	21%
Every year	12%	26%
	Cover crop barriers	
Cost	5%	10%
Limited space	15%	15%
Questions about logistics	16%	11%
Limited time	23%	29%
Concern about performance of next crop	6%	2%
Other	35%	33%

Despite these differences in cover crop adoption between high tunnels and open fields, barriers were nearly identical across environments (Table 3). In fields, farmer participants were slightly more likely to cite costs and time constraints as a barriers. In high tunnels, they were slightly more likely to cite performance about the next crop and questions about logistics as barriers. The choice most frequently selected by the farmers for both fields and high tunnels when asked about barriers to cover crop use was "Other", suggesting that a key factor (or factors) was missed in the survey questionnaire. There were 52 write-in responses, some of which included multiple barriers: 28 responses related to a lack of time, not necessarily in terms of labor, but in terms of being able to fit cover crops into existing crop rotations. Of these 28 responses, 12 were related to fields, and 16 were related to high tunnels. Two people mentioned struggles incorporating cover crops into their perennial plantings, and seven mentioned weather as a barrier, specifically a lack of rainfall at the right time. Seven people mentioned lacking experience or being unsure of where to find seed. Two people mentioned labor and cost, and five discussed equipment barriers. Among the equipment barriers, two felt that the equipment needed to plant and terminate cover crops was too costly, and the other three shared that cover crops do not work with their system due to the use of plastic mulch or landscape fabric.

Of the demographic variables surveyed, only organic certification was correlated with cover crop use ($\chi^2 = 21.149$, $p = 0.012$). Overall, certified organic farmers were the most likely group to plant a cover crop every year in both environments, and the least likely group to have never planted a cover crop (Table 4). In high tunnels, farmers claiming to use mostly or exclusively organic practices without certification were slightly more likely to plant a cover crop than farmers who identified as conventional, though, in fields, the results were more mixed (Table 4).

Table 4. Frequency of cover crop use across environments according to farmer's organic certification status based on farmer survey responses at 100 Minnesota vegetable farms. Percentages reflect the percentage of participants who selected each option on a multiple-choice survey.

	Never	Occasionally	Every 2–3 Years	Every Year
	High tunnels			
Conventional (n = 7)	86%	14%	0%	0%
Using mostly organic practices (n = 17)	71%	6%	24%	0%

Table 4. *Cont.*

	Never	Occasionally	Every 2–3 Years	Every Year
Using exclusively organic practices but not certified (n = 43)	58%	26%	2%	14%
Certified organic (n = 15)	33%	20%	20%	27%
	Fields			
Conventional (n = 11)	27%	9%	45%	18%
Using mostly organic practices (n = 17)	35%	18%	18%	29%
Using exclusively organic practices but not certified (n = 44)	32%	34%	16%	18%
Certified organic (n = 14)	7%	21%	21%	50%

Farmer experience (χ^2 = 5.287, p = 0.5075), land access arrangement (χ^2 = 28.194, p = 0.2519), and race (χ^2 = 24.046, p = 0.1535) did not significantly impact cover crop use. In fields, total acreage was not correlated with cover crop use (χ^2 = 34.345, p = 0.356).

3.5. Tillage Practices

Overall, tillage intensity, as determined by tillage score (Equation (1)), was slightly lower in high tunnels than in fields (Figure 7). This difference was borderline significant according to a chi-square test (χ^2 = 3.1817, p = 0.0745).

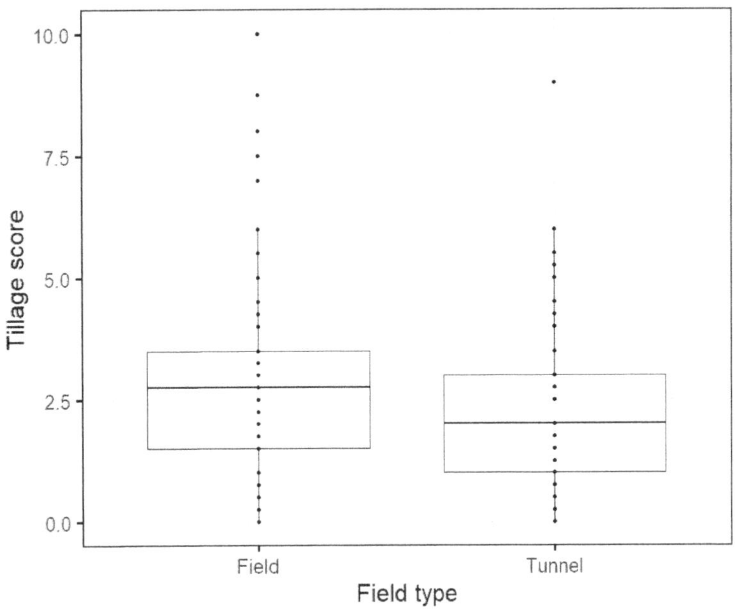

Figure 7. Tillage score (according to Equation (1)) in 100 fields vs. 100 high tunnels on Minnesota vegetable farms. Box plots indicate the median, 1st, and 3rd quartile for each group.

Organic certification and land access did not significantly correlate to tillage score (χ^2 = 6.5276, p = 0.1631 and, χ^2 = 8.9553, p = 0.3461, respectively) but more experience was significantly correlated with tillage score (χ^2 = 8.8471, p = 0.0120). In general, experienced farmers were more likely to use more intensive tillage than their less experienced counterparts (Figure 8). More experienced farmers did, on average, have larger farms than those with less experience (Figure 9).

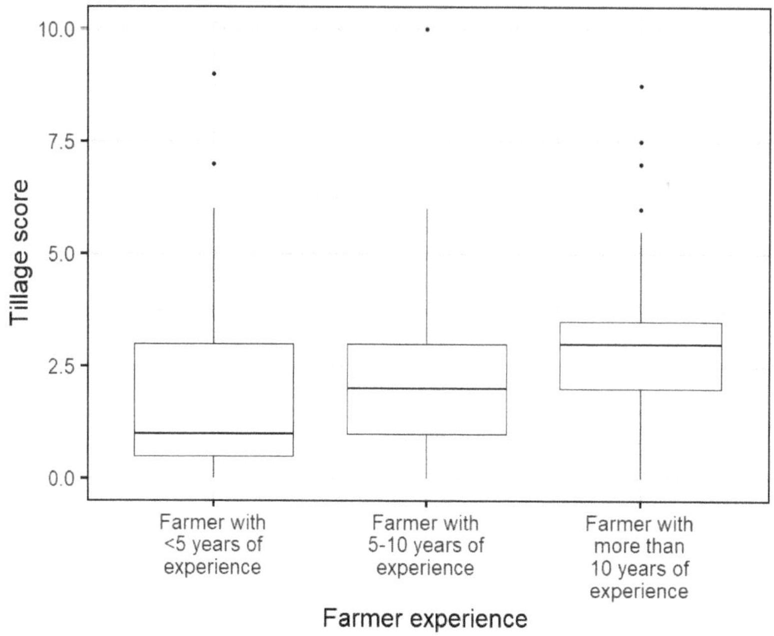

Figure 8. Tillage score (according to Equation (1)) in 100 fields and 100 high tunnels (data aggregated across sites) based on farmer experience level. Box plots indicate the median, 1st, and 3rd quartile for each group.

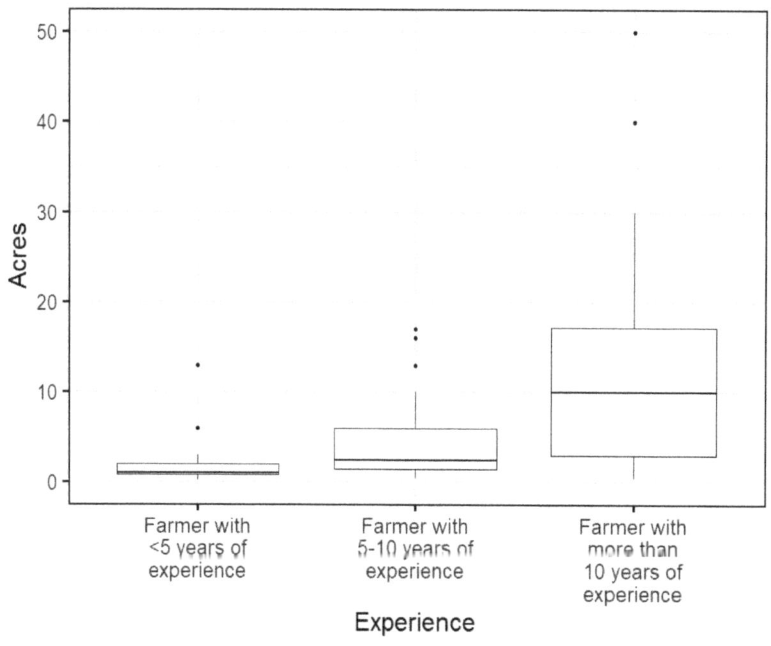

Figure 9. Farm size (acres in production) vs. farmer experience at 100 Minnesota vegetable farms. Box plots indicate the median, 1st, and 3rd quartile for each group.

When we looked at types of tillage rather than the overall tillage score, we saw differences between farms of different scales. The smallest farms used less frequent deep tillage (rototillers or moldboard plows) and less shallow tillage (harrows, tilthers) than larger farms, but used manual tillage techniques like broadforks much more frequently. Farmers with 1–3 acres were more likely to use shallow tillage techniques, whereas farmers with more than 3 acres were more likely to use more deep tillage techniques. Tillage techniques in high tunnels fell somewhere in the middle (Figure 10).

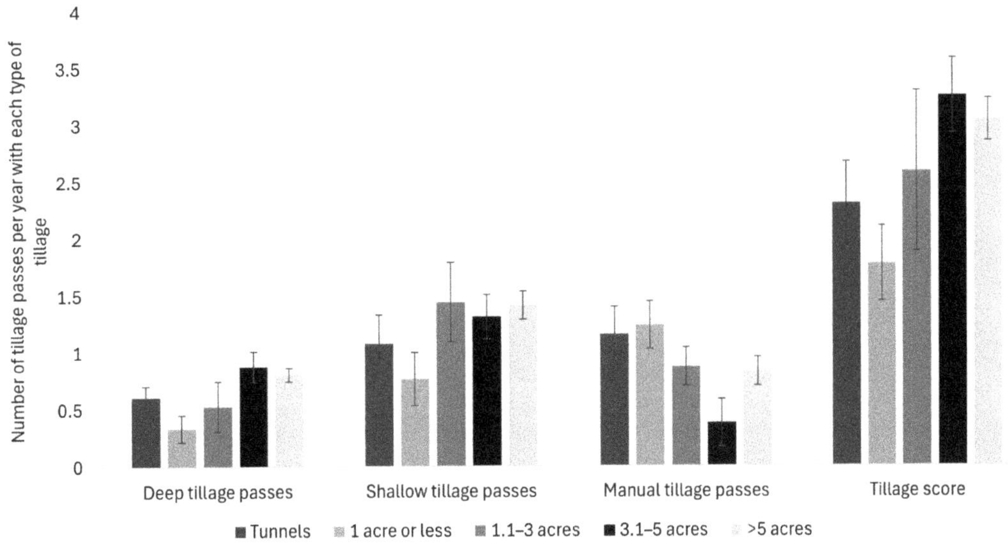

Figure 10. Types of tillage and frequency of tillage passes at 100 Minnesota vegetable farms based on farm size and production environment (high tunnel vs. field). Tillage score is the cumulative tillage intensity based on Equation (1). Bars represent means, error bars represent standard error.

There was no significant association between cover crop use and tillage ($\chi^2 = 0.1795$, $p = 0.9808$).

4. Discussion

4.1. Nutrient Management Decision-Making

More experienced farmers and certified organic farmers in our study were more likely to utilize soil testing than their less experienced or non-certified organic peers. The National Organic Program requires that certified organic producers manage inputs in a manner that does not contribute to contamination of crops, soil, or water by plant nutrients [39]. Rule §205.601(j) prevents farmers from adding micronutrients to their soil without first documenting a deficiency by testing the soil [39]. These rules, along with support provided by certifiers likely contribute to the higher rates of soil testing among certified organic farmers. While there is limited research exploring the links between organic certification and soil testing, organic certification has been linked to improved soil health in a variety of studies and contexts [40–42].

This finding suggests a need for more outreach and education targeted to beginning farmers about how to correctly collect soil samples for soil testing and also how to interpret laboratory soil test results, particularly for farmers who are not certified organic or pursuing certification. Extension agents have been identified as crucial for promoting organic farming practices and other agroecological practices [42]. As such, Extension programs should invest in promoting soil testing and soil test interpretation with this audience and also consider promoting organic certification to connect farmers with additional support for sustainable nutrient management practices.

Even when farmer participants tested their soil regularly, this did not translate to meaningful differences in soil nutrient concentrations, and most fields and high tunnels in the study had excessive soil phosphorus. Excess phosphorus in agricultural soils is correlated with runoff and leaching, resulting in eutrophication of freshwater ecosystems [43,44]. A 2022 soil health-related needs assessment of emerging and beginning farmers in Minnesota identified that even when farmers test their soil, they often struggle to interpret soil test results and make nutrient management decisions accordingly [32]. These results support the finding that soil testing does not always translate into nutrient management decision-making and further emphasizes the need for additional education about soil test interpretation among small-scale vegetable farmers. This is consistent with farmers in both Michigan and Ghana, whose observations of physical and biological attributes of soil generally aligned well with soil health assays, but whose perceptions of chemical properties of soil did not consistently align with soil test results [45,46].

The farmers in our study based fertility decisions more on their own observations of previous crop performance and other farmers' recommendations than on consultant or Extension recommendations. This is consistent with findings from studies on farmer decision-making across the region and the world, in which farmers prefer to learn from one another versus seeking advice from agricultural professionals [26,47–49]. By leaning into this finding and using train-the-trainer models, peer learning cohorts, and other methods to engage communities of farmers in learning together, educators can have more impact when sharing best practices for soil health and nutrient management.

4.2. Compost Confusion

This study highlights a particular need for compost-specific and composted-manure-specific education among diversified vegetable farmers. Composted manure was the most commonly used input among farmers in the study. They applied it primarily to add fertility and yet treated it very differently from other nutrient sources. They tended to apply as much as possible or a set amount each year rather than calculating compost and composted manure inputs based on soil test values. A previous qualitative study identified similar issues among beginning and emerging farmers in Minnesota, specifically, over-application of composts including composted manure, and conceptualizing composted products separately from nutrient sources [32]. Our findings here show that these challenges do not just apply to emerging farmers and that even experienced diversified vegetable farmers struggle to apply nutrient management principles to the use of composted manure and vegetative compost.

Compost applications can enhance the sustainability of a farm by improving root growth and nutrient uptake, improving crop yields and quality, and improving soil characteristics including porosity, aggregate stability, moisture and nutrient contents, and organic matter [50]. Mismanagement of synthetic fertilizers and pesticides has led to pollution and the degradation of soil on a global scale [51], and supplementing fertility programs with compost can mitigate some of these challenges. The use of compost to improve soil conditions has implications for the ability of a farm to remain sustainable, including the ability to withstand unpredictable weather conditions, manage pathogens, reduce soil erosion and nutrient runoff, and reduce the need for chemical fertilizers and irrigation [50]. Composting also improves the sustainability of global waste management and phosphorus management. A significant portion of human municipal waste is compostable, with estimates of 50% in Europe [52], and 40–70% globally [53]. This is particularly true in urban environments [54]. By diverting this waste from landfills and recycling it into high-quality compost, there are sustainability implications for greenhouse gas reductions and soil organic matter improvement at a global scale [52,53]. A 2011 analysis of the Minnesota Twin Cities Capitol Region Watershed estimated that food consumed and wasted by humans was the largest source of phosphorus inputs to the watershed and that diverting food and yard waste from landfills would reduce the storage of phosphorus in the watershed landscape by 66% [55].

Using compost to supplement soil fertility programs contributes to improving the sustainability of agriculture by improving nutrient use efficiency and productivity [56]. However, to promote more informed use of composts and mitigate externalities, there is a need for more research to develop accurate recommendations about the right volumes and types of compost needed for substituting commercial or synthetic fertilizers across different soils, crops, and growing conditions [53,56]. Compost is a highly variable product, referring to any organic matter that has been derived from plants, animals, or people and decomposed under aerobic conditions [50]. The final product depends on composting temperature, moisture, carbon-to-nitrogen ratio, source, and particle size of the feedstock, the presence of microorganisms, and the amount of oxygen and aeration available during the process [50].

If farmers are not using soil tests to determine compost application rates, but using compost and composted manure as primary fertility sources, soil tests have limited utility. This, combined with the excessively high soil nutrient concentrations on the farms in our study, represents a need for more targeted education about how to translate soil test results for farming systems primarily relying on composted vegetable scraps and manure for fertility.

4.3. Cover Crops

The majority of farmers in this study (72%) used cover crops at least occasionally in their fields, while the majority (58.5%) had never done so in their high tunnels. Adoption of cover crops was significantly higher in this group overall compared to estimates of the average Minnesota farmer. According to the 2022 Census of Agriculture, only 9% of Minnesota farms that participated in the census planted cover crops, representing about 3% of farmland represented in the census [6]. Of the demographic variables studied, only organic certification was significantly correlated with higher cover crop adoption. Farm size, experience, race, or land access were not significantly correlated with cover crop use. Other studies of small-scale vegetable farmers have identified that this group of farmers tends to be very conservation-minded, prioritizing soil health practices as a core motivation behind their approach to farming [10,13]. While a study of 541 organic fruit and vegetable farmers in the United States found that smaller farmers (<40 acres) were more likely to use cover crops and other conservation practices than their larger counterparts, almost all of the farmers in our study fell into the "small" farm category [41].

Despite the lower adoption of cover crops in high tunnels, we did not identify any barriers that were more significant in high tunnels than in open fields. Overall, the most significant barriers for cover crop use were a lack of time, and not being able to fit cover crops into crop rotations. This is consistent with the characterization that small-scale vegetable farms tend to plant very intensively, using season extension and succession planting to efficiently produce high yields on a small scale, leaving limited windows for cover crops and fallow periods [19].

The cost was one of the least significant barriers to cover crop adoption. This is important, as many of the conservation payment programs that incentivize cover crop use cover the cost of seed, but not of farmers' time. A 3-year study on economic tradeoffs for high tunnel cover crops in Minnesota, Kansas, and Kentucky found that seed made up only 1% or less of the cost of planting cover crops, and that labor accounted for the vast majority of the cost [57]. Our results support the finding that incentive programs may be more effective if they can offset the costs of labor related to planting cover crops versus simply covering the cost of seed [57].

Only 16% of high tunnel participants and 11% of field participants identified educational barriers (questions about logistics). This suggests that the primary barriers to cover crop use on small-scale vegetable farms are not educational in nature. Nonetheless, targeted educational outreach could still be beneficial for improving cover crop adoption among the farmers who identified it as a barrier.

4.4. Tillage

More experienced farmers in our study were more likely to use more intensive tillage than their less experienced counterparts. While we do not have enough data to investigate the reasons behind this trend, this finding is consistent with a study of 96 organic (including those using organic practices but not certified) field crop and vegetable producers in Michigan. The survey found that farmers with fewer years of experience were more likely to be interested in reduced tillage methods and that vegetable farmers were more likely to be interested in reduced tillage methods than field crop farmers [58]. While this study did not report on why, they cited that smaller, newer farms generally have less capital to invest in equipment like reduced tillage machinery and that the return on investment for equipment may be lower on highly diversified vegetable farms where equipment may only be suitable for a subset of the crops grown. Small beginning farmers are also likely to be debt-averse [10]. Thus, smaller vegetable farms may be more motivated to find alternatives [58]. Our breakdown of tillage types among different farm scales supports this hypothesis. Our results also show that while reduced tillage methods are common among the smallest-scale vegetable farmers (1 acre or less), there is a need for education, as well as research into methods for reducing tillage at the 3–50-acre scale. This is especially important for organic systems; conventional no-till approaches rely heavily on herbicides [59–61].

While organic certification was significantly correlated with soil testing and cover crop use, it was not significantly correlated with tillage score.

Finally, a commonly cited tradeoff of using cover crops, particularly in systems that do not use herbicides, is that tillage is typically required to terminate them [62]. We did not find a correlation between the frequency of cover crop use and tillage score to support this. This may be partially explained by the fact that if well timed, the weed suppression benefits of cover crops may offset additional tillage passes for weed management [62].

4.5. Limitations and Future Research

Because of the number of participants in our study and the complexity of their agricultural systems (i.e., many crops per farm), our analysis lacks a detailed accounting of inputs. While our general categorization of input frequency (e.g., number of times per year participants applied a variety of inputs), the lack of quantification limits our ability to draw detailed conclusions about nutrient management practices. We can infer that the use of compost, particularly composted manure, is contributing to excessive nutrient accumulation based on soil nutrient levels and qualitative survey data, but without a full cost accounting of inputs, these conclusions are speculative in nature.

We also acknowledge the limited soil data presented in this paper. Due to the volume of data produced in this project, a follow-up paper will document more detailed soil test data from the 100 farms. Future research should address both agronomic challenges addressed here, including the development of better guidance for using compost products for fertility in a diversified agricultural system, as well as the social dynamics of how to effectively communicate soil health practices to farmers.

5. Conclusions

Small-scale vegetable farmers face unique challenges related to managing their soils due to intensive planting windows, labor demands, insufficient access to capital and equipment, and heavy reliance on inputs like compost and composted manure with hard to calculate nutrient concentrations.

Extension and other educational programs should develop targeted resources and education to support farmers with nutrient management in these systems, particularly related to the application of compost and avoiding over-application of nutrients like phosphorus as well as soil testing and test interpretation. Farmers in our study used composted manure and vegetative compost far more often than synthetic fertilizers (58% of farmers used composted manure at least once per year, 34% used vegetative compost compared to 18% using synthetic complete fertilizers, and 11% using synthetic nitrogen at least once per year in fields), and yet the majority (87% of high tunnels and 84% of fields) had soil phosphorus levels that were "very high", compromising the sustainability of these systems. Farmer participants did not treat vegetative compost and composted manure like other fertility inputs: rather than relying on soil tests to make decisions about application rates, they tended to apply as much as possible or a set amount each year, likely contributing to the over-application of phosphorus. More research is needed to help farmers make informed decisions about organic inputs, and educators may be most effective by tapping into existing farmer networks and leveraging peer learning.

Cover crops are particularly challenging for this audience due to short planting windows, labor availability, and equipment requirements. Despite this, small-scale vegetable farmers in this study far surpassed the average Minnesota farmer in their adoption of cover crops with 72% of farmers in our study using cover crops at least occasionally, compared to 9% of Minnesota farmers in the 2022 agriculture census. Organic certification was the only significant driver of cover crop adoption; farm size, farmer experience, race, and land access were not significantly correlated with cover crop adoption. Conservation incentive programs that compensate labor in addition to seed costs in combination with more targeted educational outreach may significantly benefit small-scale vegetable farmers and other farmers.

The use of tillage was related to farm size (or at least land in vegetable production), with smaller areas (high tunnels and open fields < 1 acre) relying mostly on manual tillage and larger areas using mechanized tillage. Other demographic variables including organic certification, farmer experience, race, and land access were not significantly correlated with tillage score. More investment in reduced tillage practices and methods for small-scale growers in the 3–50-acre range is critical for reducing tillage among small-scale vegetable farmers.

Overall, farmers in our study were likely to use practices widely considered sustainable including the use of organic inputs, cover crops, and reduced tillage. However, they need more targeted information and educational resources to support their continued sustainability.

Author Contributions: Conceptualization, formal analysis, data curation, writing—original draft preparation, visualization, project administration, and funding acquisition, N.H.; methodology, N.H. and P.H.P.; investigation, N.H., S.M.B. and E.L.; writing—review and editing, N.H., S.M.B., E.L. and P.H.P.; supervision, P.H.P. All authors have read and agreed to the published version of the manuscript.

Funding: This research was funded by the University of Minnesota Agricultural Experiment Station Rapid Response Fund, which is supported by the Minnesota state legislature.

Institutional Review Board Statement: Ethical review and approval were waived for this study by the University of Minnesota Institutional Review Board because it was determined to be "Not human research" (study #00018630).

Informed Consent Statement: Informed consent was obtained from all subjects involved in the study.

Data Availability Statement: The data presented in this study are available on request from the corresponding author.

Acknowledgments: We are thankful to our team of educators who participated in soil testing for this project: Anthony Adams, Katie Hagen, Katie Drewitz, Claire LaCanne, Randy Nelson, Jennifer Hahn, Sarah Waddle, Colleen Carlson, Mercedes Moffet, Troy Salzer, Tarah Young, Kaitlyn Albers, and Erik Heimark. We are also grateful to Julie Grossman, Adria Fernandez, and Carl Rosen for their input on survey design and analysis.

Conflicts of Interest: The authors declare no conflicts of interest.

References

1. Tahat, M.M.; Alananbeh, K.M.; Othman, Y.A.; Leskovar, D.I. Soil Health and Sustainable Agriculture. *Sustainability* **2020**, *12*, 4859. [CrossRef]
2. Lichtfouse, E.; Navarrete, M.; Debaeke, P.; Souchère, V.; Alberola, C.; Ménassieu, J. Agronomy for Sustainable Agriculture—A Review. *Agron. Sustain. Dev.* **2009**, *29*, 1–6. [CrossRef]
3. Biswas, S.; Ali, M.D.N.; Goswami, R.; Chakraborty, S. Soil Health Sustainability and Organic Farming: A Review. *J. Food Agric. Environ.* **2014**, *12*, 237–243.
4. United States. *U.S. Census of Agriculture*; Library of Congress: Washington, DC, USA, 2012.
5. United States. *U.S. Census of Agriculture*; Library of Congress: Washington, DC, USA, 2017.
6. United States. *U.S. Census of Agriculture*; Library of Congress: Washington, DC, USA, 2022.
7. Bailey, P.; Kagan, A. *Emerging Farmers in Minnesota*; Minnesota Department of Agriculture: St. Paul, MN, USA, 2020.
8. Ferguson, R.S.; Lovell, S.T. Livelihoods and Production Diversity on U.S. Permaculture Farms. *Agroecol. Sustain. Food Syst.* **2017**, *41*, 588–613. [CrossRef]
9. Inwood, S.; Clark, J.K.; Bean, M. The Differing Values of Multigeneration and First-Generation Farmers: Their Influence on the Structure of Agriculture at the Rural-Urban Interface. *Rural. Sociol.* **2013**, *78*, 346–370. [CrossRef]
10. Rissing, A. Alternative Economic Strategies and the Technology Treadmill: Beginning Vegetable Farmers in Iowa. *Econ. Anthropol.* **2016**, *3*, 304–314. [CrossRef]
11. Lamont, W.J. Overview of the Use of High Tunnels Worldwide. *HortTechnology* **2009**, *19*, 25–29. [CrossRef]
12. Pierre, J.F.; Jacobsen, K.L.; Wszelaki, A.; Butler, D.; Velandia, M.; Woods, T.; Sideman, R.; Grossman, J.; Coolong, T.; Hoskins, B.; et al. Sustaining Soil Health in High Tunnels: A Paradigm Shift toward Soil-Centered Management. *Hortte* **2024**, *34*, 594–603. [CrossRef]
13. DeLong, A.; Swisher, M.E.; Chase, C.A.; Irani, T.; Ruiz-Menjivar, J. The Roots of First-Generation Farmers: The Role of Inspiration in Starting an Organic Farm. *Land* **2023**, *12*, 1169. [CrossRef]
14. Natural Resource Conservation Service Soil Health Starter Kit 2022. Available online: https://www.nrcs.usda.gov/sites/default/files/2022-10/Soil-Health-Starter-Kit-reduced.pdf (accessed on 19 December 2024).
15. Sahu, P.K.; Singh, D.P.; Prabha, R.; Meena, K.K.; Abhilash, P.C. Connecting Microbial Capabilities with the Soil and Plant Health: Options for Agricultural Sustainability. *Ecol. Indic.* **2019**, *105*, 601–612. [CrossRef]
16. Gathala, M.K.; Timsina, J.; Islam, M.S.; Rahman, M.M.; Hossain, M.I.; Harun-Ar-Rashid, M.; Ghosh, A.K.; Krupnik, T.J.; Tiwari, T.P.; McDonald, A. Conservation Agriculture Based Tillage and Crop Establishment Options Can Maintain Farmers' Yields and Increase Profits in South Asia's Rice–Maize Systems: Evidence from Bangladesh. *Field Crops Res.* **2015**, *172*, 85–98. [CrossRef]
17. Gozubuyuk, Z.; Sahin, U.; Ozturk, I.; Celik, A.; Adiguzel, M.C. Tillage Effects on Certain Physical and Hydraulic Properties of a Loamy Soil under a Crop Rotation in a Semi-Arid Region with a Cool Climate. *CATENA* **2014**, *118*, 195–205. [CrossRef]
18. Cannell, R.Q.; Hawes, J.D. Trends in Tillage Practices in Relation to Sustainable Crop Production with Special Reference to Temperate Climates. *Soil Tillage Res.* **1994**, *30*, 245–282. [CrossRef]

19. Maughan, N.; Pipart, N.; Dyck, B.V.; Visser, M. The Potential of Bio-Intensive Market Gardening Models for a Transformative Urban Agriculture. In *Resourcing an Agroecological Urbanism*; Tornaghi, C., Dehaene, M., Eds.; Routledge: London, UK, 2021; pp. 144–165, ISBN 978-0-429-43356-6.
20. Waldman, K.B.; Conner, D.S.; Biernbaum, J.A.; Hamm, M.W.; Montri, A.D. Determinants of Hoophouse Profitability: A Case Study of 12 Novice Michigan Farmers. *HortTechnology* **2012**, *22*, 215–223. [CrossRef]
21. Bloom, D.; Lelekacs, J.; Dunning, R.; Piner, A.; Brinkmeyer, E. Local Food Systems Course for Extension Educators in North Carolina: Summary of an Innovative Program. *JOE* **2017**, *55*, 22. [CrossRef]
22. Niewolny, K.; Bendfeldt, E.; Latimer, J.; MacAuley, L. Model of Community, Local, and Regional Food Systems Extension Programming. *JOE* **2022**, *60*. [CrossRef]
23. Raison, B. Educators or Facilitators? Clarifying Extension's Role in the Emerging Local Food Systems Movement. *J. Ext.* **2010**, *48*, v48-3comm1.
24. Klodd, A.; Hoidal, N. *Needs Assessment of Minnesota Fruit and Vegetable Producers*; University Digital Conservancy: St. Paul, MN, USA, 2019.
25. McCamant, T. *Educational Interests, Needs and Learning Preferences of Immigrant Farmers*; Center for Rural Policy and Development: Mankato, MN, USA, 2014.
26. Rust, N.A.; Stankovics, P.; Jarvis, R.M.; Morris-Trainor, Z.; De Vries, J.R.; Ingram, J.; Mills, J.; Glikman, J.A.; Parkinson, J.; Toth, Z.; et al. Have Farmers Had Enough of Experts? *Environ. Manag.* **2022**, *69*, 31–44. [CrossRef]
27. Ruch, B.; Hefner, M.; Sradnick, A. Excessive Nitrate Limits the Sustainability of Deep Compost Mulch in Organic Market Gardening. *Agriculture* **2023**, *13*, 1080. [CrossRef]
28. Perkins, R. *Regenerative Agriculture: A Practical Whole Systems Guide to Making Small Farms Work*; RP 59°N: Sunne, Sweden, 2019; ISBN 978-91-519-1051-2.
29. Frost, J. *The Living Soil Handbook: The No-Till Grower's Guide to Ecological Market Gardening*; Chelsea Green Publishing: White River Junction, Vermont, 2021; ISBN 978-1-64502-027-1.
30. Möller, K. Soil Fertility Status and Nutrient Input–Output Flows of Specialised Organic Cropping Systems: A Review. *Nutr. Cycl. Agroecosystems* **2018**, *112*, 147–164. [CrossRef]
31. Pagliari, P.H.; Laboski, C.A.M. Dairy Manure Treatment Effects on Manure Phosphorus Fractionation and Changes in Soil Test Phosphorus. *Biol. Fertil. Soils* **2013**, *49*, 987–999. [CrossRef]
32. Hoidal, N.; Fernandez, A.; Kubovcik, K.; Berry, K.; Andreasen, A.; Ispache, D.; Grossman, J. Emerging Specialty Crop Farmer Perspectives and Educational Needs Related to Soil Health and Nutrient Management in the Upper Midwest. *Renew. Agric. Food Syst.* **2024**, *39*, e25. [CrossRef]
33. University of Minnesota. *University of Minnesota Research Analytical Laboratory: Major Equipment*; University of Minnesota: Minneapolis, MN, USA, 2024.
34. University of Minnesota. *University of Minnesota Research Analytical Laboratory: Method References*; University of Minnesota: Minneapolis, MN, USA, 2024.
35. Kucera, M. Soil Tillage Intensity Rating STIR. 2023. Available online: https://www.nrcs.usda.gov/sites/default/files/2023-01/Soil-Tillage-Intensity-Rating-Fact-Sheet3-27-2020.pdf (accessed on 19 December 2024).
36. Büchi, L.; Georges, F.; Walder, F.; Banerjee, S.; Keller, T.; Six, J.; Van Der Heijden, M.; Charles, R. Potential of Indicators to Unveil the Hidden Side of Cropping System Classification: Differences and Similarities in Cropping Practices between Conventional, No-till and Organic Systems. *Eur. J. Agron.* **2019**, *109*, 125920. [CrossRef]
37. R Core Team. *R: A Language and Environment for Statistical Computing*; R Core Team: Vienna, Austria, 2022.
38. Rosen, C.; Eliason, R. *Nutrient Management for Commercial Fruit & Vegetable Crops in Minnesota*; University of Minnesota Extension Service: Minneapolis, MN, USA, 2005.
39. National Organic Program (NOP). *Soil Fertility and Crop Nutrient Management Practice Standard*; 7 CFR §205.203; National Organic Program: Washington, DC, USA, 2011.
40. Kamau, J.W.; Schader, C.; Biber-Freudenberger, L.; Stellmacher, T.; Amudavi, D.M.; Landert, J.; Blockeel, J.; Whitney, C.; Borgemeister, C. A Holistic Sustainability Assessment of Organic (Certified and Non-Certified) and Non-Organic Smallholder Farms in Kenya. *Environ. Dev. Sustain.* **2022**, *24*, 6984–7021. [CrossRef]
41. Liebert, J.; Benner, R.; Bezner Kerr, R.; Björkman, T.; De Master, K.T.; Gennet, S.; Gómez, M.I.; Hart, A.K.; Kremen, C.; Power, A.G.; et al. Farm Size Affects the Use of Agroecological Practices on Organic Farms in the United States. *Nat. Plants* **2022**, *8*, 897–905. [CrossRef]
42. Sapbamrer, R.; Thammachai, A. A Systematic Review of Factors Influencing Farmers' Adoption of Organic Farming. *Sustainability* **2021**, *13*, 3842. [CrossRef]
43. Wang, Y.T.; Zhang, T.Q.; Hu, Q.C.; Tan, C.S.; Halloran, I.P.O.; Drury, C.F.; Reid, D.K.; Ma, B.L.; Ball-Coelho, B.; Lauzon, J.D.; et al. Estimating Dissolved Reactive Phosphorus Concentration in Surface Runoff Water from Major Ontario Soils. *J. Environ. Qual.* **2010**, *39*, 1771–1781. [CrossRef]

44. Mekonnen, M.M.; Hoekstra, A.Y. Global Anthropogenic Phosphorus Loads to Freshwater and Associated Grey Water Footprints and Water Pollution Levels: A High-Resolution Global Study. *Water Resour. Res.* **2018**, *54*, 345–358. [CrossRef]
45. O'Neill, B.; Sprunger, C.D.; Robertson, G.P. Do Soil Health Tests Match Farmer Experience? Assessing Biological, Physical, and Chemical Indicators in the Upper Midwest United States. *Soil Sci. Soc. Am. J.* **2021**, *85*, 903–918. [CrossRef]
46. Obour, P.B.; Dadzie, F.A.; Arthur, E.; Munkholm, L.J.; Saba, C.K.S.; Rubæk, G.H.; Owusu, K. Integration of Farmers' Knowledge and Science-Based Assessment of Soil Quality for Peri-Urban Vegetable Production in Ghana. *Renew. Agric. Food Syst.* **2020**, *35*, 128–139. [CrossRef]
47. DeDecker, J.; Malone, T.; Snapp, S.; Thelen, M.; Anderson, E.; Tollini, C.; Davis, A. The Relationship between Farmer Demographics, Social Identity and Tillage Behavior: Evidence from Michigan Soybean Producers. *J. Rural. Stud.* **2022**, *89*, 378–386. [CrossRef]
48. Morgan, S.L. Social Learning among Organic Farmers and the Application of the Communities of Practice Framework. *J. Agric. Educ. Ext.* **2011**, *17*, 99–112. [CrossRef]
49. Manono, B.O. New Zealand Dairy Farm Effluent, Irrigation and Soil Biota Management for Sustainability: Farmer Priorities and Monitoring. *Cogent Food Agric.* **2016**, *2*. [CrossRef]
50. Ho, T.T.K.; Tra, V.T.; Le, T.H.; Nguyen, N.-K.-Q.; Tran, C.-S.; Nguyen, P.-T.; Vo, T.-D.-H.; Thai, V.-N.; Bui, X.-T. Compost to Improve Sustainable Soil Cultivation and Crop Productivity. *Case Stud. Chem. Environ. Eng.* **2022**, *6*, 100211. [CrossRef]
51. Devarinti, S.R. Natural Farming: Eco-Friendly and Sustainable? *Agrotechnol* **2016**, *5*, 2. [CrossRef]
52. Razza, F.; D'Avino, L.; L'Abate, G.; Lazzeri, L. The Role of Compost in Bio-Waste Management and Circular Economy. In *Designing Sustainable Technologies, Products and Policies*; Benetto, E., Gericke, K., Guiton, M., Eds.; Springer International Publishing: Cham, Germany, 2018; pp. 133–143. ISBN 978-3-319-66980-9.
53. Manea, E.E.; Bumbac, C.; Dinu, L.R.; Bumbac, M.; Nicolescu, C.M. Composting as a Sustainable Solution for Organic Solid Waste Management: Current Practices and Potential Improvements. *Sustainability* **2024**, *16*, 6329. [CrossRef]
54. Childers, D.L.; Corman, J.; Edwards, M.; Elser, J.J. Sustainability Challenges of Phosphorus and Food: Solutions from Closing the Human Phosphorus Cycle. *BioScience* **2011**, *61*, 117–124. [CrossRef]
55. Baker, L.A. Can Urban P Conservation Help to Prevent the Brown Devolution? *Chemosphere* **2011**, *84*, 779–784. [CrossRef]
56. Oyetunji, O.; Bolan, N.; Hancock, G. A Comprehensive Review on Enhancing Nutrient Use Efficiency and Productivity of Broadacre (Arable) Crops with the Combined Utilization of Compost and Fertilizers. *J. Environ. Manag.* **2022**, *317*, 115395. [CrossRef]
57. DiGiacomo, G.; Gieske, M.; Grossman, J.; Jacobsen, K.; Peterson, H.; Rivard, C. Economic Trade-Offs: Analysis of Hairy Vetch (*Vicia villosa*) Cover Crop Use in Organic Tomato (*Solanum Lycopersicum* L.) High Tunnel Systems across Multiple Regions. *Renew. Agric. Food Syst.* **2023**, *38*, e10. [CrossRef]
58. Lowry, C.J.; Brainard, D.C. Organic Farmer Perceptions of Reduced Tillage: A Michigan Farmer Survey. *Renew. Agric. Food Syst.* **2019**, *34*, 103–115. [CrossRef]
59. Friedrich, T. Does No-Till Farming Require More Herbicides? *OutLook Pest Manag.* **2005**, *16*, 188–191. [CrossRef]
60. Cordeau, S.; Baudron, A.; Adeux, G. Is Tillage a Suitable Option for Weed Management in Conservation Agriculture? *Agronomy* **2020**, *10*, 1746. [CrossRef]
61. Colbach, N.; Cordeau, S. Are No-Till Herbicide-Free Systems Possible? A Simulation Study. *Front. Agron.* **2022**, *4*, 823069. [CrossRef]
62. Osterholz, W.R.; Culman, S.W.; Herms, C.; de Oliveira, F.J.; Robinson, A.; Doohan, D. Knowledge Gaps in Organic Research: Understanding Interactions of Cover Crops and Tillage for Weed Control and Soil Health. *Org. Agric.* **2021**, *11*, 13–25. [CrossRef]

sustainability

MDPI

Article

Organic Management and Intercropping of Fruit Perennials Increase Soil Microbial Diversity and Activity in Arid Zone Orchard Cropping Systems

Rhonda R. Janke [1,2,*], Daniel Menezes-Blackburn [3], Asma Al Hamdi [3] and Abdul Rehman [4]

1 Department of Plant Sciences, Sultan Qaboos University, P.O. Box 34, Al-Khoud 123, Oman
2 Department of Horticulture, Kansas State University, Manhattan, KS 66506, USA
3 Department of Soils, Water and Agricultural Engineering, Sultan Qaboos University, P.O. Box 34, Al-Khoud 123, Oman; danielblac@squ.edu.om (D.M.-B.)
4 Department of Agronomy, The Islamia University of Bahawalpur, Bahawalpur 63100, Pakistan; abdurehmanuaf@gmail.com
* Correspondence: rrjanke@ksu.edu; Tel.: +1-785-236-1994

Abstract: Organic farming encourages soil management practices that can improve soil health and fertility by increasing soil organic matter inputs and system sustainability. This study evaluated the effect of three years of continuous organic farming and intercropping orchard treatments on soil microbial diversity, microbial enumeration, respiration, soil fertility and fruit yields. Organic management resulted in higher soil organic matter content, Olsen P, and water holding capacity, but did not affect soil pH, electrical conductivity (EC), K, or Na levels. Growth parameters measured on all fruit trees were not significantly different among treatments. The enumeration of bacteria was significantly higher in organic plots when compared to conventionally managed plots. Soil respiration and substrate-induced respiration were significantly higher in the organic diverse plots in comparison to both conventional systems. The genomic analysis of prokaryotes (16S rRNA) and eukaryotes/fungi (ITS) revealed a significantly higher number of taxa, Shannon H index, and Equitability index in the organic systems for both prokaryotes and eukaryotes, in comparison to conventional farming, all of which are indicators of system sustainability. The relative abundance of Operational Taxonomic Units (OTUs) previously reported as diazotrophs, denitrifiers, or involved in the sulfur cycle, as well as Arbuscular Mychorrizae Fungi (AMF)/glomeromycotan, were highest in the organically managed soils than in the conventional plots. A multivariate correlation network clustering revealed that the microbial communities within the organic and conventional soils had strong dissimilarities regarding soil microbial niches. Our work provides evidence that organic management can be used for increasing soil microbial diversity and soil health, leading to higher levels of sustainability in fruit orchard systems.

Keywords: bacteria; archaea; fungi; soil fertility; organic farming; microbial metagenomics

1. Introduction

Long-term cropping systems studies have made a significant contribution to our understanding of organic agriculture, especially regarding the impact of organic farming soil management on nutrient and carbon cycling [1], yields and pest management [2]. In this regard, the recent literature has emphasized the need to use a systems approach to understand organic farming's environmental, human and economic aspects compared to conventional farming systems [3–6].

A number of related studies have also looked at the relationship between soil properties and microbial communities. A meta-analysis of organic farming systems comparisons from both field and greenhouse trials ranging from 3 years to 100 years, in four climatic zones, found that the organic systems had increased microbial biomass carbon by up to

41%, total nitrogen by 51%, and showed increased activities of protease, urease and dehydrogenases as compared to conventional farming systems [7]. Interestingly, the microbial metabolic quotient (qCO_2) was unaffected by the type of farming system for the same meta-analysis. These authors attribute the observed results to greater rates of organic matter inputs in organic farms, including manures, composts and legume cover crops. Several other studies, both from temperate and tropical regions, have highlighted that organic systems tend to be higher in soil organic carbon content [8–11], have higher microbial biomass [12–14] and soil respiration rates [13,15,16].

In addition to documenting changes in soil quality, many studies are using a range of methods to quantify and identify soil microbial taxa and, when possible, their function. These methods have drastically changed over the past 20 years from culture techniques to "next-generation DNA sequencing" (NGS) [17], dramatically increasing in sophistication and ability to quantify the presence and abundance of different taxa. Drawbacks of NGS include artifacts introduced by different DNA extraction protocols, bias on PCR steps, sequencing errors [18], limited DNA libraries and difficulty of linking taxonomic assignation to function [19]. However, the reduced cost allows for more samples to be tested, and for a better representation of soil bacteria [20] and fungal [21] diversity.

Identifying "keystone species" by NGS diversity data is advantageous in understanding soil ecological systems [18]. Some studies have compared the ratios of dominant phyla within the bacterial and fungal kingdoms, as indicators of nutrient availability or other aspects of soil health, while others look at the presence or absence of plant disease or disease antagonists [22]. For example, the ratio between the number of proteobacteria and acidobacteria was suggested to be indicative of the nutritional status of the soil and would be highest in organic soils with high inputs [19]. On occasion, the soil nutrient levels in organic and conventional systems were similar, but with different chemical speciation, different bioavailability and recalcitrance. Geisseler and Scow [23] found in a review of 64 long-term trials that the effect of mineral fertilizers (as compared to non-fertilized controls) led to a 15% increase in microbial biomass and 12.8% increase in soil organic carbon levels; these results appeared to be pH-dependent. Nitrogen inputs did not significantly affect soil microbial biomass, but the change in soil pH from NH_4^+ addition may have inhibitory effects on the soil microbiota. However, this review did not include high organic matter treatments in its analysis which may have affected its conclusions. Organic farming and its associated higher organic matter inputs also leads to an increase in water soil holding capacity [24] and soil porosity [11], which may also be associated with soil microbial abundance and diversity. Although the available literature showed the use of a wide range of different methods for assessing the effect of organic farming on increasing soil microbial diversity, they have come to similar conclusions.

Although different farming systems in desert climates have been compared [22,25], and AMF have been documented to be abundant and diverse in these organic farming systems [26,27], there is still a knowledge gap for the following: (a) the effect of organic and conventional soil nutrient management on soil microbial diversity for perennial fruit crops in arid climates; and (b) the effect of diverse intercropped systems on soil microbial diversity and soil health. High organic matter inputs and intercropping (or crop rotation) are common in organic farming systems, and their effect on soil microbes are often integrated as the 'organic farming' effect. The rhizodeposition of different carbon compounds and their effect on stimulating soil microbial biomass and changing the soil microbial community composition is known to be plant-specific [28].

Our general objective was to test the effect of three cumulative years of organic farming on plant growth, soil fertility levels, and soil microbial diversity and activity. In addition, the effect of crop diversity (citrus vs. citrus/mango/banana) within both systems (conventional and organic) was of interest. We hypothesized that both components of organic farming systems in this experiment (organic matter inputs and crop diversity) will stimulate soil microbial biomass, activity and diversity, and, therefore, contribute to better soil health and long-term cropping system sustainability.

2. Materials and Methods

2.1. Site Description and Experimental Design

Experiments were conducted in a fan-cooled greenhouse at the Sultan Qaboos University Agriculture Experiment Station, Muscat, Oman at 23°36′00.3″ N, 58°10′02.3″ E. The greenhouse (dimensions 9 m × 39 m) is covered with rigid polycarbonate, with a height of 3.5 m at the side, and 5.5 m at the roof, allowing for tall crops like fruit trees to be grown. Though greenhouse fruit production is not common in Oman, this production system allowed for year-round temperature control (range of approximately 22 to 32 °C) and isolated the plots from external contaminants such as pesticides.

The original greenhouse soil was coarse gravel with small boulders. Before planting, a 30 cm topsoil (mix of 2 parts sand and 1 part silt loam) layer was brought in to create a suitable growth medium. Initial soil fertility was low, and to our knowledge, no prior soil amendments or fertilizers were used in the brought-in soil substrate. Annual sesbania (*Sesbania sesban*) was seeded in October 2017 to enrich the soil and add organic matter. Though *Rhizobium* sp. was not added, the roots were nodulated. In March 2018, the cover crop was pulled and added as mulch to the organic plots into which trees had been previously planted (December 2017). The data included in this paper were collected in 2000, three years after the experiment was established. The experiment is ongoing, with plans to publish additional data after 6 and 9 years.

Fruit tree seedlings, previously grown in mixed media in pots for 1 to 2 years, were transplanted into the plots with 2 kg of local compost added to each hole. A drip irrigation system was installed, and subsequent fertility inputs were documented (Table S1). Only organic amendments were applied to the organic farming plots (compost, fish fertilizer and potassium sulfate), and only conventional soluble fertilizer was applied to the conventional plots after transplanting.

In both organic and conventional farming plots, macronutrient inputs were calculated for approximate equivalency, and USDA recommended rates were used as a guide. However, organic fertilizer sources are slow-release (mineralization) and contain micronutrients. The total nitrogen added was higher in the organic plots to compensate for the lower efficiency of compost as compared to the soluble N-fertilizer used in the conventional plots (588 and 358 g per tree over three years for organic and conventional, respectively). Phosphorus was applied at 306 and 292 g P_2O_5 per tree, while potassium was applied at 574 and 430 g K_2O per tree, respectively.

Regarding the carbon inputs, sesbania mulch (year 1 only) and banana leaf mulch (after each harvest, unquantified) were applied to the organic plots annually, while the conventional plots remained as bare soil. All plots received 1.6 kg of compost per tree at the time of planting, and then the organic plots received an additional 27.5 kg of compost per tree over the first 3 years, while the conventional plots received no additional carbon. In summary, the carbon budget in Table S1 showed that organic plots received 15.7 kg of carbon, and the conventional plots only 0.9 kg of carbon at the start of the experiment. Only organic pest management practices were applied to both systems as needed to avoid cross-contamination, including Neem Oil for scale insects on mango, and 'Spinosad' for leaf miners on citrus.

The experimental design included two organic treatments. One was planted with four citrus trees, the "monoculture," with a mix of varieties that included grapefruit, orange, lime and lemon, and the other being "diverse" with one citrus, two mangoes (var. 'Saffron' and 'Ros') and one banana (3–5 leads of local var. 'Nagal'). Two conventional treatments with the same species mix were also planted, and each of the four treatments was replicated three times in a randomized complete block design. The plot size was 4 m × 4 m, with 2 m between trees allowing for root intermingling.

Trees were measured annually for height, stem diameter (50 cm above the ground, or above the site of the graft) and rated for vigor and pest pressure. The banana harvest began in December 2018, with the weight and number in each bunch continuously recorded, since bananas fruit year-round. Soil samples (0–20 cm) were collected annually in October for soil fertility analysis from a composite sample of 4 subsamples per plot. The pH and EC were determined in 1:2 soil to distilled water. The same soil water extract was used for sodium and potassium quantification using Horiba hand-held sensors and available phosphorus extracts were determined colormetrically using the Olsen method [29]. Soil organic matter content was estimated by the loss of mass during ignition in a muffle furnace at 550 °C for 2 h. For the quantification of the soil water holding capacity, 30 mL of fresh soil was saturated in a funnel, and the gravimetric soil moisture content was assayed after 4 h of free drainage.

2.2. Enumeration of Culturable Heterotrophic Bacteria, Actinomycetes and Fungi

For microbial analysis, fresh soil samples were collected during March 2020, at a distance of 20–30 cm from the base of each of the four plants (at two opposing points) from each plot, at a depth of 0–5 cm. Therefore, each of the 12 plot samples corresponded to a composite of eight subsamples from within each plot.

Soil culturable bacterial, actinomycetes and fungal colony enumeration were assayed from soil matrices by suspending 10 g of freshly collected soil samples in 95 mL of saline solution (0.85% $NaCl_2$). Samples were shaken for 10 min at 100 rpm. Each soil suspension was then subjected to five sequential 10× dilutions. Soil suspensions at increasing dilutions were plated in the following agar growth media: Peptone yeast agar (for general bacteria), Casein glycerol agar (for actinomycetes) and Rose Bengal with streptomycin (for fungi) media. Agar plates were incubated at room temperature (20 to 25 °C) for 72 h. For each sample, six analytical replicates were averaged to obtain the number of colony-forming units (CFUs) per gram of soil on a dry weight basis.

2.3. Microbial Respiration

Microbial respiration was measured using the MicroResp™ system [30]. Each well of the deep-well microplate (96 wells, 1.2 mL per well) was filled with 0.46 ± 0.02 g of fresh soil samples, with four replicate samples/wells for each soil sample and each carbon substrate. The gravimetric moisture content for each soil sample was independently identified via oven drying at 105 °C. Before incubation, soils were moistened with either 25 μL sterile deionized water (for respiration) or a glucose solution (30 μg per well) to assess the substrate-induced respiration (SIR). A second microplate holding a CO_2 detection gel (12.5 mg kg^{-1} cresol red, 150 mM KCl, and 2.5 mM $NaHCO_3$ in 1% purified agar) was assembled on top of the soil microplate using an air-tight sealing system provided by the manufacturer. The system was then incubated in the dark at room temperature, and detection plates were read after 20 h. The absorbance of the indicator plates was measured at 570 nm using a Microplate reader before and after incubation with soils. The CO_2 released from soils (%CO_2) was converted to respiration rate (μg CO_2-C g^{-1} dry soil h^{-1}) [30].

2.4. Microbial Diversity

The DNA from the freshly collected soil samples was extracted using a PurelinkTM microbiome DNA purification kit (Thermo Fisher Scientific, Waltham, MA, USA). The DNA extracts were then sent to Molecular Research (MR-DNA, Shallowater, TX, USA) for V4 16S rRNA and ITS NGS diversity analysis. The 16S rRNA gene V4 variable region, and the ITS 1–2 PCR primers 515/806 were used with a barcode on the forward primer in a 35 cycle PCR using the HotStarTaq Plus Master Mix Kit (Qiagen, USA) under the following conditions: 95 °C for 5 min, followed by 35 cycles of 95 °C for 30 s, 53 °C for 40 s and 72 °C for 1 min, after which a final elongation step at 72 °C for 10 min was performed. After amplification, PCR products were checked in 2% agarose gel to determine the amplification's success and the bands' relative intensity. Samples were purified using calibrated SPRI beads. The pooled and purified PCR products were used to prepare the Illumina DNA library. The DNA sequencing was performed at MR DNA (https://www.mrdnalab.com/, accessed on 1 June 2020, Shallowater, TX, USA) on a MiSeq following the manufacturer's guidelines.

The sequence data derived from the sequencing process were processed using the MR DNA ribosomal and functional gene analysis pipeline and in-house built software (http://www.mrdnafreesoftware.com, accessed on 1 June 2020, https://www.mrdnalab.com/, accessed on 1 June 2020, MR DNA, Shallowater, TX). In this process, sequences were joined, depleted of barcodes and primers and then sequences with <150 bp or ambiguous base calls were removed. The sequences were filtered for quality using a maximum expected error threshold of 1.0 and dereplicated. Sequences were then denoised, and operational taxonomic units were defined as clustering at 3% divergence (97% similarity) followed by removal of singleton sequences and chimaeras. The final OTUs were taxonomically classified using BLASTn against a curated database derived from NCBI (www.ncbi.nlm.nih.gov, accessed on 1 June 2020).

2.5. Statistical Analyses

Data calculations, manipulation, average, standard deviation and correlation analysis were performed using Microsoft Office Excel 2016. Analysis of variance was performed using Minitab (version 20.0), and the Fisher test was used for mean separation. The Krona Excel template [31] was used to construct HTML hierarchical microbial diversity graphics to visualize and summarize differences in microbial community composition within the Prokaryotes and Eurkarotes. The PAST4 software (version 4.15) [32] was used to calculate the Shannon-H and Equitability microbial diversity indices. Graphia 2.0 software was used for correlation network analysis of OTUs, with Pearson's correlation coefficients above 0.95 [33], clustered by Markov Cluster Algorithm (MCL, granularity 1.1). The taxa of known functional microbial groups, such as mycorrhiza, diazotrophs, nitrifiers and denitrifiers, were grouped to represent the effect of the treatments on soil functional microbial groups.

3. Results

3.1. Soil Physicochemical Properties, Plant Growth and Yield

Though different levels of total N and C had been added to the experimental plots over the three years of the experiment (Table S1), the results showed nonsignificant differences in the soil pH, concentration of K and Na (Table 1). Soil salinity (EC) was also unaffected by our treatments leading to nonsignificant changes in electrical conductivity. Soil organic matter significantly increased, more than doubling in value from 2.55% in conventional plots to 5.49% in the organic plots. Olsen P and water holding capacity (WHC) also significantly increased in the organic plots, though there was no significant difference comparing monoculture and diverse treatments.

Table 1. Soil test data from October 2019. Letters denote means separation statistical significance using Fisher pairwise comparison.

	pH	**EC**	**K**	**Na**	**Water Holding Capacity**	**Soil Organic Matter**	**Olsen Phosphorus**
		dS m^{-1}	**ppm**	**ppm**	**%**	**%**	**mg Kg^{-1}**
Conventional Mono	8.1 a	1757 a	35 a	283 a	41.86 a	2.69 b	161 ab
Conventional Diverse	8.1 a	1973 a	37 a	213 a	42.41 a	2.40 b	142 b
Organic Mono	8.1 a	1818 a	65 a	287 a	51.51 a	5.81 a	231 a
Organic diverse	8.1 a	2234 a	46 a	277 a	52.00 a	5.18 a	205 ab
Conventional	8.1 a	1865 a	36 a	248 a	42.14 b	2.55 b	151 b
Organic	8.1 a	2026 a	55 a	282 a	51.76 a	5.49 a	218 a
Monoculture	8.1 a	1787 a	50 a	285 a	46.69 a	4.25 a	196 a
Polyculture	8.1 a	2104 a	42 a	245 a	47.21a	3.79 a	173 a
average all	8.1	1946	46	265	46.95	4.02	185
p-values two-way anovas:							

Table 1. *Cont.*

	pH	EC	K	Na	Water Holding Capacity	Soil Organic Matter	Olsen Phosphorus
		dS m^{-1}	ppm	ppm	%	%	mg Kg^{-1}
org vs. conv	0.780	0.685	0.076	0.598	0.018	0.000	0.023
mono vs. diverse	0.412	0.432	0.381	0.529	0.876	0.330	0.366
interaction	0.780	0.801	0.287	0.634	0.994	0.712	0.896

By the end of the first three years of experimentation, the mango and citrus trees were still young and not yet fruiting, but the banana plants started producing fruit 12 months after transplanting. The total number of bunches per plant was 11 for organic and 8.3 for conventional. The number of hands per bunch was the only significantly different (p = 0.043) banana yield parameter, and was higher for conventional farming with 5.76, compared to 4.85 for organic farming plots (Table S2). However, the total weight per bunch was similar and was not statistically different between conventional and organic farming treatments. Mango and citrus stem diameter, height, and vigor ratings showed no statistically significant differences (Table S3). Due to pruning management, citrus showed a decline in height in some treatments from 2020 to 2021.

3.2. Enumeration of Culturable Heterotrophs and Soil Microbial Respiration

Of the three-culture media used for microbial enumeration, only the nutrient agar (bacteria) showed statistically significant differences, likely due to the high variability of the data (Table 2). The organic diverse plots had the highest bacteria CFU g^{-1} soil and was significantly higher than conventional diverse. The enumeration of actinomycetes in glycerol casein agar and fungi in Rose Bengal agar was highly variable and imprecise, preventing conclusions about the effect of organic farming and intercropping. For the culturable fungi, the highest levels were in the organic monoculture, but were not statistically different than the other treatments (Table 3).

Table 2. Enumeration of culturable heterotrophic bacteria, actinomycetes and fungi. Soil respiration, substrate induced respiration (SIR) and metabolic quotient (MQ) estimation (based on bacteria abundance). Letters designate mean separation statistical significance by Fisher pair-wise comparison.

		Bacteria	Actinomycetes	Fungi	Resp. DW	SIR DW	MQ Resp.
		CFU/g	CFU/g	CFU/g	(µg CO2-C /g/h)	(µg CO2-C /g/h)	(µg CO2-C /10^{7} CFU/h)
Conv. Mono	**Mean**	1.63×10^{7} ab	0.00 a	3.48×10^{4} a	0.604 b	0.804 b	0.384 b
	Stnd Dev.	4.27×10^{6}	0.00	3.90×10^{4}	0.001	0.078	0.085
Conv. Diverse	**Mean**	6.93×10^{6} b	2.74×10^{6} a	2.43×10^{4} a	0.557 b	1.295 ab	0.869 a
	Stnd Dev.	3.49×10^{6}	5.42×10^{6}	2.38×10^{4}	0.017	0.75	0.264
Organic mono	**Mean**	1.65×10^{7} ab	4.50×10^{5} a	1.32×10^{5} a	0.578 b	1.329 ab	0.535 ab
	Stnd Dev.	2.11×10^{7}	6.51×10^{5}	2.63×10^{5}	0.076	0.319	0.341
Organic diverse	**Mean**	3.65×10^{7} a	4.40×10^{4} a	1.99×10^{4} a	0.760 a	1.963 a	0.268 b
	Stnd Dev.	3.64×10^{7}	1.08×10^{5}	3.17×10^{4}	0.137	0.753	0.177
p-value - one way	by trt	0.137	0.282	0.442	0.048	0.167	0.064
p-value two way	org vs conv	0.101	0.326	0.409	0.087	0.101	0.140
	mono vs diverse	0.548	0.308	0.279	0.175	0.118	0.450
	interaction	0.106	0.174	0.367	0.036	0.829	0.025

Table 3. Number of taxa, sequences, Shannon H and Equitability indices for Prokaryotes (16S) and Eukaryotes (ITS). Letters designate mean separation statistical significance based on 95% confidence interval (CI).

Prokaryotes 16S	**# OTU (Taxa)**	**# Sequences**	**Shannon H**	**CI**	**Equitability**	**CI**
Conventional Mono	2193	22,588	6.62	0.021 c	0.86	0.003 a
Conventional Diverse	501	46,371	4.98	0.015 d	0.80	0.003 c
Organic Mono	7084	384,697	7.10	0.006 a	0.80	0.001 c
Organic diverse	4820	81,646	6.84	0.013 b	0.81	0.001 b
Conventional	2453	68,959	6.03	0.015 b	0.77	0.002 b
Organic	7388	466,343	7.15	0.005 a	0.80	0.001 a
Monoculture	7556	407,285	7.19	0.006 a	0.81	0.001 a
Polyculture	5064	128,017	6.75	0.010 b	0.79	0.001 b
All	7842	535,302	7.27	0.005	0.81	0.001
Eukaryotes ITS	# OTU (Taxa)	# Sequences	Shannon H	CI	Equitability	CI
Conventional Mono	361	44,518	3.71	0.020 c	0.63	0.003 c
Conventional Diverse	54	14,900	2.01	0.030 d	0.50	0.007 d
Organic Mono	410	20,601	4.57	0.020 b	0.76	0.003 a
Organic diverse	433	20,672	4.63	0.023 a	0.76	0.004 a
Conventional	413	59,418	3.85	0.017 b	0.64	0.003 b
Organic	581	41,273	5.01	0.016 a	0.79	0.002 a
Monoculture	645	65,119	4.51	0.016 a	0.70	0.002 a
Polyculture	485	35,572	4.21	0.022 b	0.68	0.004 b
All	813	100,691	4.83	0.013	0.72	0.002

Soil respiration rate was highest in the organic diverse plots, and significantly different from the other three treatments (Table 2). Substrate-induced respiration was also significantly higher in the diverse organic plots compared to the conventional monoculture; and the conventional-diverse and organic-monoculture plots were not significantly different from other treatments.

The metabolic quotient (qCO_2) corresponds to the respiration rate per unit of soil microbial biomass. The qCO_2 was highest in the conventional-diverse plots, and significantly different than the conventional-monoculture and the organic-diverse (Table 2). For this experiment, the bacteria enumeration was also lowest in the conventional-diverse treatment, explaining its high qCO_2. The organic-diverse plots, on the other hand, showed the highest respiration rates, the highest culturable bacteria enumeration and the lowest qCO_2.

3.3. Soil Microbial Taxonomic Diversity Using 16S and ITS Gene Sequencing

The number of OTUs, the Shannon diversity index (H) and the equitability (EH) indexes are presented for both the prokaryotes (16S rRNA V4 regions) and fungi (ITS) derived from the NGS microbial diversity analysis are presented in Table 3. Together, these indices represent the number of taxa, abundance, and the evenness of population distribution for both culturable and unculturable soil microbes. This analysis resulted in a more sensitive way to represent the differences between the treatments, with statistically significant differences among all treatments. For the prokaryotes 16S diversity (bacteria and archaea), the number of OTUs was highest in the organic-monoculture, followed by the organic-diverse plots, and the conventional-diverse treatment was the lowest (Table 3). A similar pattern was observed for the Shannon H, with the highest values in both the monoculture and diverse organic treatments, compared to the conventionally farmed plots. Species equitability was highest in the conventional monoculture plots, followed by the organic diverse plots. Overall, when analyzed together, the organic plots showed higher equitability than the conventionally farmed soils. For fungi, the highest number of OTUs were found for both the diverse and monoculture organic plots. The Shannon H and equitability indices were also higher for fungi in both organic plots and for the organic treatments overall, and these differences were highly significant.

3.4. Functional Microbes and Correlation Networks

The hierarchical taxonomic abundance from kingdom to genus illustrates some of the data described above, plus reveals additional information regarding the abundance of certain specialized taxa. For example, the difference between the prokaryotes for the averaged organic and conventional systems (Figure 1a,b) show the relative proportions of bacteria and archaea, with higher archaea in the organic vs. conventional plots (4% vs. 1% of total prokaryotes, respectively). The ratio of proteobacteria to acidobacteria has been suggested as an indicator of readily available nutrients in soil [19], and was higher in the conventional plots. The proteobacteria made up 55% and 61% of the total bacteria in the organic and conventional systems, respectively, and the acidobacteria were 5% and 3% of bacterial in the two systems, respectively. Thus, the resulting ratio was 11 for the organic systems and 20.3 for the conventional.

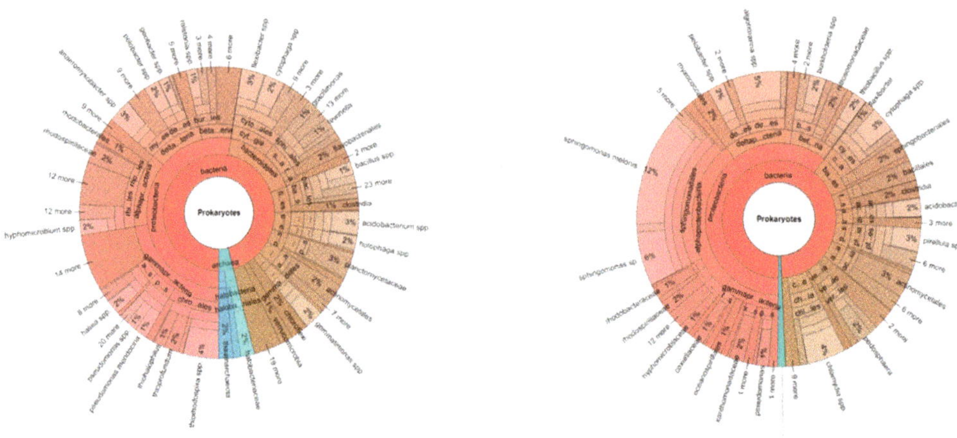

(**a**) (**b**)

Figure 1. Hierarchical taxonomic abundance (from kingdom to genus) of prokaryotes based on 16S ribosomal RNA for organic (**a**) and conventional (**b**) soils (averaged across monoculture and diverse treatments). Bacteria are represented in pink-brown sections of the chart, and archaea in the blue section.

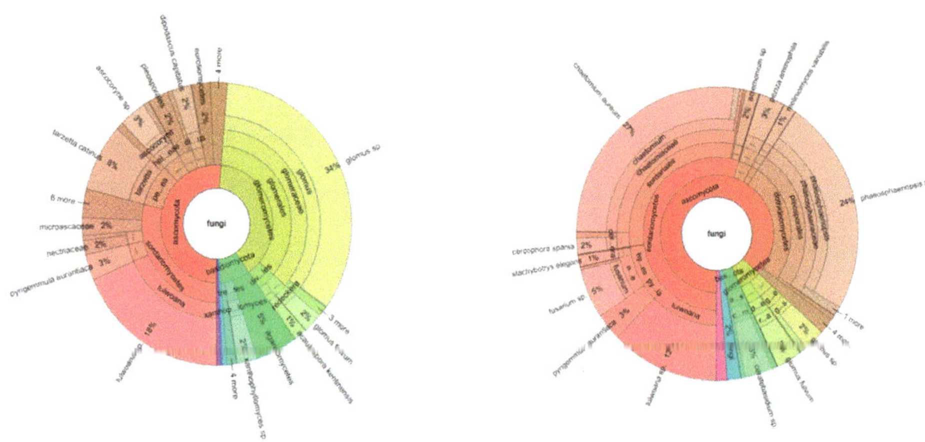

(a) (b)

Figure 2. Hierarchical taxonomic abundance (from kingdom to genus) of eukaryotes based on ITS ribosomal RNA for organic (**a**) and conventional (**b**) soils (averaged across monoculture and diverse treatments).

For fungi, the plot of the eukaryotes (Figure 2a,b) clearly shows that the proportion of fungi in the phyla glomeromycota are higher (39% of fungi) in organic as compared to conventional (only 6% of fungi). The relative abundance of ascomycota and basidomycota is also different, with 51% and 9% of these two phyla in organic, and 87% and 4% in the conventional, respectively, resulting in ratios of ascomycota to basidomycota of 5.9 and 20.5 for the organic and conventional systems, respectively.

Graphia 2.0 was used to build a correlation network of OTUs (includes both prokaryotes and eukaryotes) with Pearson's correlation coefficients above 0.95 (Figure 3). Cluster 1 is dominated by OTUs found in monoculture organic (72%), followed by the diverse organic system (15%). Cluster 2 is also dominated by monoculture organic (93%). Cluster 3 is made of 50% diverse organic OTU, followed by monoculture organic with 36%. Cluster 4 is dominated by monoculture conventional (65%) and cluster 5 by diverse conventional (77%). Thus, we see that the microbial communities/associations were different between the two conventional systems, and were also different between the conventional and the organic systems. The first two clusters show differences between organic monoculture and organic diverse cropping systems (Figure 3).

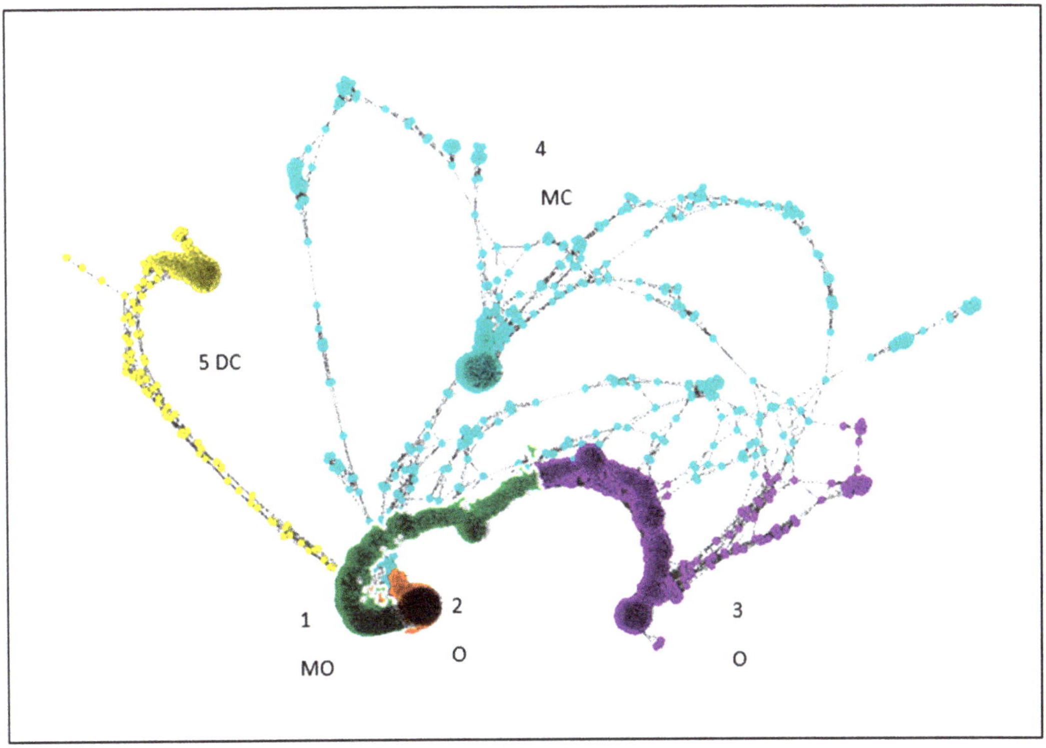

Figure 3. Graphia 2.0 Pearson correlation network, positive correlation polarity, minimum correlation value 0.95. 1-MO: monoculture organic treatment cluster, 2-O and 3-O: organic (both monoculture and diverse) clusters, 4-MC monoculture conventional cluster, and 5-DC: diverse conventional treatment cluster.

When we look at the correlations of the clusters with specific groups of organisms, clusters 1, 2 and 3 were significantly positively correlated with archaea and bacteria, while cluster 4 was more significantly correlated with fungi. Cluster 5 was negatively correlated with archaea and bacteria, with no significant correlation with fungi (Table S5). Within the kingdom of bacteria, clusters 1, 2 and 3 were strongly and positively correlated with most of the phyla, except tenericutes, which was only positively correlated with cluster 5, thermotogae (only cluster 3), chlamydiae (only cluster 5), cyanobacteria (clusters 1, 2 and 4) and lentisphaerae (only cluster 4) (Table S5). Within the fungi, clusters 1, 2 and 3 were significantly and positively correlated with glomeromycota and cryptomycota. Cluster 4 was significantly and positively correlated with ascomycota, chytridiomycota and basidiomycota, while cluster 5 was either not correlated or negatively correlated with all phyla.

When we look at the soil data, clusters 1, 2 and 3 were significantly and positively correlated with all soil variables except pH, including the soil organic matter, water holding capacity, and the Olsen phosphorus test. However, the correlations between archaea, bacteria and fungi with the soil test data show that only the archaea and bacteria are positively correlated with organic matter, water holding capacity and phosphorus, and fungi does not show a significant correlation. Within the kingdom of fungi, the phyla of glomeromycota and cryptomycota are significantly and positively correlated with organic matter, water holding capacity and phosphorus level; the phyla of ascomycota and chytridiomycota are negatively correlated; and basidiomycota have a weak but positive correlation (Table S5).

When the taxa are sorted by name to indicate their function, the percentage of diazatrophs, denitrifiers and those involved in the sulfur cycle are highest in the organic monoculture system, followed by the organic diverse system (Table S4). The percentage classified as AMF was also higher in the two organic soils/treatments.

4. Discussion

In the present study, we observed an increase in soil organic matter under organic practices as compared to conventional farming. Our findings are consistent with previous studies that have found increases in soil organic matter in organic cropping systems receiving manures, composts, or both, in both tropical and temperate environments [8–11,14–16,24,34]. The organic system also resulted in an increase in soil water holding capacity compared to the conventional farming treatment. The increase in soil water holding capacity can be attributed to several factors associated with organic farming practices, including the addition of organic matter and the promotion of soil micro-organisms [11]. Organic matter improves soil structure [11,24], which enhances water infiltration and retention. Additionally, the presence of beneficial microorganisms in organic soils can enhance soil structure, leading to increased water holding capacity as was evident from the positive correlation of soil organic matter (SOM) with soil microbes in this study (Table S5). While the positive effects of organic matter on soil structure and water holding capacity are clear, our study did not find significant differences in soil EC, pH, K, and Na between organic and conventional farming systems. This finding may be due to the complex and context-specific nature of nutrient levels in soils, as they can vary depending on a variety of factors, including the specific experiment conducted, the duration of the study, the type of cropping system or rotation, and the rates of fertilizers applied [8,35,36]. However, Olsen P was higher in the organic farming compared to the conventional farming system. Organic farming systems receiving organic fertilizer often have higher P level in soils [9,37].

The organic farming system resulted in a higher abundance and diversity of soil bacteria, archaea and fungi in organic farming systems and can be attributed to the presence of organic matter inputs such as manures or composts [11], which were strongly and positively correlated with soil organic matter (Table S5). This is consistent with findings from other studies [8,11,24,34,38,39]. Additionally, the dominance of clusters 1, 2 and 3 by organic farming systems (Figure 3) also contributed to the increased abundance and diversity of soil microorganisms. However, the richness and diversity of species in some systems cannot be solely attributed to whether a system is defined as organic. Factors such as the amount of organic matter inputs and pH can also play significant role [8,10,15,40]; in some cases, soil pH has been negatively correlated with microbial abundance and diversity [8,15]. Some studies have shown different results when comparing microbial diversity in control plots with no fertility amendments, with some finding the lowest diversity in the control [11], others finding the lowest diversity in the fertilizer plot [38], and others finding no significant difference between the no fertilizer and conventional fertilizer treatments, but both being lower than the organic treatments [39].

The composition of bacterial communities in soil can be influenced by agricultural practices. Specifically, the ratio of proteobacteria and acidobacteria has been shown to differ between organic and conventional systems, with higher ratios in conventional systems [10], which is attributed to the more readily available nutrients in conventional systems promoting the rapid growth of proteobacteria, as compared to the slower growing acidobacteria [19]. In our study, the ratio was 11.2 in the organic system, and 18.2 in conventional. These results are consistent with the findings of Suyal et al. [10] in organic and conventional chickpea cropping systems in India, where ratios of 10.3 and 22.2 were observed. In contrast, cowpea and citrus cropping systems in India had ratios of 20.0 and 60.0 for organic and conventional, respectively [24], while a barley–beet crop rotation in Germany had ratios of 2.9 in rotations receiving manure or NPK fertilizer, and 2.2 where no fertilizer or manure was used [34].

Culture-based methods have limitations in detecting soil microbes due to the inability to culture a large percentage of soil microbes. In our study, only the nutrient agar medium for bacteria showed significant differences for the field treatments, likely due to the high variability of the data. This finding is consistent with the limitations of plate count methods for bacteria, fungi and actinomycetes that have been found by others [36]. The differences in results from the NGS methods and plate count methods are likely due to the extremely small percentage of soil microbes that can be cultured, as compared to those that are detected using DNA extraction/genomic methods [36,41]. Our study also found a higher respiration rate and SIR in the organic systems, which is consistent with other studies [13,15,16,36] and has been correlated with the soil organic matter levels in these studies.

In recent years, few studies have compared the farming systems to "natural" ecosystems, which are either desert, or dryland-grown wild plants in Oman. Khan et al. [42], using Next Generation Sequencing (NGS), found 121 fungal and 3662 unique bacterial OTUs in the rhizosphere of 3 native plants growing in arid conditions in Oman. In comparison, our study found 581 and 413 fungal OTUs in the organic and conventional systems, respectively. Our prokaryote OTUs were 7388 and 2453 in organic and conventional plots, respectively. The higher numbers in our study may be due to the fertility supplements and/or regular watering regime. In Egypt, a desert soil was compared to an organic vegetable farm for rhizosphere bacteria, and found higher diversity in the organic farm, with a Shannon index of 11.21 compared to 7.90 for the desert soil [22]. The study also identified more pathogen antagonists in the organic farm, but more N-fixing bacteria in the desert soil [22].

Kazerooni et al. [17] compared conventional tomato farming and desert soils in Oman, and found similar fungal diversity between the two systems, but with slightly more species in the farmed soil. The Shannon index was slightly higher in the desert soil (1.9 vs. 1.4). In another study, however, that included an organic farming system, the fungal diversity was highest for organic tomato and cucumber, intermediate for date palm and lime farmed with traditional methods that included manures, and lowest in conventional cucumber due to lower use of organic matter inputs plus the use of fungicide to control disease [43]. In experiments that compare organic and conventional farming to other ecosystems in an environment with more rainfall (West Java, Indonesia), the highest organic matter (OM) and % total N were in forests, followed by organic cropping, and lower in conventional. Microbial biomass also followed this same pattern [14].

In our experiment, the organic cropping systems showed a higher percentage of fungi in the Glomeracota, which is similar to the findings of Mullath et al. [26], who reported a higher quantity and diversity of arbuscular mycorrhizal fungi (AMF) in organic cropping systems in the UAE. The desert plants sampled in their study did not contain AMF. All fruit species used in this study are known hosts of AMF, which has been widely documented in citrus [44–47], including acid lime in Oman [48], mango [49,50] and banana [51]. The benefits of AMF root colonization on fruit tree seedlings have also been documented [47,49,52,53]. It has been reported that fertilizer application may reduce AMF [54,55], that manure and compost application can decrease root colonization by [13], or that higher levels of soil P may reduce the abundance of AMF [55] at low soil Olsen-P (10 mg kg^{-1}); the root colonization and pi transporter genes are significantly upregulated [9]. However, other studies show the positive correlation of AMF colonization with soil properties such as organic matter and available P in citrus orchards [39,56] or no correlation with soil P content [50]. In one study, available phosphorus was negatively correlated with spore density of AMF, but positively correlated with root colonization of papaya [57]. Another study with banana found that P and soil carbon negatively affect AMF colonization, but soil carbon was positively correlated with spore abundance and species richness [51]. In our study, the P levels measured were higher in the organic systems though similar rates were applied, which did not seem to negatively impact AMF.

A study with annual cropping systems demonstrated an increase in diversity of AMF communities in more diverse polycultures as compared to monoculture [58], and it would seem likely to have the same effect in perennial systems as well. In our experiment the monoculture citrus had a higher Shannon H index and higher equitability index for both prokaryotes and eukaryotes than the diverse system that also included banana and mango. When only the glomeromycota are looked at, they were more abundant in the organic systems as compared to conventional, but also more abundant in the monoculture as compared to diverse. This finding should be followed up with more research.

The Pearson correlations and multivariate analysis in our study are consistent with previous research by Liao et al. [59] which reported higher abundance of genera associated with organic farming systems. We observed a positive correlation between archaea and bacteria with clusters 1, 2 and 3 representing organic farming systems (Figure 3), suggesting that these clusters are dominated by prokaryotic organisms, and may have similar ecological roles in the environment. It could also indicate that these clusters share similar environmental preferences or require similar resources to thrive. This positive impact on the microbial community may be attributed to the use of organic fertilizers and soil management practices in organic farming systems [11,59]. In contrast, cluster 4, representing the mono conventional farming system, showed a higher correlation with fungi, indicating a greater abundance of eukaryotic organisms in these soils. This aligns with previous studies indicating a negative impact on soil health and microbial diversity in conventional farming systems [55]. Conversely, the significant correlation between fungi and conventional farming systems suggests that these systems are dominated by eukaryotic organisms, particularly fungi, which may have different ecological roles compared to organic farming systems.

That organic farming systems support a diverse microbial community dominated by prokaryotic organisms is consistent with previous work by Liao et al. [59] who reported a higher abundance of genera associated with organic farming systems. The positive correlation with archaea and bacteria in the present study may indicate that these microorganisms play important roles in nutrient cycling and other ecosystem services in organic farming systems. This suggests that organic farming systems may have a positive impact on the microbial communities in the soil, possibly due to the use of organic fertilizers and reduced tillage practices. Cluster 4, which represents the mono conventional farming system, was more significantly correlated with fungi, indicating a higher abundance of eukaryotic organisms in these soils. This finding is consistent with previous studies that have found that conventional farming systems tend to have a negative impact on soil health and microbial diversity.

The positive correlation between archaea and bacteria with organic farming systems suggests that these clusters are dominated by prokaryotic organisms, and may have similar ecological roles in the environment. It could also indicate that these clusters share similar environmental preferences or require similar resources to thrive. On the other hand, the significant correlation between fungi and conventional farming systems suggests that these systems are dominated by eukaryotic organisms, specifically fungi, and may have a different ecological role compared to organic farming systems. Fungi are known to play important roles in nutrient cycling and decomposition, so this cluster may be more involved in these processes.

5. Conclusions

Organic farming practices can significantly improve soil health and fertility by increasing soil organic matter inputs, soil biodiversity and microbial diversity. The organic farming system resulted in higher soil organic matter content, Olsen P and water holding capacity, but did not affect soil pH, EC, K or Na levels. The enumeration of bacteria, soil respiration and substrate-induced respiration were higher in the organic and diverse plots compared to conventionally-managed and monoculture plots. The genomic analysis of prokaryotes and eukaryotes/fungi revealed a significantly higher number of taxa, Shannon H index, and Equitability index in the organic systems for both prokaryotes and eukaryotes, in comparison to conventional farming. Limitations of this study are that only the first 3 years of what is anticipated to be a longer-term study are included. The study so far shows that organic farming practices and intercropping can increase soil microbial diversity and health, which can have positive impacts on crop yields and overall ecosystem health and sustainability.

Supplementary Materials: The following supporting information can be downloaded from the following website: https://www.mdpi.com/article/10.3390/su16219391/s1, Table S1: Soil fertility inputs from the inception of the experiment to fall 2020; Table S2: Banana stem count per plant and yield from 2018 to present (August 2021); Table S3: Citrus and mango stem diameter, height, and vigor ratings for organic and conventional trts; 2020 and 2021; Table S4: Functional microbial taxa; mycorrhiza (AMF), diasotrophs, denitrifying and sulfur-cycle by number and percent of sequences; Table S5: Correlation matrix for soil fertility measurements and selected taxonomic data.

Author Contributions: Conceptualization, D.M.-B. and R.R.J.; methodology, D.M.-B. and R.R.J.; software, D.M.-B. and A.A.H.; validation, D.M.-B., A.A.H. and R.R.J.; formal analysis, A.A.H.; investigation, A.A.H.; resources, R.R.J. and D.M.-B.; data curation, D.M.-B.; writing—original draft preparation, A.A.H.; writing—review and editing, A.R., R.R.J. and D.M.-B.; visualization, D.M.-B.; supervision, D.M.-B. and R.R.J.; project administration, R.R.J.; funding acquisition, R.R.J. and D.M.-B. All authors have read and agreed to the published version of the manuscript.

Funding: This research was funded by His Majesty Grant Fund, grant number SR/AGR/CROP/19/02.

Institutional Review Board Statement: Not applicable.

Informed Consent Statement: Not applicable.

Data Availability Statement: The original contributions presented in the study are included in the article, further inquiries can be directed to the corresponding author.

Acknowledgments: The authors gratefully acknowledge in-kind support from the College of Agricultural and Marine Sciences, Sultan Qaboos University, for use of laboratory and greenhouse facilities, and contributions by the support staff.

Conflicts of Interest: The authors declare no conflict of interest.

References

1. Drinkwater, L.E.; Wagoner, P.; Sarrantonio, M. Legume-based cropping systems have reduced carbon and nitrogen losses. *Nature* **1998**, *396*, 262–265. [CrossRef]
2. Liebhardt, W.C.; Andrews, R.W.; Culik, M.N.; Harwood, R.R.; Janke, R.R.; Radke, J.K.; Riegerschartz, L.L. Crop production during conversion from conventional to low-input methods. *Agron. J.* **1989**, *81*, 150–159. [CrossRef]
3. Barker, A.V. *Science and Technology of Organic Farming*; CRC Press: Boca Raton, FL, USA, 2010.
4. Francis, C. (Ed.) *Organic Farming: The Ecological System. Agronomy Monograph*; American Society of Agronomy: Madison, WI, USA, 2009; Volume 54.
5. Lockeretz, W. (Ed.) *Organic Farming: An International History*; CAB International Press: Wallingford, UK, 2007.
6. Pacini, C.; Wossink, A.; Giesen, G.; Huirne, R.J. Ecological-economic modelling to support multi-objective policy making: A farming systems approach implemented for Tuscany. *Agric. Ecosyst. Environ. Health Perspect.* **2004**, *102*, 349–364. [CrossRef]
7. Lori, M.; Symnaczik, S.; Mäder, P.; De Deyn, G.; Gattinger, A. Organic farming enhances soil microbial abundance and activity—A meta-analysis and meta-regression. *PLoS ONE* **2017**, *12*, e0180442. [CrossRef]
8. Chen, X.; Henriksen, T.M.; Svensson, K.; Korsaeth, A. Long-term effects of agricultural production systems on structure and function of the soil microbial community. *Appl. Soil Ecol.* **2020**, *147*, 103387. [CrossRef]

9. Ding, G.-C.; Bai, M.; Han, H.; Li, H.; Ding, X.; Yang, H.; Xu, T.; Li, J. Microbial taxonomic, nitrogen cycling and phosphorus recycling community composition during long-term organic greenhouse farming. *Microbiol. Ecol.* **2019**, *95*, fiz042. [CrossRef]
10. Suyal, D.C.; Soni, R.; Singh, D.K.; Goel, R. Microbiome change of agricultural soil under organic farming practice. *Biologia* **2021**, *76*, 1315–1325. [CrossRef]
11. Tang, H.; Li, C.; Xiao, X.; Shi, L.; Chen, K.; Wen, L.; Li, W. Effects of short-term manure nitrogen input on soil microbial community structure and diversity in a double-cropping paddy field of southern China. *Sci. Rep.* **2020**, *10*, 13540. [CrossRef]
12. Araújo, A.S.F.; Melo, W.J. Soil microbial biomass in organic farming system. *Ciência Rural* **2010**, *40*, 2419–2426. [CrossRef]
13. Jannoura, R.; Bruns, C.; Joergensen, R.G. Organic fertilizer effects on pea yield, nutrient uptake, microbial root colonization and soil microbial biomass indices in organic farming systems. *Eur. J. Agron.* **2013**, *49*, 32–41. [CrossRef]
14. Moeskops, B.; Sukristiyonubowo; Buchan, D.; Sleutel, S.; Herawaty, L.; Husen, E.; Saraswati, R.; Setyorini, D.; De Neve, S. Soil microbial communities and activities under intensive organic and conventional vegetable farming in West Java, Indonesia. *Appl. Soil Ecol.* **2010**, *45*, 112–120. [CrossRef]
15. Ge, T.; Chen, X.; Yuan, H.; Li, B.; Zhu, H.; Peng, P.; Li, K.; Jones, D.L.; Wu, J. Microbial biomass, activity, and community structure in horticultural soils under conventional and organic management strategies. *Eur. J. Soil Biol.* **2013**, *58*, 122–128. [CrossRef]
16. Velmourougane, K. Impact of Organic and Conventional Systems of Coffee Farming on Soil Properties and Culturable Microbial Diversity. *Scientifica* **2016**, *2016*, 3604026. [CrossRef]
17. Van Dijk, E.L.; Auger, H.; Jaszczyszyn, Y.; Thermes, C. Ten years of next-generation sequencing technology. *Trends Genet.* **2014**, *30*, 418–426. [CrossRef]
18. Schloter, M.; Nannipieri, P.; Sørensen, S.J.; van Elsas, J.D. Microbial indicators for soil quality. *Biol. Fertil. Soils* **2018**, *54*, 1–10. [CrossRef]
19. Torsvik, V.; Øvreås, L. Microbial diversity and function in soil: From genes to ecosystems. *Curr. Opin. Microbiol.* **2002**, *5*, 240–245. [CrossRef]
20. Delgado-Baquerizo, M.; Oliverio, A.M.; Brewer, T.E.; Benavent-González, A.; Eldridge, D.J.; Bardgett, R.D.; Maestre, F.T.; Singh, B.K.; Fierer, N. A global atlas of the dominant bacterial found in soil. *Science* **2018**, *359*, 320–325. [CrossRef]
21. Tedersoo, L.; Bahram, M.; Põlme, S.; Kõljalg, U.; Yorou, N.S.; Wijesundera, R.; Ruiz, L.V.; Vasco-Palacios, A.M.; Thu, P.Q.; Suija, A.; et al. Global diversity and geography of soil fungi. *Science* **2014**, *346*, 1256688. [CrossRef]
22. Köberl, M.; Müller, H.; Ramadan, E.M.; Berg, G. Desert farming benefits from microbial potential in arid soils and promotes diversity and plant health. *PLoS ONE* **2011**, *6*, e24452. [CrossRef]
23. Geisseler, D.; Scow, K.M. Long-term effects of mineral fertilizers on soil microorganisms—A review. *Soil Biol. Biochem.* **2014**, *75*, 54–63. [CrossRef]
24. Aparna, K.; Pasha, M.A.; Rao, D.L.N.; Krishnaraj, P.U. Organic amendments as ecosystem engineers: Microbial, biochemical and genomic evidence of soil health improvement in a tropical arid zone field site. *Ecol. Eng.* **2014**, *71*, 268–277. [CrossRef]
25. Kazerooni, E.A.; Maharachchikumbura, S.N.; Rethinasamy, V.; Al-Mahrouqi, H.; Al-Sadi, A.M. Fungal diversity in tomato rhizosphere soil under conventional and desert farming systems. *Front. Microbiol.* **2017**, *8*, 1462. [CrossRef] [PubMed]
26. Mullath, S.K.; Błaszkowski, J.; Govindan, B.N.; Al Dhaheri, L.; Symanczik, S.; Al-Yahya'ei, M.N. Organic farming practices in a desert habitat increased the abundance, richness and diversity of arbuscular mycorrhizal fungi. *Emir. J. Food Agric.* **2019**, *31*, 969–979. [CrossRef]
27. Symanczik, S.; Błaszkowski, J.; Koegel, S.; Boller, T.; Wiemken, A.; Al-Yahya'ei, M.N. Isolation and identification of desert habituated arbuscular mycorrhizal fungi newly reported from the Arabian Peninsula. *J. Arid. Land* **2014**, *6*, 488–497. [CrossRef]
28. Jones, D.L. Organic acids in the rhizosphere—A critical review. *Plant Soil* **1998**, *205*, 25–44. [CrossRef]
29. Olsen, S.R. *Estimation of Available Phosphorus in Soils by Extraction with Sodium Bicarbonate*; US Department of Agriculture: Washington, DC, USA, 1954; Volume 939.
30. Campbell, C.D.; Chapman, S.J.; Cameron, C.M.; Davidson, M.S.; Potts, J.M. A rapid microtiter plate method to measure carbon dioxide evolved from carbon substrate amendments so as to determine the physiological profiles of soil microbial communities by using whole soil. *Appl. Environ. Microbiol.* **2003**, *69*, 3593–3599. [CrossRef]
31. Ondov, B.D.; Bergman, N.H.; Phillippy, A.M. Interactive metagenomic visualization in a web browser. *BMC Bioinform.* **2011**, *12*, 385. [CrossRef]
32. Hammer, O.; Harper, D.A.; Ryan, P.D. PAST: Paleontological statistics software package for education and data analysis. *Palaeontol Electron* **2001**, *4*, 9.
33. Köhler, J.; Baumbach, J.; Taubert, J.; Specht, M.; Skusa, A.; Rüegg, A.; Rawlings, C.; Verrier, P.; Philippi, S. Graph-based analysis and visualization of experimental results with ONDEX. *Bioinformatics* **2006**, *22*, 1383–1390. [CrossRef]
34. Francioli, D.; Schulz, E.; Lentendu, G.; Wubet, T.; Buscot, F.; Reitz, T. Mineral vs. organic amendments: Microbial community structure, activity and abundance of agriculturally relevant microbes are driven by long-term fertilization strategies. *Front. Microbiol.* **2016**, *7*, 1446. [CrossRef]
35. Orr, C.H.; Leifert, C.; Cooper, J.M.; Cummings, S.P. The effect of organic and conventional farm management practices on the soil microbial community at the phylum level. *Asp. Appl. Biol.* **2013**, *120*, 1–7.
36. Bettiol, W.; Ghini, R.; Galvao, J.A.H.; Ligo, M.A.V.; Mineiro, J.L.C. Soil organisms in organic and conventional cropping systems. *Sci. Agric.* **2002**, *59*, 565–572. [CrossRef]

37. Norgaard, A.E.; Lewis, D.; Borden, K.A.; Krzic, M.; Carrillo, J.; Smukler, S.M. Trade-offs in organic nutrient management strategies across mixed vegetable farms in Southwest British Columbia. *Front. Sustain. Food Syst.* **2022**, *6*, 706271. [CrossRef]
38. Fadiji, A.E.; Ayangbenro, A.S.; Babalola, O.O. Organic farming enhances the diversity and community structure of endophytic archaea and fungi in maize plant: A shotgun approach. *J. Soil Sci. Plant Nutr.* **2020**, *20*, 2587–2599. [CrossRef]
39. Zhang, Q.-C.; Shamsi, I.H.; Xu, D.-T.; Wang, G.-H.; Lin, X.-Y.; Jilani, G.; Hussain, N.; Chaudhry, A.N. Chemical fertilizer and organic manure inputs in soil exhibit a vice versa pattern of microbial community structure. *Appl. Soil Ecol.* **2012**, *57*, 1–8. [CrossRef]
40. Hartmann, M.; Frey, B.; Mayer, J.; Mäder, P.; Widmer, F. Distinct soil microbial diversity under long-term organic and conventional farming. *Int. Soc. Microb. Ecol. J.* **2015**, *9*, 1177–1194. [CrossRef]
41. Al-Sadi, A.M.; Al-Mazroui, S.S.; Phillips, A.J.L. Evaluation of culture-based techniques and 454 pyrosequencing for the analysis of fungal diversity in potting media and organic fertilizers. *J. Appl. Microbiol.* **2015**, *119*, 500–509. [CrossRef]
42. Khan, A.L.; Asaf, S.; Abed, R.M.M.; Chai, Y.N.; Al-Rawahi, A.N.; Mohanta, T.K.; Al-Rawahi, A.; Schachtman, D.P.; Al-Harrasi, A. Rhizosphere microbiome of arid land medicinal plants and extra cellular enzymes contribute to their abundance. *Microorganisms* **2020**, *8*, 213. [CrossRef]
43. Kazeeroni, E.A.; Al-Sadi, A.M. 454-Pyrosequencing Reveals Variable Fungal Diversity Across Farming Systems. *Front. Plant Sci.* **2016**, *7*, 314. [CrossRef]
44. Nemec, S.; Menge, J.A.; Platt, R.G.; Johson, E.L.V. Vesicular-arbuscular mycorrhizal fungi associated with citrus in Florida and California and notes on their distribution and ecology. *Mycologia* **1981**, *73*, 112–127. [CrossRef]
45. Ortas, I. Mycorrhiza in citrus: Growth and nutrition. In *Advances in Citrus Nutrition*; Srivastava, A.K., Ed.; Springer Science+Business Media: Berlin/Heidelberg, Germany, 2012; pp. 333–351. [CrossRef]
46. Wang, P.; Zhang, J.J.; Shu, B.; Xia, R.X. Arbuscular mycorrhizal fungi associated with citrus orchards under different types of soil management, southern China. *Plant Soil Environ.* **2012**, *58*, 302–308. [CrossRef]
47. Bourazza, M.; Chetto, O.; Talha, A.; Farih, A.; Douira, A.; Benyahia, H. The influence of arbuscular mycorrhizal colonization on key growth parameters of five citrus rootstock cultivars under salt stress. *Plant Cell Biotechnol. Mol. Biol.* **2021**, *22*, 125–138.
48. Al-Sadi, A.M.; Kazerooni, E.A. Illumina-Mi-Seq analysis of fungi in acid lime roots reveals dominance of Fusarium and variation in fungal taxa. *Sci. Rep.* **2018**, *8*, 17388. [CrossRef] [PubMed]
49. Mohandas, S. Arbuscular mycorrhizal fungi benefit mango (*Mangifera indica* L.) plant growth in the field. *Sci. Hortic.* **2012**, *143*, 43–48. [CrossRef]
50. Govindan, M.; Rajeshkumar, P.P.; Varma, C.K.Y.; Anees, M.M.; Rashmi, C.R.; Nair, A.B. Arbuscular mycorrhizal fungi status of Mango (Mangifera indica) cultivars grown in typic quartzipsamments soil. *Agric. Res.* **2020**, *9*, 188–196. [CrossRef]
51. Jefwa, J.M.; Kahangi, E.; Losenge, T.; Mung'atu, J.; Ngului, W.; Ichami, S.M.; Sanginga, N.; Vanluawe, B. Arbuscular mycorrhizal fungi in the rhizosphere of banana and plantain and the growth of tissue culture cultivars. *Agric. Ecosyst. Environ.* **2012**, *157*, 24–31. [CrossRef]
52. Chebet, D.K.; Wanzala, F.K.; Kariuki, W. Versicular arbuscular mycorrhizal inoculation influences growth, nutrient absorption and hyphae colonization of rough lemon (*Citrus limon*) seedlings. *Afr. J. Educ. Sci. Technol.* **2021**, *6*, 75–85.
53. Menge, J.A.; Johnson, E.L.V.; Platt, R.G. Mycorrhizal dependency of several citrus cultivars under three nutrient regimes. *New Phytol.* **1978**, *81*, 553–559. [CrossRef]
54. Chen, K.; Liu, W.; Guo, S.; Liu, R.; Li, M. Diversity of arbuscular mycorrhizal fungi in continuous cropping soils used for pepper production. *Afr. J. Microbiol. Res.* **2012**, *6*, 2469–2474.
55. Liao, J.; Xu, Q.; Xu, H.; Huang, D. Natural farming improves soil quality and alters microbial diversity in a cabbage field in Japan. *Sustainability* **2019**, *11*, 3131. [CrossRef]
56. Wang, P.; Wang, Y.; Shu, B.; Liu, J.-F.; Xia, R.-X. Relationships between arbuscular mycorrhizal symbiosis and soil fertility factors in citrus orchards along an altitudinal gradient. *Pedosphere* **2015**, *25*, 160–168. [CrossRef]
57. Khade, S.W.; Rodrigues, B.F. Arbuscular mycorrhizal fungi associated with varieties of *Carica papaya* L. in tropical agro-based ecosystem of Goa, India. *Trop. Subtrop. Agroecosystems* **2009**, *10*, 369–381.
58. Guzman, A.; Montes, M.; Hutchins, L.; DeLaCerda, G.; Yang, P.; Kakouridid, A.; Dahlquist-Willard, R.M.; Firestone, M.K.; Bowles, T.; Kremen, C. Crop diversity enriches arbuscular mycorrhizal fungal communities in an intensive agricultural landscape. *New Phytol.* **2021**, *231*, 447–459. [CrossRef] [PubMed]
59. Liao, J.; Liang, Y.; Huang, D. Organic farming improves soil microbial abundance and diversity under greenhouse condition: A case study in Shanghai (Eastern China). *Sustainability* **2018**, *10*, 3825. [CrossRef]

sustainability

MDPI

Article

The Integration of Phosphorus-Solubilizing Rhizobacteria, *Eisenia fetida* and Phosphorus Rock Improves the Availability of Assimilable Phosphorus in the Vermicompost

Alfonso Andrade-Sifuentes, Gabriel de Jesús Peña-Uribe, Jorge Sáenz-Mata *, Jesús Josafath Quezada-Rivera *, Rubén Palacio-Rodríguez and Gisela Muro-Pérez

Facultad de Ciencias Biológicas, Universidad Juárez del Estado de Durango, Gómez Palacio 35010, Mexico; ing.andrade_85@hotmail.com (A.A.-S.); biol.mar.gabriel@gmail.com (G.d.J.P.-U.); ruben.palacio@ujed.mx (R.P.-R.); giselamuro@ujed.mx (G.M.-P.)

* Correspondence: jsaenz_mata@ujed.mx (J.S.-M.); josafath.quezada@ujed.mx (J.J.Q.-R.); Tel.: +52-871-709-6999 (J.S.-M.); +52-871-118-1838 (J.J.Q.-R.)

Abstract: Due to increasing soil degradation caused by unsustainable agricultural practices and the continued demand for quality food for the human population, it is imperative to find sustainable strategies for high-quality food production. For this reason, the objective of the present study was to evaluate the interaction between the factors of consortium of phosphorus-solubilizing rhizobacteria, addition of phosphate rock and worm load in horse manure to produce an organic fertilizer fortified with phosphorus. For this, consortia of phosphate-solubilizing rhizobacteria of the genus *Bacillus* (*Bacillus aryabhattai*, *Bacillus subtilis* and *Bacillus cereus*) isolated from the rhizosphere of *Distichlis spicata* were inoculated. Igneous phosphate rock (0 and 2%) was added in the vermicomposting process (with 25 and 50 g of *E. fetida* worms per kg of horse manure). The results obtained show that there is a significant interaction between the factors of inoculation with bacterial consortia (1×10^8 CFU mL^{-1}), phosphate rock (2%) and earthworm biomass (50 g kg^{-1} of manure), and that this interaction promotes the production of assimilable forms of phosphorus for plants (such as monobasic phosphate ions $H_2PO_4{}^{-1}$ or dibasic phosphate ions $HPO_4{}^{-2}$) within the vermicomposting process, having as a product an organic substrate supplemented with the optimal nutritional requirements for the development and growth of crops. This work can serve as a basis to produce high-quality organic fertilizer. However, field studies are required in order to observe the impact of vermicompost on the yield and quality of the fruits, and it can be compared with other types of fertilizers and the relevance of their use in different types of climates.

Keywords: bacterial consortia; sustainable agriculture; phosphate; vermidegradation

1. Introduction

The current food demand due to the accelerated growth of the human population is a worrying issue [1]. To satisfy this demand, excessive doses of agrochemicals are commonly applied to increase and accelerate crop yields, resulting in soil degradation [2,3]. Therefore, sustainable strategies are sought to replace these practices in order to obtain organic substrates rich in nutrients that satisfy the requirements of plants [4].

One of these strategies is the use of functional microorganisms for agriculture, such as plant-growth-promoting rhizobacteria (PGPR), which have demonstrated increases in crop yields [5]. The beneficial effect of these bacteria is that they are naturally part of the biological cycles of nutrients, facilitating their availability to plants in ionic forms. Among the promoter microorganisms are phosphate-solubilizing bacteria. These have an important role in the conversion to inorganic phosphorus from phosphate rock, thus increasing its bioavailability for the plant [6].

Phosphorus is considered an essential nutrient for the development of organisms (mainly in the form of phosphate). It is also part of the structure of nucleic acids, participates in different metabolic pathways and in energy storage, regulates enzymatic activity and signaling pathways through protein phosphorylation/dephosphorylation cycles and constitutes 0.2% of the dry weight of plants [7,8]. However, phosphates are considered one of the scarcest nutrients in soil intended for agricultural activities. To cover this deficit, one strategy is the addition of phosphate rock (PR) to manure-based organic substrates, allowing the microbiota to solubilize the phosphate and convert it into plant-available forms. This process can be carried out by different microorganisms, including bacteria belonging to the genera *Azospirillum*, *Acinetobacter*, *Alcaligenes*, *Arthrobacter*, *Bacillus*, *Pseudomonas*, *Burkholderia*, *Enterobacter*, *Erwinia*, *Flavobacterium*, *Proteus*, *Rhizobium*, *Serratia*, and *Xanthomonas*, among others [9,10], which transform inorganic and insoluble organic phosphorus compounds into soluble forms [11,12].

Vermicomposting can convert polluting organic waste, such as raw manure, into value-added products [13]. These products can be used in different agricultural production systems (open field or under protected systems) as an option to produce higher-quality organic foods. From this perspective, the objective of the present study was to evaluate the interaction between the factors of consortium of phosphorus-solubilizing rhizobacteria, addition of phosphate rock and worm load in horse manure to produce an organic fertilizer fortified with phosphorus.

2. Materials and Methods

2.1. Experiment Preparation

We carried out the experiment in the greenhouses of the Facultad de Ciencias Biológicas of the Universidad Juárez del Estado de Durango (FCB-UJED) (25°35′14.08″ N 103°30′2.43″ W). Three factors were evaluated: phosphate rock (PR) (0 and 2% per kg of manure), consortium of phosphorus-solubilizing bacteria (PSB) (at a concentration of 0 and 1×10^8 CFU mL^{-1}) and earthworm biomass (*E. fetida*) (25 and 50 g of worms kg^{-1} of manure). We used plastic containers (0.55 × 0.40 × 0.20 m, length × width × height). The containers had plastic lids to protect the worms and maintain the moisture of the substrate, since conditions that are too humid can cause anaerobic environments that are harmful to earthworms [14] (Figure 1).

Figure 1. Establishment of the experiment and raw material used. (**a**) Horse manure, (**b**) earthworm (*E. fetida*), (**c**) phosphorus-solubilizing bacteria grown in LB nutrient broth. (**d**) Phosphorus rock and (**e**) experimental units (plastic containers).

In each plastic container, 3 kg of solarized top-quality alfalfa fodder-based horse manure (dry weight) obtained from a stable in the region of Coahuila was mixed (Figure 1a). Depending on the treatment, some containers were mixed with 2% PR (Brow Depot®) per kg of manure [15], commercial product obtained from the La Negra mine of Fosforita de México, S.A. de C.V. located in Pachuca, Hidalgo, at 20°57′55″ N 99°20′45″ W 1820 m. PR is characterized as phosphorite with high amounts of phosphate minerals of dark brown to ochre color, a density of 3.12 gr cm^{-3} and an insoluble solubility of (>0.01%). It also has a melting point of 1400 °C, a hardness of 5 Mohs, a pH of 7.25 and a phosphorus (P_2O_5) content of 29.07% [16].

The plastic containers were placed onto well-ventilated metal shelves. Also, the containers were inclined to leach the substrate solution and maintain the required moisture, which was monitored with a hygrometer. We placed juvenile *E. fetida* worms in plastic containers filled with horse manure with population densities of 25 and 50 g of worms per kg^{-1} of manure [17]. Moisture was maintained at 70% [18] during the vermicomposting period (60 days).

The bacteria used in this project were supplied by the Microbial Ecology Laboratory of the Faculty of Biological Sciences of the Universidad Juárez del Estado de Durango. As a bacterial consortium, we used strains of *Bacillus aryabhattai* (Cryrizos1), *Bacillus subtilis* (CR7) and *Bacillus cereus* (CR5). The activation of bacteria was carried out in liquid Luria Bertani (LB) culture medium; they were grown in a shaking incubator at 37 °C and 120 rpm for a period of 24 h. Once grown, we used a cell density of 1×10^8 CFU mL^{-1}, with a Neubauer chamber for counting.

2.2. Experimental Design

We used a factorial design where the biomasses of *E. fetida* (25 and 50 g kg^{-1} of manure), (Figure 1b) bacterial consortia (0 and 1×10^8 CFU mL^{-1}) (Figure 1c) and phosphate rock (0 and 2%) (Figure 1d) were considered as factors. As a result of the combination of these factors, a total of 8 treatments were obtained (Table 1) with three repetitions, for a total of 24 experimental units (Figure 1e).

Table 1. Treatments established in solarized horse manure.

Treatments	FACTOR 1. Phosphorus Rock (PR)	FACTOR 2. Biomass *E. fetida*	FACTOR 3. Cell Density of Bacterial Consortia
	(%)	g Worms per kg^{-1} Manure	CFU mL^{-1}
T1	2	25	1×10^8
T2	2	25	0
T3	0	25	1×10^8
T4	0	25	0
T5	2	50	1×10^8
T6	2	50	0
T7	0	50	1×10^8
T8	0	50	0

2.3. Sampling

In each experimental unit, sampling was carried out at three depth levels (approx. 6, 12 and 18 cm). All samples were mixed to obtain composite samples of 150 g of vermicompost per treatment. This was carried out at 30 and 60 days. The composite samples were analyzed at the Laboratorio Nacional de Servicios de Análisis de Aguas, Suelos, Plantas y Medio Ambiente del Centro Nacional de Investigación Disciplinaria en Relación Agua, Suelo, Planta, Atmósfera (CENID-RASPA) in Gómez Palacio, Durango, México.

2.4. Physicochemical Analysis

The samples were dried in an oven at a temperature of 45 °C until they reached a constant weight. They were subsequently ground (<2 mm) to obtain a homogeneous sample for the analysis of pH, electrical conductivity (EC), total nitrogen (TN), total phosphorus (TP), soluble phosphorus (SP), nitrate (NO_3^-) and ammonium (NH_4^+).

Electrical conductivity and pH were determined from a vermicompost–water suspension (1:10). This was shaken at 230 rpm for 30 min and allowed to settle for one hour before measuring pH and EC. A pocket pH/conductivity meter, model HI98129 (HANNA Instruments®), was used as described in [19]. Total nitrogen (TN) was measured by the standard digestion and distillation method with Kjeldahl equipment (Kelplus) using reagents (magnesium oxide for the determination of NH_4^+ and Devarda alloy for NO_3^- [20] to subsequently titrate with HCL (0.005 N) according to the NOM-021-RECNAT-2000 standard.

Total phosphorus (TP) was determined by digesting 0.5 g of vermicompost sample using hot-plate crude-fiber digestion equipment according to NOM-F-90-S-1978 and the Hedley 1982 method modified by Tiessen in 1993 [21,22].

2.5. *E. fetida* Egg and Worm Count

Once the experiment was finished (60 days), the total biomass of the worms was weighed using a digital scale (Santul 5927) and the average weight of a single individual was obtained. The eggs were counted manually with a 20 cm straight stainless steel spatula.

2.6. Statistical Analysis

The data obtained were analyzed with the Anderson Darling normality test. For data corresponding to a normal distribution, we performed analysis of variance using the Statistical Analysis System software, version 9.4 (SAS. 2016). Tukey was used as a post hoc test ($p \leq 0.05$). For data that did not have a normal distribution, we performed Kruskal–Wallis tests.

3. Results and Discussion

3.1. Vermicompost Physicochemical Characteristics

The results obtained show that there is a significant interaction between the factors of inoculation with bacterial consortia (1×10^8 CFU mL^{-1}), phosphate rock (2%) and earthworm biomass (50 g kg^{-1} of manure), and that this interaction promotes the production of assimilable forms of phosphorus for plants (such as monobasic phosphate ions $(H_2PO_4)^{-1}$ or dibasic phosphate ions $(HPO_4)^{-2}$) within the vermicomposting process, having as a product an organic substrate supplemented with the optimal nutritional requirements for the development and growth of crops. The physicochemical characteristics of the vermicompost treatments at 60 days can be seen in Table 2. During this process, the moisture in the vermireactors was between 70 and 80%, providing a favorable environment for the development and multiplication of the *E. fetida* earthworms. Some authors indicate that the best conditions for the survival of *Eisenia fetida* range between 50 and 90% moisture in the substrate [23]. This is important because it is a crucial factor that would put the vermidegradation of the organic fertilizer at risk and alter the physical and chemical characteristics throughout the process [24].

Table 2. Physicochemical parameters analyzed based on the interaction of phosphorus rock (PR), consortium of phosphorus-solubilizing bacteria (PSB) and quantity of earthworms (*E. fetida*) with the vermidegradation of nutrients in the substrate at 60 days.

	pH	EC	TN	NO_3^-	NH_4^+	TP	SP
		(dS m^{-1})	%	ppm	ppm	%	ppm
Initial Manure Values	10	2.8	0.65	Nd	Nd	0.31	Nd
T1 (2% + 25 g + 1 × 10^8)	9.10 ± 0.04 a	4.2 ± 0.06 c	1.22 ± 0.08 a	481.66 ± 53.30 a	112.12 ± 25.41 ba	1.59 ± 0.08 a	350.26 ± 28.37 b
T2 (2% + 25 g)	8.83 ± 0.04 a	4.8 ± 0.69 b	1.43 ± 0.08 a	427.33 ± 48.78 a	148.61 ± 47.76 ba	1.95 ± 0.08 a	329.99 ± 35.34 cb
T3 (25 g + 1 × 10^8)	9.00 ± 0.19 a	8.3 ± 0.80 a	1.34 ± 0.05 a	390.33 ± 38.37 a	82.39 ± 15.85 b	0.51 ± 0.01 b	252.31 ± 11.66 d
T4 (25 g)	9.03 ± 0.22 a	8.0 ± 0.60 a	1.32 ± 0.14 a	397.00 ± 110.26 a	108.28 ± 15.01 ba	0.53 ± 0.04 b	269.12 ± 28.37 cd
T5 (2% + 50 g + 1 × 10^8)	9.14 ± 0.04 a	6.6 ± 1.02 bc	1.21 ± 0.07 a	472.66 ± 75.19 a	160.27 ± 30.29 a	1.62 ± 0.09 a	442.44 ± 17.29 a
T6 (2% + 50 g)	9.15 ± 0.03 a	4.1 ± 0.76 c	1.22 ± 0.07 a	481.00 ± 79.68 a	108.87 ± 31.73 ba	1.47 ± 0.06 a	325.70 ± 20.27 cb
T7 (50 g + 1 × 10^8)	9.00 ± 0.21 a	7.1 ± 0.65 a	1.19 ± 0.18 a	435.33 ± 115.31 a	101.51 ± 16.35 ba	0.55 ± 0.07 b	346.21 ± 24.32 b
T8 (50 g)	8.87 ± 0.04 a	7.5 ± 0.00 a	1.21 ± 0.13 a	368.33 ± 112.26 a	112.69 ± 13.96 ba	0.56 ± 0.04 b	236.73 ± 28.38 d

Means and standard deviations. Same letters in the column for each factor represent non-significant differences (Tukey, $p < 0.05$). Nd = not determined.

3.1.1. pH

The pH values did not show significant differences between treatments ($p \leq 0.05$) at 60 days (Table 2). However, the highest values were the treatments containing PR, T6 (2% + 50 g, 0 CFU), T5 (2% + 50g + 1 × 10^8 CFU) and T1 (2% + 25 g + 1 × 10^8 CFU), with 9.15 ± 0.03, 9.14 ± 0.04 and 9.10 ± 0.04 respectively. These values were high with respect to the NMX-FF-109-SCFI-2007 standard; however, this work was carried out under the standards and parameters established by the USDA [25]. This specifies the parameters of a quality vermicompost, and the pH values should vary from 5.5 to 8.5. The probable cause of high pH values may be due to higher pH values of the pre-composted horse manure, which registered an initial alkaline value of 10. Different authors have mentioned that the pH of horse manure can vary depending on the type of feed, having a more acidic pH (5.7~) when the feed is enriched with grains and a more neutral pH (6.7~) when the horses are fed with grass and forage [26]. In addition, no studies have been reported with pH values higher than 8, so the high pH values in all treatments may be due to the initial pH of the horse manure, which was 10. However, it should be noted that the pH variable decreased from the beginning of the experiment. After 30 days of sampling, it decreased to 9.8, and at 60 days, the pH value was 8.83, slightly alkaline. This may be because the bacteria and number of worms contributed to the decrease in the pH of the medium towards either neutrality or acidity throughout the process [27,28]. Some authors [19,29] have mentioned that the metabolism of earthworms decreases the pH in the surface layer of the soil (Horizon A) or in the vermicompost due to the processes of bioturbation and/or vermidegradation (mineralization of nitrogen into nitrates and nitrites and phosphorus into orthophosphates) and the bioconversion of organic matter into other organic acids. In addition, the metabolism of phosphorus-solubilizing bacteria released organic acids, causing decreases in the pH of the substrate and greater permeability [30,31]. Also, it was demonstrated in [32] that microbes become additives for the reduction in solid waste in the composting process. This confirms that the vermidegradation process was successful, significantly reducing the pH noted in Figure 2 and supporting the Table 3 data that show the interactions among pH and the factors. There were significant differences according to the factors and their interactions. Regarding the separate analyses of the factors, for the worm gram factor (25 and 50 g), no significant differences were found ($p > 0.05$) in the two periods evaluated (30 and 60 days). For the phosphorus rock factor (0 and 2%), significant differences were recorded ($p \leq 0.05$) at 30 days, with a higher pH value for 2% (9.79), but not at 60 days. For the bacterial consortia factor (0 and 1 × 10^8 CFU), significant differences ($p \leq 0.05$) were recorded at 30 days, registering values of 9.62 and 9.70, respectively; on the contrary, no significant differences were recorded for this factor at 60 days. Regarding the interaction between the factors, an interaction between factors was recorded 60 days after the start of the experiment (Table 3, Figure 3a). On the contrary, there was no interaction between factors at the first sampling time (30 days). The values closest to neutral resulted from treatments T2 (2% + 25 g + 0 CFU) and T8 (0% + 50 g + 0 CFU), with 8.83 ± 0.04 and 8.87 ± 0.04. However, if the process of vermidegradation were carried out for 90 days, we could obtain more desirable pH values between 7.8 and 8.5, which are reported to be optimal for worm growth and microbial activity [33].

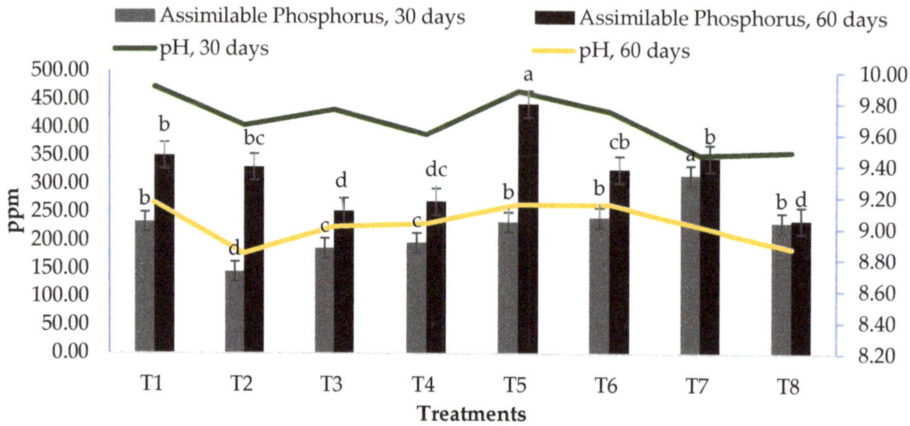

Figure 2. Statistical analysis of the behavior of the assimilable phosphorus and pH in the treatments. Bars with different letters are statistically different according to Tukey's test ($p \leq 0.05$).

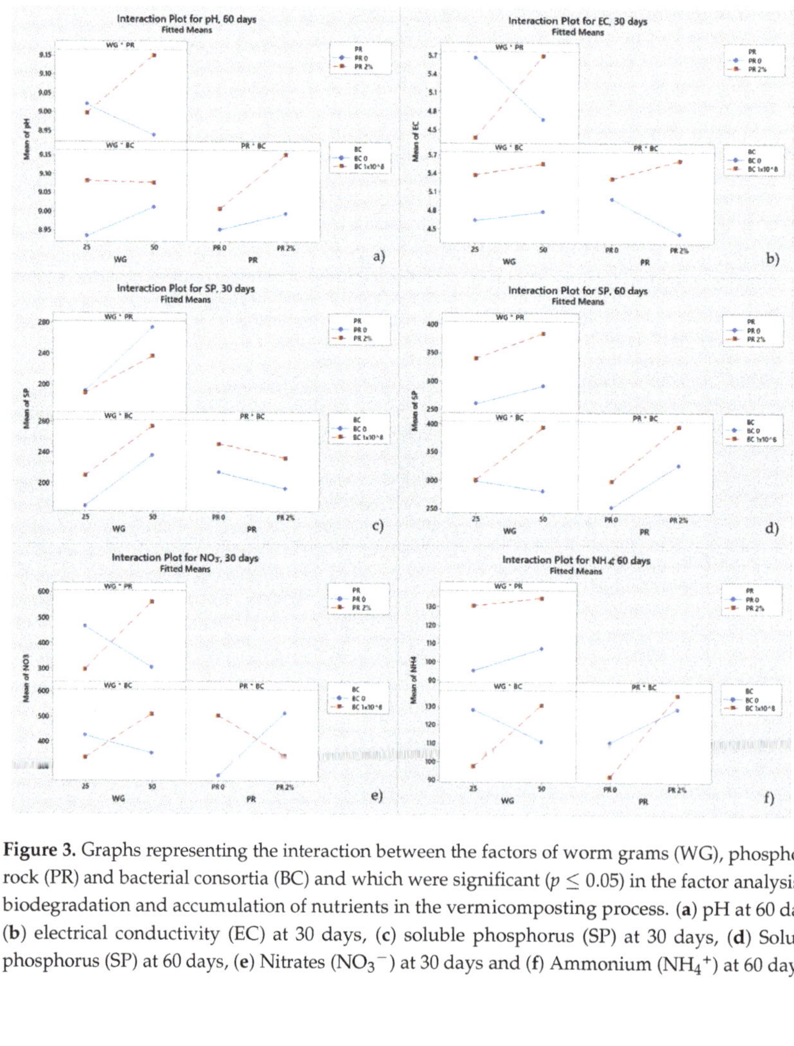

Figure 3. Graphs representing the interaction between the factors of worm grams (WG), phosphoric rock (PR) and bacterial consortia (BC) and which were significant ($p \leq 0.05$) in the factor analysis of biodegradation and accumulation of nutrients in the vermicomposting process. (**a**) pH at 60 days, (**b**) electrical conductivity (EC) at 30 days, (**c**) soluble phosphorus (SP) at 30 days, (**d**) Soluble phosphorus (SP) at 60 days, (**e**) Nitrates (NO_3^-) at 30 days and (**f**) Ammonium (NH_4^+) at 60 days.

Table 3. Factor analysis of the effects of phosphorus rock (PR), consortium of phosphorus-solubilizing bacteria (PSB) and amount of earthworms (*E. fetida*) and their interaction (WG × PR and BC) on the vermidegradation of nutrients in the substrate.

Factor	pH		EC		TN		NO_3^-		NH_4^+		TP		SP	
			(dS m^{-1})		%		ppm		ppm		%		ppm	
Worm grams (WG)														
	30 days	60 days	30 days	60 days	30 days	60 days	30 days	60 days	30 days	60 days	30 days	60 days	30 days	60 days
25 g	9.72 ± 0.04 a	9.0 ± 0.0 a	5.09 ± 0.04 b	6.3 ± 0.0 a	1.2 ± 0.05 a	1.3 ± 0.05 a	381 ± 26 b	439 ± 7.5 a	77 ± 11 b	121 ± 4.0 a	0.84 ± 0.0 a	1.2 ± 0.1 a	190 ± 33 b	300 ± 15 b
50 g	9.64 ± 2.96 a	9.0 ± 0.0 a	5.16 ± 0.09 a	6.3 ± 0.0 a	1.1 ± 0.05 a	1.2 ± 0.00 b	434 ± 27 a	424 ± 7.5 a	99 ± 11 a	113 ± 4.0 a	0.81 ± 0.0 a	1.0 ± 0.1 a	255 ± 33 a	308 ± 16 a
Phosphorus rock (PR)														
	30 days	60 days	30 days	60 days	30 days	60 days	30 days	60 days	30 days	60 days	30 days	60 days	30 days	60 days
0%	9.58 ± 0.0 b	8.9 ± 0.0 a	5.15 ± 0.0 a	5.9 ± 0.4 a	1.1 ± 0.0 a	1.3 ± 0.0 a	386 ± 22 b	397 ± 34 a	106 ± 18 a	101 ± 15 b	0.46 ± 0.4 b	0.5 ± 0.6 b	232 ± 9.5 a	276 ± 43 b
2%	9.79 ± 0.0 a	9.0 ± 0.0 a	5.02 ± 0.0 a	6.7 ± 0.4 a	1.2 ± 0.0 a	1.3 ± 0.0 a	430 ± 22 a	465 ± 34 a	70 ± 18 b	132 ± 15 a	1.19 ± 0.4 a	1.7 ± 0.6 a	213 ± 9.5 a	362 ± 43 a
Bacterial consortia (BC)														
	30 days	60 days	30 days	60 days	30 days	60 days	30 days	60 days	30 days	60 days	30 days	60 days	30 days	60 days
0 CFU	9.62 ± 0.0 b	8.9 ± 0.0 a	4.71 ± 0.3 b	6.1 ± 0.2 a	1.2 ±0.0 a	1.2 ± 0.0 a	391 ± 17 b	418 ± 13 a	74 ± 14 b	114 ± 2.5 a	0.83 ± 0.0 a	1.0 ± 0.1 a	203 ± 19 b	290 ± 28 b
1×10^8 CFU	9.70 ± 0.0 a	9.0 ± 0.0 a	5.46 ± 0.3 a	6.5 ± 0.2 a	1.1 ± 0.0 a	1.3 ± 0.0 a	425 ± 17 a	445 ± 13 a	102 ±14 a	119 ± 2.5 a	0.81 ± 0.0 a	1.1 ± 0.1 a	242 ± 19 a	348 ± 29 a
WG × PR × BC														
	ns	*	*	ns	*	ns	*	ns	ns	*	*	*	*	ns

Means and standard deviation. Equal letters for each factor represent non-significant differences. * = significant difference; ns = non-significant difference according to Tukey test ($p \leq 0.05$).

3.1.2. Electric Conductivity (EC)

For electrical conductivity, significant differences were recorded between treatments ($p \leq 0.05$) at 60 days, with treatments T3 (0% + 25 g + 1 × 10^8 CFU), T4 (0% + 25 g + 0 CFU), T8 (0% + 50 g + 0 CFU) and T7 (0% + 50 g + 1 × 10^8 CFU) showing the highest EC values (8.3, 8.0, 7.5 and 7.1 dS m^{-1}, respectively) (Table 2). Like the pH values, the EC values were high according to NMX-FF-109-SCFI-2007 (regarding the quality specifications of worm humus produced in Mexico), which establishes that the permissible EC value is less than 4 (dS m^{-1}), while for other countries, this value may be less than 8.2 (dS m^{-1}) [34,35]. Nevertheless, the initial electrical conductivity value was 2.8 dS m^{-1}, showing an increase in all treatments towards the end of the experiment, with treatment T3 (0% + 25 g + 1 × 10^8 CFU) being the one with the highest EC value. These variations could be attributed to the mineralization of organic matter produced by *E. fetida* organisms and inoculations with bacterial consortia [36]. In addition, some authors have mentioned that EC is undoubtedly the most impactful factor that favors the acceptability of the final product, considering that it is a good indicator of safety and suitability for agricultural purposes [37]. Regarding the separate analyses of factors, it was found that for the worm grams factor (25 and 50 g), significant differences were recorded ($p \leq 0.05$) at 30 days, with the highest value being 5.16 dS m^{-1}; however, no statistical differences were recorded at 60 days. For the phosphate rock factor (0 and 2%), no significant differences were recorded ($p > 0.05$) at either of the two sampling times. For the bacterial consortium factor (0 and 1 × 10^8 CFU), significant differences were recorded ($p \leq 0.05$) at 30 days, recording values of 4.71 and 5.46, respectively; on the contrary, no significant differences were recorded for this factor at 60 days. Regarding the interaction between factors, it was found that there was an interaction between factors at the first sampling time (30 days), but on the contrary, no interaction between factors was recorded 60 days after the experiment began (Table 3, Figure 3b).

3.1.3. Total Nitrogen, NO_3^- and NH_4^+

For the total nitrogen (TN), there were no significant differences between treatments ($p > 0.05$). However, it is important to highlight that the simple vermicomposting process increased the TN values towards the end of the experiment, since the initial TN value of the horse manure was 0.65% and at the end of the experiment, it ranged between 1.19 and 1.43% (Table 2). This could be because, as some researchers have stated, nitrogen concentrations at the end of vermicomposting are determined by the amount of nitrogen provided by the substrate material used [38,39]. Other factors that can contribute to the mineralization of the substrate are the feeding of microorganisms and worms, the release of mucus, other nitrogenous excretions and the death of worms [40]. Regarding the separate analyses of the factors, it was determined that, for the worm grams factor (25 and 50 g), no significant differences were recorded ($p > 0.05$) in the first period evaluated (30 days). On the contrary, significant differences ($p \leq 0.05$) were determined at the 60-day sampling, registering 1.3% for the 25 g. For the percentage of phosphate rock (0 and 2%) and bacterial consortium (0 and 1 × 10^8 CFU) factors, no significant differences were recorded ($p \leq 0.05$) in either of the periods evaluated (30 and 60 days) (Table 3). Specifically, for nitrate (NO_3^-) there were no significant differences between treatments ($p > 0.05$). The range of concentrations between treatments ranged from 368.3 to 481.6 ppm (Table 2). Regarding the separate analyses of the factors, it was determined that, for the worm grams factor (25 and 50 g), significant differences were recorded ($p \leq 0.05$) in the first period evaluated (30 days), with the 50 g load registering the highest concentration of NO_3^- (434 ppm). For the factors of percentage of phosphate rock (0 and 2%) and bacterial consortia (0 and 1 × 10^8 CFU), significant differences were recorded ($p \leq 0.05$) at 30 days. Regarding the interaction between factors, it was found that there was an interaction between factors at the first sampling time (30 days), but on the contrary, no interaction between factors was recorded 60 days after the experiment began (Table 3, Figure 3e). Regarding ammonium NH_4^+, according to the analysis of variance, there were significant differences between the treatments ($p \leq 0.05$), with treatment 5 being (2% + 50 g + 1 × 10^8) the one that registered the highest concentration of ammonium (160 ppm), while treatment 3 (25 g + 1 $\times$ 10^8) was the one that registered the lowest concentration of this ion (82.39 ppm) (Table 2). This may be proof of the ability of these bacterial consortia to transform ammonia, a product of worms' metabolism, into ammonium. Regarding the separate analyses of the factors, it was determined that, for worm grams factor (25 and 50 g), there were significant differences ($p \leq 0.05$) in the first period evaluated (30 days), with the load of 50 g being the one that registered a higher concentration of NH_4^+ (99 ppm). At 60 days, there were no significant differences. For the percentage of phosphate rock (0 and 2%) and bacterial consortia (0 and 1 $\times$ 10^8 CFU) factors, significant differences ($p \leq 0.05$) were recorded in both periods evaluated. Regarding the interaction between factors, it was found that there was an interaction between factors at the second sampling time (60 days), but on the contrary, no interaction between the factors was recorded 30 days after the start of the experiment (Table 3, Figure 3f).

For the total nitrogen (TN), there were no significant differences between treatments ($p > 0.05$). However, it is important to highlight that the simple vermicomposting process increased the TN values towards the end of the experiment, since the initial TN value of the horse manure was 0.65% and at the end of the experiment, it ranged between 1.19 and 1.43% (Table 2). This could be because, as some researchers have stated, nitrogen concentrations at the end of vermicomposting are determined by the amount of nitrogen provided by the substrate material used [38,39]. Other factors that can contribute to the mineralization of the substrate are the feeding of microorganisms and worms, the release of mucus, other nitrogenous excretions and the death of worms [40]. Regarding the separate analyses of the factors, it was determined that, for the worm grams factor (25 and 50 g), no significant differences were recorded ($p > 0.05$) in the first period evaluated (30 days). On the contrary, significant differences ($p \leq 0.05$) were determined at the 60-day sampling, registering 1.3% for the 25 g. For the percentage of phosphate rock (0 and 2%) and bacterial consortium (0 and 1 $\times$ 10^8 CFU) factors, no significant differences were recorded ($p \leq 0.05$) in either of the periods evaluated (30 and 60 days) (Table 3). Specifically, for nitrate (NO_3^-) there were no significant differences between treatments ($p > 0.05$). The range of concentrations between treatments ranged from 368.3 to 481.6 ppm (Table 2). Regarding the separate analyses of the factors, it was determined that, for the worm grams factor (25 and 50 g), significant differences were recorded ($p \leq 0.05$) in the first period evaluated (30 days), with the 50 g load registering the highest concentration of NO_3^- (434 ppm). For the factors of percentage of phosphate rock (0 and 2%) and bacterial consortia (0 and 1 $\times$ 10^8 CFU), significant differences were recorded ($p \leq 0.05$) at 30 days. Regarding the interaction between factors, it was found that there was an interaction between factors at the first sampling time (30 days), but on the contrary, no interaction between factors was recorded 60 days after the experiment began (Table 3, Figure 3e). Regarding ammonium NH_4^+, according to the analysis of variance, there were significant differences between the treatments ($p \leq 0.05$), with treatment 5 being (2% + 50 g + 1 $\times$ 10^8) the one that registered the highest concentration of ammonium (160 ppm), while treatment 3 (25 g + 1 $\times$ 10^8) was the one that registered the lowest concentration of this ion (82.39 ppm) (Table 2). This may be proof of the ability of these bacterial consortia to transform ammonia, a product of worms' metabolism, into ammonium. Regarding the separate analyses of the factors, it was determined that, for worm grams factor (25 and 50 g), there were significant differences ($p \leq 0.05$) in the first period evaluated (30 days), with the load of 50 g being the one that registered a higher concentration of NH_4^+ (99 ppm). At 60 days, there were no significant differences. For the percentage of phosphate rock (0 and 2%) and bacterial consortia (0 and 1 $\times$ 10^8 CFU) factors, significant differences ($p \leq 0.05$) were recorded in both periods evaluated. Regarding the interaction between factors, it was found that there was an interaction between factors at the second sampling time (60 days), but on the contrary, no interaction between the factors was recorded 30 days after the start of the experiment (Table 3, Figure 3f).

3.1.4. Total Phosphorus and Assimilable Soluble Phosphorus

As time goes by, it becomes more important to integrate strategies to improve crop nutrition and yield in a more sustainable way. In this sense, phosphate rock (PR) has been of great importance, since it is among the few inorganic compost amendments that are allowed in organic agriculture [41]. Regarding the total phosphorus variable (TP), significant differences were obtained between treatments ($p \leq 0.05$), resulting in a higher percentage in those to which phosphate rock was added (T1 (2% + 25 g + 1 × 10^8 CFU), T2 (2% + 25 g + 0 CFU), T5 (2% + 50 g + 1 × 10^8 CFU) and T6 (2% + 50 g + 0 CFU)). It is important to mention that the simple addition of phosphate rock (2%) to horse manure drastically increased the content of this element. The initial TP value recorded in horse manure was 0.31%, and this increased to 1.91% for T2 (2% + 25 g + 0 CFU) (Table 2). This confirms that increases in total phosphorus in the vermicompost are directly related to the supply of phosphate rock in the treatments. According to the separate analyses of the factors, it was found that for the worm grams factor (25 and 50 g), there were no significant differences ($p > 0.05$) in any of the periods evaluated (30 and 60 days); however, it is important to mention that it increased, on average, from 0.82% to 1.1% in the TP content towards the end of the experiment. However, it is important to highlight that this phosphorus is not completely assimilated by crops. Phosphorus assimilated by plants occurs in the forms of dihydrogen phosphate ($H_2PO_4{}^{-1}$) and hydrogen phosphate ($HPO_4{}^{-2}$). For the percentage of phosphate rock factor (0 and 2%), obvious significant differences were recorded in both periods evaluated. For the treatments with 2% of phosphate rock, a value of 1.19 was reached at 30 days and a value of 1.7 was reached at 60 days. The bacterial consortia factor (0 and 1 × 10^8 CFU) behaved in a similar way to the grams of worm factor. Regarding the interaction between factors, it was found that there was an interaction between factors at both evaluation times (Table 3). One of the most important variables and objects of study in the present investigation was the presence of assimilable phosphorus in the designed vermicomposts. According to the ANOVA, significant differences ($p \leq 0.05$) were found between the treatments; the T5 treatment (2% + 50 g + 1 × 10^8) had the highest quantified concentration of soluble phosphorus with 442.44 ppm. This was mainly due to the biodegradation and solubilization processes carried out by the worms in interaction with the bacterial consortium in the substrate [42]. Regarding the separate analyses of the factors, it was determined that for the grams of earthworm factor (25 and 50 g), significant differences were recorded ($p \leq 0.05$) for both periods evaluated (30 and 60 days), highlighting that the highest values were quantified for the 50 g load. This is supported by studies that have shown that the joint action between bacteria and worms increases the solubilization of phosphorus due to the intestinal phosphatases of the worms [43]. This alkaline phosphatase (esterase group phosphatase) from worm feces is directly involved in the phosphorus cycle [44]. For the percentage of phosphate rock factor (0 and 2%), significant differences were recorded only for the 60-day period, reaching an average value of 362 ppm for the 2% load. For the bacterial consortia factor (0 and 1 × 10^8 CFU), significant differences were recorded in both periods, highlighting that for the 30-day period, the treatments with bacterial load reached 242 ppm, and for the 60-day period, a concentration of 348 ppm was reached. Also, phosphorus-solubilizing bacteria release organic acids that reduce pH and improve permeability, since these organic acids are byproducts of microbial fermentation produced by oxidative respiration, while glucose is used as a carbon source [30]. This is why a positive relationship was confirmed between the release of organic acids and the segregation of phosphatases, with an effect on the solubilization of phosphate rock. Regarding the interaction between factors, it was determined that if there was an interaction at 30 days, this was not the case for 60-day evaluation between factors evaluated at both times (Table 3, Figure 3c,d).

3.1.5. *E. fetida* Earthworm Biomass and Number of Eggs

According to the Kruskal–Wallis test, significant differences were recorded in the two variables analyzed, i.e., earthworm biomass (H-Value = 18.05, p = 0.012) and number of egg capsules (H-Value = 21.26, p = 0.003). The treatment that led to a greater biomass of *E. fetida* was treatment T6 (2% + 50 g + 0 CFU), with an average weight per individual of 1.03 ± 0.3 g (Table 4), while treatment T1 (2% + 25 g + 1 × 10^8 CFU) was the one that presented the lowest biomass, with 0.23 ± 0.06 g. For the number of eggs variable, the treatment that had the greatest impact was treatment T8 (0% + 50 g + 0 CFU), with an average of 384 ± 65.29, while the treatment that recorded the lowest number of eggs was T5 (2% + 50 g + 1 × 10^8), with an average of 132. Curiously, the treatments to which bacterial consortia were not added were those that presented a higher biomass and number of eggs. This is because, as some authors [45,46] have mentioned, the metabolic activity of worms decreases with the presence of bacterial communities. However, the survival of the organisms is not affected. In this research, it is shown that the presence of the bacterial consortium, beyond having a limiting effect on the reproduction of the worms, participates efficiently in the process of nutrient bioavailability.

Table 4. Results of the KruskalWallis test determining the differences in biomass and *E. fetida* eggs between treatments.

	Worm Biomass H-Value = 18.05 p = 0.012				Egg Capsules H-Value = 21.26 p = 0.003			
Treatment	Mean (g)	Median	Mean Rank	Z-Value	Mean	Median	Mean Rank	Z-Value
T1 (2% + 25 g + 1 × 10^8)	0.23 ± 0.06	0.23529	2.0	−2.75	160 ± 37.70	156	14.5	−1.87
T2 (2% + 25 g)	0.83 ± 0.08	0.83333	20.3	2.05	216 ± 53.70	216	27.2	0.5
T3 (25 g + 1 × 10^8)	0.52 ± 0.11	0.58333	10.3	−0.57	168 ± 77.40	144	16.5	−1.5
T4 (25 g)	0.51 ± 0.12	0.43750	10.7	−0.48	221.33 ± 26.85	220	27.8	0.62
T5 (2% + 50 g + 1 × 10^8)	0.47 ± 0.13	0.38462	8.8	−0.96	180 ± 103.51	132	17.2	−1.37
T6 (2% + 50 g)	1.03 ± 0.3 *	1.00000	22.7 *	2.66	192 ± 42.93	192	20.8	−0.69
T7 (50 g + 1 × 10^8)	0.54 ± 0.13	0.54000	11.2	−0.35	214 ± 43.04	210	26.5	0.37
T8 (50 g)	0.62 ± 0.08	0.58974	14.0	0.39	384 ± 65.29	348	45.5 *	3.93
Overall			12.5				24.5	

* = significant difference.

4. Conclusions

This study evaluated the potential of phosphorus-solubilizing rhizobacteria to enhance the degradation and nutrient release from phosphate rocks in vermicompost based on horse manure, aiming to create an organic fertilizer enriched with soluble phosphorus. While total nitrogen and NO_3^- values did not show significant differences across treatments, NH_4^+ levels correlated with treatments containing higher worm populations (T5, T6, T7 and T8). Notably, the results underscored the importance and potential of phosphorus-solubilizing rhizobacteria and their interaction with worms in the promotion of soluble phosphorus availability in the substrate (vermicompost from horse manure). Treatments that included phosphate rock and bacterial consortia achieved high levels of $(H_2PO_4)^{-1}$. In addition to the above, the treatment with the highest worm load was the one that presented the highest amount of soluble phosphorus (T5). This positive interaction between worms and bacteria enhanced phosphorus solubilization, highlighting the potential to produce vermicompost suitable for phosphorus-deficient soils. Such vermicompost could significantly improve soil nutritional characteristics, thereby boosting crop yields. The findings of this research provide a foundation for producing high-quality organic fertilizer that contributes to environmental sustainability. Future research should address the optimal application rates of this product and involve field trials to compare yields with other fertilizers, assessing its relevance across various crops and climates.

Author Contributions: Conceptualization, J.S.-M., G.d.J.P.-U. and R.P.-R.; methodology, J.S.-M., J.J.Q.-R. and A.A.-S.; software, R.P.-R. and G.d.J.P.-U.; validation, J.S.-M. and G.d.J.P.-U.; formal analysis, R.P.-R., A.A.-S. and J.S.-M.; investigation, A.A.-S., J.J.Q.-R. and R.P.-R.; resources, R.P.-R.; A.A.-S.

and G.d.J.P.-U.; data curation, G.M.-P., G.d.J.P.-U. and J.S.-M.; writing—original draft preparation, J.S.-M., R.P.-R., A.A.-S. and J.J.Q.-R.; writing—review and editing, J.J.Q.-R.; visualization, G.M.-P.; supervision, J.J.Q.-R., A.A.-S. and J.S.-M.; project administration, G.M.-P.; funding acquisition, A.A.-S., G.M.-P. and J.J.Q.-R. All authors have read and agreed to the published version of the manuscript.

Funding: This research was supported in part by the Consejo Nacional de Humanidades Ciencia y Tecnología of Mexico (CONAHCYT), with a postdoctoral fellowship number of CVU 560948.

Institutional Review Board Statement: Not applicable.

Informed Consent Statement: Not applicable.

Data Availability Statement: Data is contained within the article.

Acknowledgments: The authors are thankful to the Consejo Estatal de Ciencia y Tecnología del Estado de Durango (COCYTED) for supporting part of the publication costs. Thank you for the technical support given by M.C. Homero Sánchez-Galvan in the greenhouse experiments of the FCB-UJED, also to M.C. Cristo Omar Puente Valenzuela for the technical assistance of the INIFAP CENID RASPA Laboratory technicians for the analysis carried out herein.

Conflicts of Interest: The authors declare that they have no conflicts of interest. The funders had no role in the design and implementation of this research or its interpretation of data, nor in the decision to write and submit this manuscript.

References

1. Sarfraz, S.; Alí, F.; Hameed, A.; Ahmad, Z.; Riaz, K. Sustainable Agriculture Through Technological Innovations. In *Sustainable Agriculture in the Era of the OMICs Revolution*; Prakash, C.S., Fiaz, S., Nadeem, M.A., Baloch, F.S., Qayyum, A., Eds.; Springer: Cham, Switzerland, 2023; pp. 223–239. [CrossRef]
2. Mesmar, A.K.; Albedwawi, S.T.; Alsalami, A.K.; Alshemeili, A.R.; Abu-Elsaoud, A.M.; El-Tarabily, K.A.; Al Raish, S.M. The Effect of Recycled Spent Coffee Grounds Fertilizer, Vermicompost, and Chemical Fertilizers on the Growth and Soil Quality of Red Radish (*Raphanus sativus*) in the United Arab Emirates: A Sustainability Perspective. *Foods* **2024**, *13*, 1997. [CrossRef]
3. Savci, S. Investigation of the Effect of Chemical Fertilizers on Environment. *APCBEE Procedia* **2012**, *1*, 287–292. [CrossRef]
4. Ajibade, S.; Mupambwa, H.A.; Manyevere, A.; Mnkeni, P.N.S. Influence of Microbial Inoculation of Igneous Rock Phosphate-Amended Cow and Pig Manures on Vermidegradation and Nutrient Release. *Agronomy* **2020**, *10*, 1587. [CrossRef]
5. Backer, R.; Rokem, J.S.; Ilangumaran, G.; Lamont, J.; Praslickova, D.; Ricci, E.; Subramanian, S.; Smith, D.L. Plant Growth-Promotion Rhizobacteria: Contex, Mechanisms of Action, and Roadmap to Commercialization of Biostimulants for Sustainable Agriculture. *Front. Plant Sci.* **2018**, *9*, 1473. [CrossRef]
6. Timofeeva, T.A.; Chebotar, V.K.; Demidov, D.V.; Gaidukova, S.E.; Yakovleva, I.V.; Kamionskaya, A.M. Effects of Apatite Concentrate in Combination with Phosphate-Solubilizing Microorganisms on the Yield of Ryegrass Cultivar Izorskiy. *Agronomy* **2023**, *13*, 1568. [CrossRef]
7. Alori, E.T.; Glick, B.R.; Babalola, O.O. Microbial Phosphorus Solubilization and Its Potential for Use in Sustainable Agriculture. *Front. Microbiol.* **2017**, *8*, 971. [CrossRef]
8. Majeed, A.; Farooq, M.; Naveed, M.; Hussain, M. Combined Application of Inorganic and Organic Phosphorous with Inoculation of Phosphorus Solubilizing Bacteria Improved Productivity, Grain Quality and Net Economic Returns of Pearl Millet (*Pennisetum glaucum* [L.] R. Br.). *Agronomy* **2022**, *12*, 2412. [CrossRef]
9. Wahid, F.; Sharif, M.; Steinkellner, S.; Khan, M.A.; Marwat, K.B.; Khan, S.A. Inoculation of arbuscular mycorrhizal fungi and phosphate solubilizing bacteria in the presence of rock phosphate improves phosphorus uptake and growth of maize. *Pak. J. Bot.* **2016**, *48*, 739–747.
10. Khoshru, B.; Sarikhani, M.R. Inoculation Effect of Phosphatic Microbial Fertilizers Containing Temperature Resistant Phosphate Solubilizing Bacteria on Nutritional Indices of *Zea mays* L. *J. Crop Prod.* **2019**, *12*, 107–122. [CrossRef]
11. Armandeh, M.; Mahmoudi, N.; Fallah Nosratabad, A.R. Screening and evaluation of phosphate-solubilizing bacteria isolated from aquaculture ponds in a step-by-step strategy as potential biofertilizer. *J. Appl. Microbiol.* **2022**, *133*, 1581–1596. [CrossRef]
12. Khoshru, B.; Mitra, D.; Khoshmanzar, E.; Myo, E.M.; Uniyal, N.; Mahakur, B.; Das Mohapatra, P.K.; Panneerselvam, P.; Boutaj, H.; Alizadeh, M.; et al. Current Scenario and Future Prospects of Plant Growth-Promoting Rhizobacteria: A Valuable Economic Resource for the Revival of Agriculture under Stress Conditions. *J. Plant Nutr.* **2020**, *43*, 3062–3092. [CrossRef]
13. Villalobos-Maldonado, J.J.; Meza-Gordillo, R.; Mancilla-Margalli, N.A. Removal of decachlorobiphenyl in vermicomposting process amended with rabbit manure and peat moss. *Water Air Soil Pollut.* **2015**, *226*, 159. [CrossRef]
14. Katiyar, R.B.; Sundaramurthy, S.; Sharma, A.K.; Arisutha, S.; Pratap-Singh, A.; Mishra, S.; Ayub, R.; Jeon, B.-H.; Khan, M.A. Vermicompost: An Eco-Friendly and Cost-Effective Alternative for Sustainable Agriculture. *Sustainability* **2023**, *15*, 14701. [CrossRef]
15. Unuofin, F.O.; Mnkeni, P.N.S. Optimization of *Eisenia fetida* stocking density for the bioconversion of rock phosphate enriched cow dung–waste paper mixtures. *Waste Manag.* **2014**, *34*, 2000–2006. [CrossRef] [PubMed]

16. By ADigital®. Grow Depot México© 2023. Roca Fosfórica—Grow Depot México. Available online: https://growdepotmexico.com/producto/roca-fosforica/ (accessed on 1 August 2024).
17. Adhami, E.; Hosseini, S.; Owliaie, H. Forms of phosphorus of vermicompost produced from leaf compost and sheep dung enriched with rock phosphate. *Int. J. Recycl. Org. Waste Agric.* **2014**, *3*, 1–6. [CrossRef]
18. Singh, S.; Singh, J.; Vig, A.P. Earthworm as ecological engineers to change the physico-chemical properties of soil: Soil vs. vermicast. *Ecol. Eng.* **2016**, *90*, 1–5. [CrossRef]
19. Ndegwa, P.M.; Thompson, S.A.; Das, K.C. Effects of stocking density and feeding rate on vermicomposting of biosolids. *Bioresour. Technol.* **2000**, *71*, 5–12. [CrossRef]
20. Tandon, H.L.S. *Methods of Analysis of Soils, Plants, Waters, Fertilizers & Organic Manures*; Fertilizer Development and Consultation Organization: New Delhi, India, 2005.
21. Tiessen, H.; Moir, O. Caracterización del P disponible por extracción secuencial. In *Toma de Muestras de Suelos y Métodos de Análisis*; Carter, M.R., Ed.; Lewis Pub: Boca Raton, FL, USA, 1993.
22. Hedley, M.J.; Steward, J.W.B.; Chauhan, B.S. Changes in Inorganic and Organic Soil Phosphorus Fractions Induced by Cultivation Practices and by Laboratory Incubations. *Soil Sci. Soc. Am. J.* **1982**, *46*, 970–976. [CrossRef]
23. Pigatin, L.B.F.; Atoloye, I.A.; Obikoya, O.A.; Borsato, A.V.; Rezende, M.O.O. Chemical study of vermicomposted agro-industrial waste. *Int. J. Recycl. Org. Waste Agric.* **2016**, *5*, 55–63. [CrossRef]
24. Fernández-Gómez, M.J.; Nogales, R.; Insam, H.; Romero, E.; Goberna, M. Continuous-feeding vermicomposting as a recycling management method to revalue tomato-fruit wastes from greenhouse crops. *Waste Manag.* **2010**, *30*, 2461–2468. [CrossRef]
25. USDA. Compost and Vermicompost in the Production of Organic Crops. USDA. NOP 7 CFR 205. 203. Available online: https://www.ams.usda.gov/rules-regulations/organic/handbook/5021 (accessed on 2 July 2024).
26. van den Berg, M.; Hoskin, S.O.; Rogers, C.W.; Grinberg, A. Fecal pH and microbial populations in thoroughbred horses during the transition from grass-fed to concentrated feeding. *J. Equine Vet. Sci.* **2013**, *33*, 215–222. [CrossRef]
27. Karmegam, N.; Jayakumar, M.; Govarthanan, M.; Kumar, P.; Ravindran, B.; Biruntha, M. Precomposting and green manure amendment for effective vermitransformation of hazardous coir industrial waste into enriched vermicompost. *Bioresour. Technol.* **2021**, *319*, 124136. [CrossRef]
28. Ramnarain, Y.I.; Ansari, A.A.; Ori, L. Vermicomposting of different organic materials using the epigeic earthworm *Eisenia foetida*. *Int. J. Recycl. Org. Waste Agric.* **2019**, *8*, 23–36. [CrossRef]
29. Desie, E.; Van Meerbeek, K.; De Wandeler, H.; Bruelheide, H.; Domisch, T.; Jaroszewicz, B.; Joly, F.; Vancampenhout, K.; Vesterdal, L.; Muys, B. Positive feedback loop between earthworms, humus form and soil pH reinforces earthworm abundance in European forests. *Funct. Ecol.* **2020**, *34*, 2598–2610. [CrossRef]
30. Satyaprakash, M.; Nikitha, T.; Reddi, E.U.B.; Sadhana, B.; Vani, S.S. Phosphorus and phosphate solubilizing bacteria and their role in plant nutrition. *Int. J. Curr. Microbiol. Appl. Sci.* **2017**, *6*, 2133–2144. [CrossRef]
31. Khoshru, B.; Nosratabad, A.F.; Mitra, D.; Chaithra, M.; Danesh, Y.R.; Boyno, G.; Chattaraj, S.; Priyadarshini, A.; Anđelković, S.; Pellegrini, M.; et al. Rock Phosphate Solubilizing Potential of Soil Microorganisms: Advances in Sustainable Crop Production. *Bacteria* **2023**, *2*, 98–115. [CrossRef]
32. Rastogi, M.; Nandal, M.; Khosla, B. Microbes as vital additives for solid waste composting. *Heliyon* **2020**, *6*, e03343. [CrossRef]
33. Katiyar, R.B.; Sundaramurthy, S.; Sharma, A.K.; Arisutha, S.; Khan, M.A.; Sillanpää, M. Optimization of Engineering and Process Parameters for Vermicomposting. *Sustainability* **2023**, *15*, 8090. [CrossRef]
34. Ministry of Urban Development, Government of India. Municipal Solid Waste Management Manual. 2016. Available online: http://swachhbharaturban.gov.in/writereaddata/Manual.pdf (accessed on 2 February 2024).
35. Lahbouki, S.; Hashem, A.; Kumar, A.; Abd_Allah, E.F.; Meddich, A. Integration of Horse Manure Vermicompost Doses and Arbuscular Mycorrhizal Fungi to Improve Fruit Quality, and Soil Fertility in Tomato Field Facing Drought Stress. *Plants* **2024**, *13*, 1449. [CrossRef]
36. Das, D.; Bhattacharyya, P.; Ghosh, B.C.; Banik, P. Bioconversion and biodynamics of *Eisenia foetida* in different organic wastes through microbially enriched vermiconversion technologies. *Ecol. Eng.* **2016**, *86*, 154–161. [CrossRef]
37. Singh, C.K.; Kumar, A. Vermicomposting of terrestrial weeds *Lantana camara* L. and *Parthenium hysterophorus* L. agriculture solid waste. *Ecol. Quest.* **2017**, *28*, 63–69. [CrossRef]
38. Gusain, R.; Suthar, S. Vermicomposting of invasive weed *Ageratum conyzoids*: Assessment of nutrient mineralization, enzymatic activities, and microbial properties. *Bioresour. Technol.* **2020**, *312*, 123537. [CrossRef] [PubMed]
39. Balachandar, R.; Biruntha, M.; Yuvaraj, A.; Thangaraj, R.; Subbaiya, R.; Govarthanan, M.; Karmegam, N. Earthworm intervened nutrient recovery and greener production of vermicompost from Ipomoea staphylina–An invasive weed with emerging environmental challenges. *Chemosphere* **2021**, *263*, 128080. [CrossRef] [PubMed]
40. Devi, C.; Khwairakpam, M. Management of invasive weed Parthenium hysterophorus through vermicomposting using a polyculture of *Eisenia fetida* and *Eudrilus eugeniae*. *Environ. Sci. Pollut. Res.* **2021**, *28*, 29710–29719. [CrossRef]
41. Cicek, H.; Bhullar, G.S.; Mandloi, L.S.; Andres, C.; Riar, A.S. Partial Acidulation of Rock Phosphate for Increased Productivity in Organic and Smallholder Farming. *Sustainability* **2020**, *12*, 607. [CrossRef]
42. Suthar, S.; Singh, S. Comparison of some novel polyculture and traditional monoculture vermicomposting reactors to decompose organic wastes. *Ecol. Eng.* **2008**, *33*, 210–219. [CrossRef]

43. Dominguez, J.; Edwards, C.A.; Webster, M. Vermicomposting of sewage sludge: Effect of bulking materials on the growth and reproduction of the earthworm *Eisenia andrei*. *Pedobiologia* **2000**, *44*, 24–32. [CrossRef]
44. Aira, M.; Monroy, F.; Domínguez, J. Earthworms strongly modify microbial biomass and activity triggering enzymatic activities during vermicomposting independently of the application rates of pig slurry. *Sci. Total Environ.* **2007**, *385*, 252–261. [CrossRef]
45. Domínguez, J.; Aira, M.; Gómez-Brandón, M. El papel de las lombrices de tierra en la descomposición de la materia orgánica y el ciclo de nutrientes. *Ecosistemas* **2009**, *18*, 20–31. Available online: https://www.revistaecosistemas.net/index.php/ecosistemas/article/view/61 (accessed on 5 May 2024).
46. Cedeño Saltos, K.A. Productive, Reproductive Behavior and Morphometry of the Californian Red Worm in Organic Waste Vermicomposting Systems, Patricia Pilar 2021. Zootechnical Engineer Thesis, Universidad Técnica Estatal de Quevedo, Quevedo, Ecuador, 2021; 72p. Available online: https://repositorio.uteq.edu.ec/server/api/core/bitstreams/4d1da33a-c18b-471c-a7df-8032f836a2e2/content (accessed on 25 July 2024).

sustainability

MDPI

Article

Effects of Nitrogen Fertilizer Application on Soil Properties and Arsenic Mobilization in Paddy Soil

Bing Han, Wei-Qing Chen, Yong-Qiang Jiao, Rui Yang, Li-Lu Niu, Xin-Ran Chen, Chen-Yang Ji and Dai-Xia Yin *

College of Resources and Environment, Anhui Agricultural University, Hefei 230036, China; hbin@stu.ahau.edu.cn (B.H.)

* Correspondence: daixia_yin@ahau.edu.cn

Abstract: Anthropogenic nitrogen (N) fertilization may substantially alter arsenic (As) behavior in the soil. However, a comprehensive understanding of how the soil As cycle responds to external N addition remains elusive. This study investigates the effects of various N fertilizers on soil properties and As mobility in paddy soil. Regardless of N sources, the concentrations of soluble As and SPLP-extractable As decreased with all N applications. Similarly, soil acidification occurred and dissolved iron (Fe) increased in most treatments, except for KNO_3 addition. However, only the KNO_3 application could reduce As desorption from soil minerals based on phosphate extraction. Also, KNO_3 enhanced both soil catalase (S-CAT) and dehydrogenase (S-DEH) activities. Other N treatments decreased S-CAT activities, but increased S-DEH activities. Principal components analysis indicated that phosphate extractable As was associated with NH_4^+-N concentration and S-DEH activity, while the concentrations of soluble As and SPLP-extractable As were associated with pH, S-CAT activity, and dissolved Fe. These results demonstrated that the soil properties induced by the N application are the main drivers of As desorption in paddy soil and that KNO_3 application is more eco-friendly than other N sources in As-contaminated paddy soil. This study shed light on the reasonable application of N-bearing fertilizers and the importance of soil properties to assess As mobility in As-contaminated paddy soil.

Keywords: nitrogen fertilizer; arsenic; soil properties; redox reactions; soil enzymatic activities

1. Introduction

Arsenic (As) is a redox-sensitive element with toxic and carcinogenic properties to organisms [1–3]. It is predominantly present in inorganic forms (As(V) and As(III)) in soils, but As(III), and As(0) also naturally occur. In general, the mobility of As depends on the redox potential, and it can be mobilized under both oxic and anoxic environments. Under oxidizing conditions, As is present in arsenate oxyanions, which prefer to be adsorbed at mineral surfaces [4,5]. In reducing soils, As occurs as uncharged arsenious acid ($H_3As^{III}O_3{}^0$) and arsenite ($H_2As^{III}O_3{}^-$) [6]. Therefore, As is substantially more mobile in reducing conditions than oxidizing conditions. Paddy soil is submerged during rice cultivation, thus forming an anaerobic environment and inducing several chemical changes in the soil, including pH and oxidation-reduction potential/redox potential (ORP/Eh) [7]. When compared to other cereals, rice is more efficient in accumulating As as a consequence of anaerobic/flooding conditions [8]. Therefore, As contamination in paddy soil and its environmental risk have been a serious global concern in recent years.

The mobility of As is responsible for its uptake by rice and is primarily affected by the physical and chemical factors of the soil like ORP, pH, soil nutrients, and adsorption/desorption processes [9,10]. Changes in pH and ORP directly affect redox transformation and As mobilization. Generally, a high pH increases negative charges at the soil surface, which may enhance the desorption of As into the aqueous solution, while a low pH induces less negative charges and promotes As adsorption to soil minerals [11]. In flooding paddy soil, ORP decreases due to the depletion of oxygen (as the electron acceptor) and the dissolution of Fe(oxy)hydroxides, promoting As desorption into the soil solution [12,13]. Microbial activities can also be regulated by physicochemical properties and play an important role in As mobility [14]. Soil enzymes are vital for biochemical processes and are often used to indicate microbial activities. In particular, Catalase (CAT) is one of the important enzymes that contributes to oxygen degradation, and dehydrogenase (DHA) acts on the biological oxidation of soil organic compounds [6]. These oxidoreductases are the catalysts of oxidation-reduction reactions and are dependent on redox fluctuation [10,15], thus affecting As mobility.

In addition, soil nutrients, such as N, P, S, and other elements, are reported to affect the mobility of As in soil. N is an essential nutrient element promoting plant growth and being widely applied to paddy fields [16]; N-contained fertilizer is predominantly present in the form of ammonia, nitrate, or urea in agriculture [17]. In the soil, N predominantly exists in inorganic forms, including nitrate-N (NO_3^--N) and ammonium-N (NH_4^+-N), which both can be utilized by rice plants. The former is prevalent under aerobic conditions and the latter often exists in anaerobic/flooding soils [18]. In flooded paddy, the main form of N is NH_4^+-N, while the NO_3^--N content is relatively low. The addition of NH_4^+-N decreases the soil pH and drives the process of ammonia oxidation coupled with Fe(III) reduction under anaerobic conditions, which may enhance As reduction [19]. However, nitrate can be used as a strong oxidant for improving the reducing environment in the paddy soil [20]. The coupled NO_3^--N reduction and Fe(II) oxidation could oxidize Fe(II) to Fe(III) in the soil solution, thus increasing As sorption and reducing its release [21]. Additionally, nitrate significantly enhanced the abundance of As oxidized microbial and stimulated anaerobic microbial arsenite oxidation, thus attenuating As mobility [22]. Therefore, fertilizing with N sources can affect the mobilization of As in paddy soil.

In recent years, fertilization with significant amounts of N is a common strategy to increase rice yields. Previous studies investigated the interactions between N fertilizers and metal(loid) fates in soils, often with a single type fertilizer [22–24]. Less attention has been paid to the effect of redox chemistry dynamics induced by various N-contained fertilizers on As' fate in paddy soil. Here, the soil pH, ORP, dissolved N, dissolved Fe, and oxidoreductase activities were measured to evaluate the soil properties and oxidation/reduction reactions. Soluble As, SPLP extractable As, and phosphate extractable As were used to assess As mobility. The objective of the present study was to investigate the variation of soil properties induced by the application of N and its effect on As mobility in paddy soil.

2. Materials and Methods

2.1. Soil Samples Collection and Pre-Treatment

The soil used in this study was collected from a paddy field in the vicinity of Shimen realgar mine in Changde City, Hunan Province, China (coordinates: 111°1′45″ E, 29°38′45″ N), with annual average temperatures between 16 and 19 °C and annual precipitation between 1200 and 1700 mm. The soil was described and classified as an Alic Umbrisol following the World Reference Base for Soil Resources [25]. After being air-dried, the soil samples were ground and sieved through a 2 mm sieve. The basic physiochemical parameters including the texture, pH, dissolved organic carbon (DOC), nutrient contents, and availability were determined (Table 1). Also, the soils were digested using the USEPA 3050B method, and the typical metals were determined via inductively coupled plasma mass spectrometry (ICP-MS, NexION 300, PerkinElmer, Shelton, USA). Their concentrations were (mg kg^{-1}) as follows: 32.78 ± 1.05 (As), 0.36 ± 0.05 (Cd), 10.41 ± 0.08 (Co), 27.62 ± 4.30 (Cr), 18.73 ± 0.61 (Cu), 801.1 ± 28.01 (Fe), 128.9 ± 5.21 (Mn), 19.87 ± 1.04 (Ni), 80.64 ± 7.13 (Zn).

Table 1. Main physical and chemical properties of the tested soils.

Soil physical-Chemical Factors	Unit	Value (Mean ± SD)
Texture		Clay loam
pH		4.69 ± 0.07
DOC	$g \cdot kg^{-1}$	9.19 ± 0.09
Total N	$g \cdot kg^{-1}$	1.26 ± 0.04
Total P	$g \cdot kg^{-1}$	0.55 ± 0.00
Total K	$g \cdot kg^{-1}$	12.0 ± 0.01
Available N	$mg \cdot kg^{-1}$	92.8 ± 2.29
Available P	$mg \cdot kg^{-1}$	5.91 ± 0.03
Available K	$mg \cdot kg^{-1}$	101 ± 0.36

2.2. Soil Incubation with Nitrogen Application

To simulate paddy conditions, a flooding soil incubation experiment was performed. Firstly, soil (230 g per pot) was weighed and pre-incubated with deionized water at a 100% water-holding capacity for 14 d. Following this, the N fertilizer was added to each pot to achieve 1, 5, and 10 mM N per kg soil, respectively. Four N fertilizer types were used in this experiment for each application rate, including KNO_3, $(NH_4)_2SO_4$, NH_4NO_3, and CH_4N_2O (urea). The soil without the N addition was referred to as the control (CK). There are 13 treatments in total, and every treatment was carried out in three replicates. The water level was maintained at ~3 cm above the soil surface via daily weighing and adding water to compensate for any losses. The experiment was conducted in a greenhouse at a controlled temperature (22/28 °C, night/day) with 60–70% relative humidity. All pots were randomly placed, and their positions were changed every week during the whole incubation.

The soil pH and ORP were measured at the end of the 30d incubation period. Deeper soil samples were collected in each pot and centrifuged to separate porewater from the soil. The separated soil was air-dried and ground to pass a 2 mm sieve for As extraction. A portion of the soil in each sample was further ground and passed through a 0.15 mm sieve for enzymatic activity analysis.

2.3. Determination of NO_3^--N, NH_4^+-N, and Fe Content in Soil Porewater

The porewater was sampled via soil centrifugation (4000 rpm for 20 min) and then passed through 0.45 μm filters. The filtrate was stored in a 4 °C refrigerator for subsequent analysis. The NO_3^--N, NH_4^+-N, and Fe content were analyzed using colorimetric techniques following previous studies [26,27]. A subsample of porewater was acidified to a pH < 2 with 0.5 M HCl immediately after being filtered, and was used to analyze the As concentration with an atomic fluorescence spectrometer (AFS, Puxi PF5, Beijing, China).

2.4. Analyses of Mobile As

The bioavailable As in each treatment was determined using the phosphate extraction method [28], which is effective in desorbing As from constituents through a ligand exchange and provides a mild extraction medium to predict the As availability to plants [29,30]. Briefly, 1 g of the air-dried sieved sample was mixed with 25 mL of 0.5 M KH_2PO_4 and placed in a 50 mL polycarbonate centrifuge tube. These tubes were shaken for 24 h at 200 rpm and then centrifuged at 4000 rpm for 20 min. The supernatants were filtered through 0.45 μm filters and acidified with 0.5 M HCl to a pH < 2. The concentrations of As in the extracts were analyzed using AFS.

To simulate an acid rain environment, the synthetic precipitation leaching procedure (SPLP, USEPA 1312, 1994a) was conducted to investigate the short-term leachability of As [31]. Briefly, the soils were extracted with a mixture of H_2SO_4 and HNO_3 with a volume ratio of 4:1, which was diluted to a pH of 4.20 ± 0.05 (the extraction was called a SPLP fluid). Furthermore, 1 g of the homogenized soil sample and 40 mL of the SPLP fluid were put into a 50 mL centrifuge tube. The tubes were capped and shaken for 18 h at 200 rpm, then centrifuged for 20 min at 4000 rpm. The supernatant leachate solutions were passed through a 0.45 μm filter, acidified to pH < 2, and analyzed for As using AFS.

2.5. Enzymatic Activity Analysis

Enzymatic activity has been considered a good indicator for assessing microorganism activity [32]. Two important intracellular enzymes, S-CAT and S-DEH, have been used to indicate the life processes of soil microbes and nutrient cycling [33]. S-CAT and S-DEH in the soil were measured to evaluate microorganism activity according to Stepniewska et al. (2009) [34]. Briefly, S-CAT was determined based on its ability to hydrolyze H_2O_2 via titration with 0.1 M $KMnO_4$. S-DEH was expressed as microgram triphenyl formazan (TPF) per gram of dry soil. Firstly, fresh soil was mixed with 0.8% TTC (2,3,5-triphenyl tetrazolium chloride) in Tris buffer (pH, 7.6) for 24 h in darkness at 30 °C. Then, triphenylformazan (TPF) extraction was conducted with a 50 mL mixture of acetone (90%, *v*/*v*) and CCl_4 (10%, *v*/*v*), followed by filtering through a 0.45 μm filter. The colorimetric intensity in the solution was measured at 485 nm.

2.6. Quality Control and Statistical Analysis

Each treatment contained triplicate samples and a reagent blank was used. During analysis, an analytical blank and calibration verification were included. The relative standard deviation (RSD) was $\leq$ 0.82%, and the recovery rate was 90–110% for As with AFS. All data are presented as the mean value of triplicates with standard error.

The linear regression analysis (Pearson's linear correlation) ($p < 0.05$) between independent variables was performed using SPSS 22.0 (IBM SPSS Statistics). Significant differences between different N treatments were determined according to a one-way analysis of variance (ANOVA), and the mean values were separated using Fisher's LSD test at $\alpha < 0.05$. The principal components analysis (PCA) of the parameters of all treatments was performed on the correlation matrix, as proposed by Andrews et al. (2002) [35] and Govaerts et al. (2006) [36]. Prior to the ANOVA analysis and PCA, dependent soil variables were tested for normality and homogeneity of variance using the Kolmogorov–Smirnov test and Leven's test, respectively. After the PCA, we extracted the variance eigenvalues and respective loadings for the first three principle components (Table S1).

3. Results and Discussion

3.1. N Fertilizers Can Acidify the Soil except for KNO_3

The original value of the pH in the control was 4.69, decreasing to 4.14 after 30 d of flooding incubation (Tables 1 and 2). All N amendments uniformly declined the soil pH, regardless of the types of N (Table 2). The pH of the soil declined by 0.20–0.94, 0.13–0.47, and 0.43–0.61 units in the $(NH_4)_2SO_4$, NH_4NO_3, and CH_4N_2O treatments, respectively. The soil pH tended to decrease with increasing N application rates, while the effect varied with N types. Compared to the control, the $(NH_4)_2SO_4$ and CH_4N_2O treatments significantly decrease the soil pH, while low-dose (1 mM) KNO_3 and NH_4NO_3 have no evident effect. This result might be attributed to how the CH_4N_2O and NH_4NO_3 addition might mainly release protons through hydrolysis under a certain threshold [37]. However, when the dose of CH_4N_2O or NH_4NO_3 reached up to 10 mM, the release of protons could be partly derived from nitrification. Also, the massive overdose might promote NH_3 production and then volatilization [38], then part of the volatilized NH_3 would re-enter the soil to produce OH^-, thus increasing the pH. In our study, 5 mM treatments of NH_4NO_3 and CH_4N_2O could release the most protons during the soil cultivation, thus having the lower pH.

Table 2. The pH and ORP were under different treatments after 30 days of incubation.

	Soil-pH				Soil-ORP (mV)			
	KNO_3	$(NH_4)_2SO_4$	NH_4NO_3	CH_4N_2O	KNO_3	$(NH_4)_2SO_4$	NH_4NO_3	CH_4N_2O
CK	4.14 ± 0.02 a	4.14 ± 0.02 a	4.14 ± 0.02 a	4.14 ± 0.02 a	−143 ± 0.42 d	−143 ± 0.42 d	−143 ± 0.42 a	−143 ± 0.42 b
1 mM	4.12 ± 0.01 a	3.94 ± 0.01 b	4.01 ± 0.01 a	3.70 ± 0.00 b	−116 ± 2.32 c	−139 ± 0.69 c	−122 ± 4.69 a	−136 ± 5.20 b
5 mM	4.02 ± 0.00 b	3.41 ± 0.01 c	3.67 ± 0.07 c	3.53 ± 0.03 c	129 ± 5.74 b	−69 ± 4.57 b	−133 ± 4.22 a	−125 ± 2.62 b
10 mM	3.94 ± 0.01 c	3.20 ± 0.01 d	3.84 ± 0.01 b	3.71 ± 0.01 b	230 ± 4.28 a	199 ± 5.78 a	−144 ± 1.78 a	−99 ± 2.92 a

Notes: Data are expressed as mean ± SD (n = 3). For each N type, numbers marked with different letters indicate significant differences between different application rates at $p < 0.05$ (LSD's test at $\alpha = 0.05$).

Soil pH is considered one of the important factors of As mobility [9], while acidification problems of soils are associated with the application of N. Our results are consistent with previous studies [39,40], which also confirmed that N fertilizer causes soil acidification. As Pan et al. (2020) [41] reported, N fertilizer applications significantly reduced the soil pH. The input of NH_4^+ strongly enhanced the H^+ production induced by nitrification [42] (Equations (1) and (2)).

$$NH_4^+ + 2O_2 = 2H^+ + NO_3^- + H_2O \quad (1)$$

$$CH_4N_2O + 4O_2 = 2H^+ + 2NO_3^- + H_2O + CO_2 \quad (2)$$

$$NO_3^- + 12H^+ + 10e^- = N_2 + 6H_2O \quad (3)$$

Under flooding cultivation, the oxygen content depleted gradually over time, and then NO_3^- replaced it as an electron acceptor to participate in microbial oxidation through the denitrification process. Denitrification consumes proton and reduces NO_3^- to N^2 (Equation (3)). Therefore, the value of pH was minimally affected by KNO_3 amendment when compared with other N fertilizers (Table 2). Similarly, previous research also reported that NH_4^+-N amendment decreased the pH of a medium loam soil more than NO_3^--N amendment [43,44]. The results indicate that soil acidification induced by the application of N is a common phenomenon, except for the appropriate dose of NO_3^--N.

3.2. Nitrogen Fertilizers Change ORP in As-Contaminated Soil

ORP is an important indicator for evaluating the oxidation-reduction status in soil [45]. The ORP values in the control soil were below −140 mV after 30 d of cultivation (Table 2), indicating a highly reducing condition and oxygen depletion via organic matter mineralization or biological/microbiological activities under flooding conditions. When compared with the control, the amendments of NH_4NO_3 and CH_4N_2O had no obvious effect on the ORP values. However, KNO_3 and $(NH_4)_2SO_4$ applications significantly increased the ORP values by ~261% and ~200%, respectively ($p < 0.05$). This is because both nitrate and sulfate could be used as electron acceptors of anaerobic bacteria after being oxygen-depleted, while the ORP scales at the same N ratio were higher in KNO_3 than in $(NH_4)_2SO_4$ treatments, which might be attributed to bacteria's preference for nitrate over sulfate [45].

Some previous research reported that ORP showed a pH-dependent behavior in paddy soils, as a consequence of microbial activities and organic matter decomposition or protons consumption for the reduction in Mn and Fe [46–48]. However, no significant correlation was observed between the pH and ORP values in the test soils (Table 3). In flooding soils, the effect of protons on ORP depends on the oxidant concentration undergoing reduction [49]. Based on the various minerals of geological interest, the redox potential fluctuated from −200 mV to 1000 mV when the pH values were around 4 in aqueous environments, but the pH range was narrow in reduced soils [50].

Table 3. Correlation coefficients among all parameters.

	pH	ORP	NH_4^+-N	NO_3^--N	S-CAT	S-DEH	Fe(II)	Total Fe	Porewater-As	SPLP-As	KH_2PO_4-As
pH	1	−0.21	−0.38 *	−0.42 **	0.77 **	−0.25	−0.04	−0.05	0.61 **	0.11	−0.18
ORP		1	0.10	0.92 **	−0.10	−0.21	−0.12	−0.15	0.18	−0.13	−0.39 *
NH_4^+-N			1	−0.17	−0.55 **	0.61 **	0.74**	0.76 **	−0.11	0.01	0.09
NO_3^--N				1	−0.33 *	−0.27	−0.25	−0.28	−0.04	−0.15	−0.32
S-CAT					1	−0.26	−0.19	−0.22	0.50 **	0.27	−0.28
S-DEH						1	0.48 **	0.51 **	−0.40 *	−0.03	0.22
Fe(II)							1	0.99 **	0.13	0.08	−0.20
Total Fe								1	0.10	0.07	−0.14
Porewater-As									1	0.33 *	−0.29
SPLP-As										1	−0.13
KH_2PO_4-As											1

Notes: The asterisks indicate the significance of correlation among the different parameters (LSD's test at $\alpha = 0.05$, *, $p < 0.05$, **, $p < 0.01$).

3.3. *Nitrogen Fertilizers Change the Solubility of Nitrogen and Iron in the Soil*

The dissolved inorganic N that occupied ~85.6% of the total inorganic N in the cultivated control soil was mainly NO_3^--N (Figure 1). Fertilization with N, including all types of fertilizers, enhanced the dissolution of NO_3^--N in the porewater, especially higher rates of KNO_3 and $(NH_4)_2SO_4$ (5 mM and 10 mM), which was a consequence of nitrification (Equations (1) and (2)). However, the dissolved NH_4^+-N contents declined with low-level N fertilization (1 mM), but increased with a high rate of $(NH_4)_2SO_4$, NH_4NO_3, and CH_4N_2O. The input of the corresponding types of N via N-bearing fertilizers was also the reason for the enhancement of NO_3^--N in the soil. The dissolved Fe was determined by the total soluble Fe in the porewater, and the results showed that N fertilization decreased Fe solubility (Figure 2). This phenomenon may be a consequence of Fe(II) oxidation coupled with nitrate reduction [51].

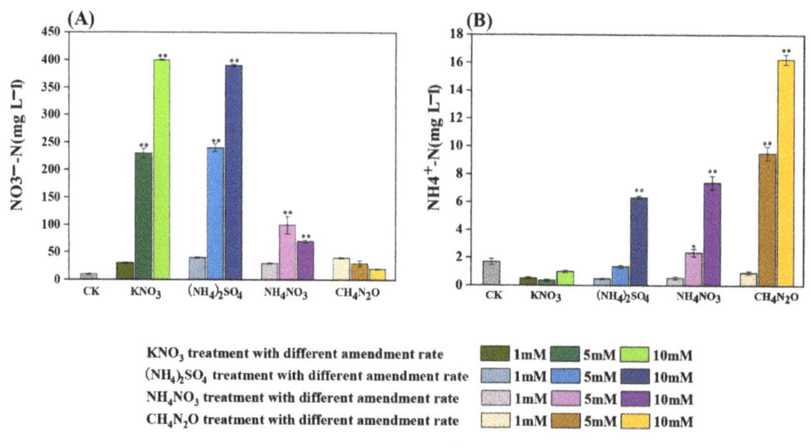

Figure 1. The contents of NO_3^--N (**A**) and NH_4^+-N (**B**) in soil porewater with different treatments. Notes: Data are expressed as mean ± SD (n = 3). The invisible error bar means that the standard deviation of the mean is small and close to zero. The asterisk indicates the significance of the difference between the control soil and different N treatments (LSD's test at $\alpha = 0.05$, *, $p < 0.05$; **, $p < 0.01$).

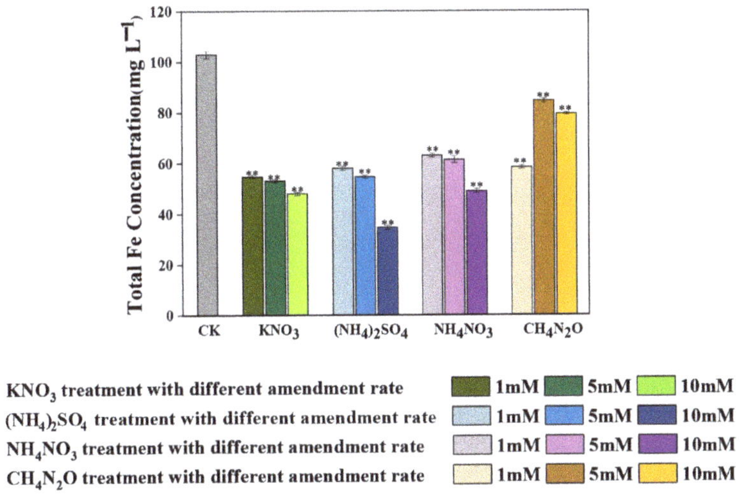

Figure 2. The contents of dissolved Fe in porewater. Notes: Data are expressed as mean ± SD (n = 3). The invisible error bar means that the standard deviation close to zero. The asterisks indicate the significance of the difference between the control soil and different N treatments (LSD's test at $\alpha = 0.05$, **, $p < 0.01$).

3.4. *Nitrogen Fertilizers Inhibit S-CAT Activities While Enhancing S-DEH Activities in the Soil*

Soil enzymes catalyze the main biochemical reactions involved in N cycles, and the enzymatic activity is sensitive to soil health [47]. S-CAT is an intracellular enzyme found in all aerobic bacteria and most facultative anaerobes, and it can split hydrogen peroxide into molecular oxygen and water, thus preventing cells from poisoning [34]. The activities of S-CAT were significantly influenced by N fertilizers, and the influences varied depending on N types and the application rates (Table 3 and Figure 3). After 30 d of cultivation, the S-CAT activities in the $(NH_4)_2SO_4$, NH_4NO_3, and CH_4N_2O treatments decreased by 18.4–71.0%, 10.3–40.1%, and 12.9–45.6%, respectively, and the mean S-CAT decreased with the increasing N fertilization rates. Dissimilarly, KNO_3 treatments increased the S-CAT activity under low and medium N additions (1 mM and 5 mM) and decreased it under high N conditions (10 mM). These results indicate that a limited dosage of KNO_3 could enhance S-CAT activities. Therefore, the application dosage of N-contained fertility needs to be addressed in paddy soil. As a previous study reported, KNO_3 amendment broke seed dormancy and improved seedling growth, while the activities of catalase and peroxidase enzymes showed an increase in concentrations of 0.2–0.4% and a strong decrease in high concentrations [52].

Correlation analysis shows a significant positive correlation between S-CAT and soil pH (Table 3, $p < 0.01$), which indicates that S-CAT activities decreased in most N fertilization treatments, which might be attributed to the reduction in the soil pH. This result is consistent with Quan et al.(2016) [53], who also found that the S-CAT activity and soil pH were significant positive correlation. However, there is no evident acidification in why the treatments with a low rate of KNO_3 obtained higher S-CAT than the control. The other reason is that medium KNO_3 applications could increase the number of aerobic microorganisms under flooding conditions [54].

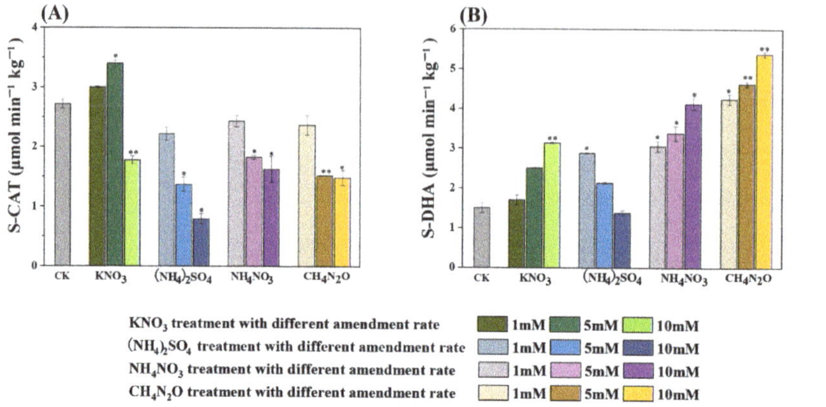

Figure 3. The activities (μmol min^{-1} kg^{-1}) of soil catalase (S-CAT) (**A**) and dehydrogenase (S-DEH) (**B**) under different treatments. Notes: Data are expressed as mean ± SD (n = 3). The invisible error bar means that the standard deviation of the mean is small and close to zero. The asterisks indicate the significance of the difference between the control soil and different N treatments (LSD's test at $\alpha = 0.05$, *, $p < 0.05$, **, $p < 0.01$).

The addition of the N fertilizer stimulated soil S-DEH activity, and it increased with the increase in the N application rate with KNO_3, NH_4NO_3, and CH_4N_2O. However, $(NH_4)_2SO_4$ application increased S-DEH activity at a low application rate (1 mM), while there was no evident effect at a higher application rate (5 mM and 10 mM). The S-DEH activity was enhanced significantly ($p < 0.05$) with the application of both NH_4NO_3 and CH_4N_2O when compared to the other treatments (Figure 3). This result indicated that N fertilization may trigger higher microbial metabolic activity than CK, except for the 10 mM $(NH_4)_2SO_4$ treatment [22]. One of the possible reasons is that the biochemical environment of the soil was less oxidative with a lower redox potential (Table 2) in the NH_4NO_3 and CH_4N_2O treatments [22,55].

3.5. The Effect of Nitrogen Amendments on As Mobility in the Soil

The concentrations of As in the porewater with KNO_3, $(NH_4)_2SO_4$, NH_4NO_3, and CH_4N_2O were 13.7–26.7%, 40.4–52.0%, 38.7–44.8%, and 30.5–50.5% lower than in the control, respectively (Figures 4A and S1A). All N applications, regardless of the applications rate, significantly decreased the dissolved As ($p < 0.05$). However, the application rate had no significant effect on the same N species, which was probably because the different nitrogen fertilizers contain different ions and have different effects on arsenic. The dissolution of As is one of the important indicators for available-As assessments. The soil pH and enzymatic activity may be the main factors as a result of the correlation analysis, demonstrating that the dissolved As was significantly positive/negative relative to the pH, S-CAT, and S-DEH activity (Table 3).

The concentrations of SPLP-extractable As were evidently reduced with 10 mM KNO_3 and 5 mM NH_4NO_3, while other treatments had no evident effect, though the concentration was reduced (Figures 4B and S1B). The water-soluble As may be the main factor of SPLP-extractable As according to the correlation analysis, which shows that the As concentration in the porewater and SPLP-extractable As were relatively significant ($p < 0.05$, Table 3). The concentrations of SPLP-extractable As were close to the water-soluble As in the control and KNO_3 treatments, but other N treatments seem to have increased acid rain leaching, demonstrated by the fact that As concentrations in SPLP extraction were evidently higher than in the porewater with $(NH_4)_2SO_4$, NH_4NO_3, and CH_4N_2O fertilization ($p < 0.05$). This result could be attributed to the acidification effect with these three N applications, while the soils were not acidified in the KNO_3 treatments (Table 2).

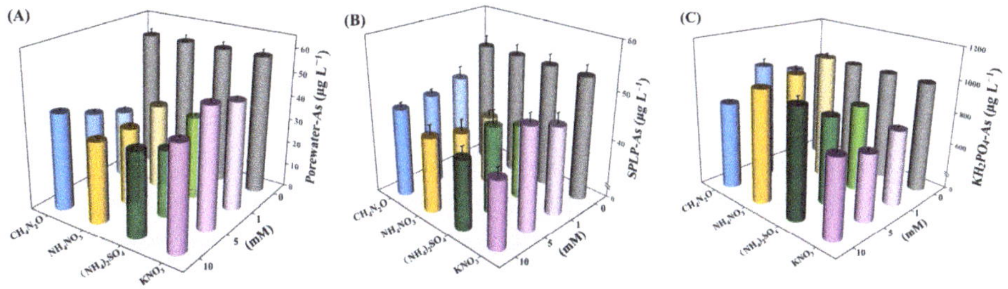

Figure 4. The porewater-As, SPLP-As, and KH_2PO_4-As concentrations under different treatments. ((**A**) Porewater-As concentration, (**B**) SPLP-As concentration, (**C**) KH_2PO_4-As concentration). Notes: Data are expressed as mean ± SD (n = 3). The invisible error bar means that the standard deviation of the mean is small and close to zero. The gray columns are the control without N application. The colors (blue, yellow, green, and purple) from light to dark represent the application rates from low to high.

The phosphate solution was found to be highly effective in extracting absorbed As from contaminated soil as a result of the ligand exchange of As and PO_4^{3-} [56]. The concentration of As in the KH_2PO_4 extraction was reduced by 20.8–27.7% in the KNO_3 treatments. Similarly, lower rates (1 mM and 5 mM) of $(NH_4)_2SO_4$ and 10 mM CH_4N_2O application also significantly decreased the KH_2PO_4 extractable As (Figures 4C and S1C). However, there was no evident decline in the extractable As in the NH_4NO_3 treatments. The results suggested that N fertilization with KNO_3 or proper application rates of $(NH_4)_2SO_4$ could significantly decrease the risk of As desorption from the contaminated soil, while the risk may be increased by NH_4NO_3 and $(NH_4)_2SO_4$ or CH_4N_2O with inappropriate application rates. The correlation analysis data revealed that KH_2PO_4 extractable As exhibited a negative correlation with the ORP in the soil, indicating that phosphate-extractable As may be dependent upon the variation of the oxidation-reduction reactions ($p < 0.01$, Table 3).

3.6. The Contribution of Soil Properties to As Mobility

For the different N treatments, PCA plots were used to visualize the key chemical parameters associated with pH, ORP, NO_3^--N, NH_4^+-N, S-CAT, S-DEH, dissolved Fe, water-soluble As, SPLP extractable As, and KH_2PO_4 extractable As (Figure 5 and Table S1). Clear associations among pH, S-CAT, and water-soluble As were found via the principal components analysis (PCA) of all N treatments. Similarly, KH_2PO_4 extractable As, NH_4^+-N, and S-DEH also presented clear associations. The PCA also highlights a clear differentiation in water-soluble As and KH_2PO_4 extractable As between the different N species. These parameters in the soil with KNO_3 applications were different from other N treatments (Figure 5B). The correlations between all parameters (Table 3) also highlight the important contribution of N fertilization on redox-sensitive changes in the soil, including the redox potential, oxidoreductase activity, and reductive dissolution of Fe, thus affecting the mobilization of As. The effects of the different parameters varied for both soluble As or extractable As.

Soil oxidation-reduction reactions play an important role in assessing As solubility, mobility, or bioavailability. Reducing soil conditions (ORP < 0 mV) greatly enhance the solubility and availability of As [57,58]. The application of KNO_3 significantly reduced water-soluble As and KH_2PO_4 extractable As, which could be attributed to the soil redox enhancement. The evident decline of ORP values with KNO_3 was observed, and the ORP values were negatively correlated with KH_2PO_4 extractable As ($p < 0.05$, Tables 2 and 3). In contrast, fertilization with other N species had fewer effects on the ORP values, especially NH_4NO_3. In addition, anaerobic microbial arsenite oxidation stimulated by nitrate also was the other reason. High rates (10 mM) of KNO_3 addition significantly decreased the S-CAT activity, but increased the S-DEH activity, demonstrating that microbe redox activities were changed in the soil. Therefore, the dosage needs to be considered during the application of nitrogenous fertilizer, as shown in a previous study concerning KNO_3 amendment to break seed fertility and improve seeding growth, where the activities of catalase and peroxidase enzymes showed an increase in concentrations of 0.2–0.4% and decreased strongly in high concentrations of the seedling [52].

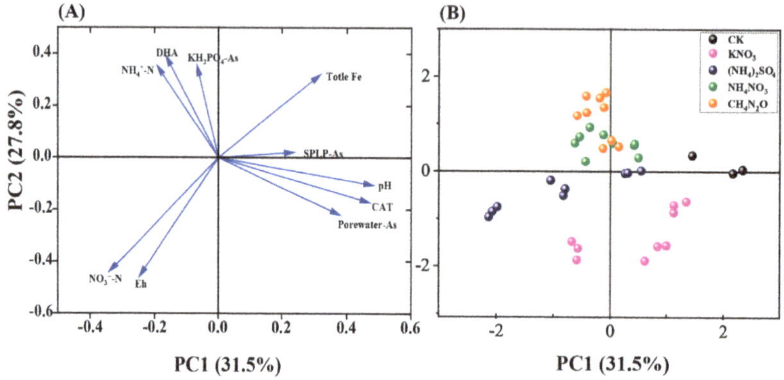

Figure 5. Factor load diagram (**A**) and score plot (**B**) of principal component analysis (the original analysis data shown in Table S1).

In paddy soil, As solubility is controlled not only by the redox potential, but also the amounts and types of the sorbent, such as Fe(III) hydroxides and As-combined minerals [39]. Soluble Fe in the porewater was also reduced with KNO_3 applications, suggesting that Fe oxidation occurred, which also adsorbed As on the mineral surface and thus decreased As mobility [59]. In general, oxidizing conditions and/or the presence of oxidants promote minerals immobilization, causing the precipitation of mobile As. Nitrate could be used as a powerful oxidant, which was responsible for oxidizing most Fe(II) under anoxic plus NO_3^--rich conditions [60]. Conversely, fertilization with NH_4^+-N ($(NH_4)_2SO_4$, NH_4NO_3, and CH_4N_2O) might increase the risks of soil acidification and As desorption from soil minerals (Table 2 and Figure 4C). This result was consistent with previous studies where NH_4^+-N could decrease the soil pH, as well as being used as a possible electron donor-induced reduction in Fe oxide minerals, thus increasing As mobility [61]. The link between NH_4^+-N and dissolved Fe ($R = 0.76$, $p < 0.01$) (Table 3) indicates the promoting effect of NH_4^+-N in Fe(III) reduction and As mobility [62,63].

4. Conclusions

In the present study, obvious changes in major redox-sensitive reactions in flooding paddy soil, such as redox potential, oxidoreductase activity, and the reductive dissolution of Fe, were caused by N fertilizers, thus affecting As mobilization. The observed phenomena suggested that N application had an important role in controlling As mobility, mainly by inducing redox-sensitive reactions in the soil. $(NH_4)_2SO_4$, NH_4NO_3, and CH_4N_2O were found to be risky for soil acidification and As desorption from soil. However, a suitable application rate of NO_3^--N fertilizer might be environmentally friendly in As-contaminated paddy soil. For example, medium rates of KNO_3 addition not only can increase soil enzymatic activities, but can also decrease As desorption with a low risk of soil acidification.

Supplementary Materials: The following supporting information can be downloaded at https://www.mdpi.com/article/10.3390/su16135565/s1, Figure S1: The Porewater-As (A), SPLP-As (B) and KH_2PO_4-As (C) concentrations under different treatments; Table S1: Principal component analysis-Linear combination coefficient and weight results.

Author Contributions: Conceptualization, B.H. and R.Y.; methodology, L.-L.N.; validation, B.H., W.-Q.C. and D.-X.Y.; formal analysis, W.-Q.C.; investigation, X.-R.C. and C.-Y.J.; writing—original draft preparation, B.H. and R.Y.; writing—review and editing, B.H. and D.-X.Y.; visualization, W.-Q.C. and Y.-Q.J.; funding acquisition, D.-X.Y. All authors have read and agreed to the published version of the manuscript.

Funding: This research was funded by the National Natural Science Foundation of China, grant number 42107403.

Institutional Review Board Statement: Not applicable.

Informed Consent Statement: Not applicable.

Data Availability Statement: The raw data supporting the conclusions of this article will be made available by the authors on request.

Conflicts of Interest: The authors declare no conflicts of interest.

References

1. Sevak, P.; Pushkar, B. Arsenic pollution cycle, toxicity and sustainable remediation technologies: A comprehensive review and bibliometric analysis. *J. Environ. Manag.* **2024**, *349*, 119504. [CrossRef]
2. Su, Q.; He, Y.; Pan, H.; Liu, H.; Mehmood, K.; Tang, Z.; Hu, L. Toxicity of inorganic arsenic to animals and its treatment strategies. *Comp. Biochem. Physiol. Part C Toxicol. Pharmacol.* **2023**, *271*, 109654. [CrossRef] [PubMed]
3. Coyte, R.M.; Vengosh, A. Factors controlling the risks of co-occurrence of the redox-sensitive elements of arsenic, chromium, vanadium, and uranium in groundwater from the Eastern United States. *Environ. Sci. Technol.* **2020**, *54*, 4367–4375. [CrossRef]
4. Ye, L.; Wang, L.; Jing, C. Biotransformation of adsorbed arsenic on iron minerals by coexisting arsenate-reducing and arsenite-oxidizing bacteria. *Environ. Pollut.* **2020**, *256*, 113471. [CrossRef]
5. Li, H.; Wang, N.; Xiao, T.; Zhang, X.; Wang, J.; Tang, J.; Kong, Q.; Fu, C.; Quan, H. Sorption of arsenate (V) to naturally occurring secondary iron minerals formed at different conditions: The relationship between sorption behavior and surface structure. *Chemosphere* **2021**, *285*, 131525. [CrossRef] [PubMed]
6. Mukherjee, A.; Coomar, P.; Sarkar, S.; Johannesson, K.H.; Fryar, A.E.; Schreiber, M.E.; Ahmed, K.M.; Alam, M.A.; Bhattacharya, P.; Bundschuh, J. Arsenic and other geogenic contaminants in global groundwater. *Nat. Rev. Earth Environ.* **2024**, *5*, 312–328. [CrossRef]
7. Azam, M.S.; Shafiquzzaman, M.; Haider, H. Arsenic release dynamics of paddy field soil during groundwater irrigation and natural flooding. *J. Environ. Manag.* **2023**, *343*, 118204. [CrossRef]
8. Zhao, F.-J.; Wang, P. Arsenic and cadmium accumulation in rice and mitigation strategies. *Plant Soil* **2020**, *446*, 1–21.
9. Hussain, M.M.; Bibi, I.; Niazi, N.K.; Shahid, M.; Iqbal, J.; Shakoor, M.B.; Ahmad, A.; Shah, N.S.; Bhattacharya, P.; Mao, K. Arsenic biogeochemical cycling in paddy soil-rice system: Interaction with various factors, amendments and mineral nutrients. *Sci. Total Environ.* **2021**, *773*, 145040. [CrossRef]
10. Kumarathilaka, P.; Seneweera, S.; Meharg, A.; Bundschuh, J. Arsenic speciation dynamics in paddy rice soil-water environment: Sources, physico-chemical, and biological factors—A review. *Water Res.* **2018**, *140*, 403–414. [CrossRef]
11. Xue, S.; Jiang, X.; Wu, C.; Hartley, W.; Qian, Z.; Luo, X.; Li, W. Microbial driven iron reduction affects arsenic transformation and transportation in soil-rice system. *Environ. Pollut.* **2020**, *260*, 114010. [CrossRef]
12. Meharg, A.A.; Rahman, M.M. Arsenic contamination of Bangladesh paddy field soils: Implications for rice contribution to arsenic consumption. *Environ. Sci. Technol.* **2003**, *37*, 229–234. [CrossRef]
13. Jung, H.-i.; Lee, J.; Chae, M.-J.; Kong, M.-S.; Lee, C.-H.; Kang, S.-S.; Kim, Y.-H. Growth-inhibition patterns and transfer-factor profiles in arsenic-stressed rice (*Oryza sativa* L.). *Environ. Monit. Assess.* **2017**, *189*, 638. [CrossRef]
14. Kumarathilaka, P.; Seneweera, S.; Meharg, A.; Bundschuh, J. Arsenic accumulation in rice (*Oryza sativa* L.) is influenced by environment and genetic factors. *Sci. Total Environ.* **2018**, *642*, 485–496. [CrossRef] [PubMed]
15. Han, Y.-S.; Park, J.-H.; Kim, S.-J.; Jeong, H.Y.; Ahn, J.S. Redox transformation of soil minerals and arsenic in arsenic-contaminated soil under cycling redox conditions. *J. Hazard. Mater.* **2019**, *378*, 120745. [CrossRef] [PubMed]
16. Suriyagoda, L.D.; Dittert, K.; Lambers, H. Mechanism of arsenic uptake, translocation and plant resistance to accumulate arsenic in rice grains. *Agric. Ecosyst. Environ.* **2018**, *253*, 23–37. [CrossRef]
17. Matyas, J.; Amonette, J.E. *Improved Moisture Management Technology for Eco-N-Control Fertilizer*; Pacific Northwest National Laboratory (PNNL): Richland, WA, USA, 2022.
18. Ma, X.; Sun, K.Y.; Dou, F.; Li, X.; Wang, X.; Sun, W. Impact of elevated nitrate and perchlorate in irrigation water on the uptake, speciation, and accumulation of arsenic in rice (*Oryza sativa* L.). *Water Air Soil Pollut.* **2020**, *231*, 309. [CrossRef]
19. Katsoyiannis, I.A.; Ruettimann, T.; Hug, S.J. pH dependence of Fenton reagent generation and As (III) oxidation and removal by corrosion of zero valent iron in aerated water. *Environ. Sci. Technol.* **2008**, *42*, 7424–7430. [CrossRef]
20. Picardal, F. Abiotic and microbial interactions during anaerobic transformations of Fe (II) and NO x. *Front. Microbiol.* **2012**, *3*, 112. [CrossRef]

21. Li, Y.; Yu, S.; Strong, J.; Wang, H. Are the biogeochemical cycles of carbon, nitrogen, sulfur, and phosphorus driven by the "Fe III–Fe II redox wheel" in dynamic redox environments? *J. Soils Sediments* **2012**, *12*, 683–693. [CrossRef]
22. Zhang, J.; Zhao, S.; Xu, Y.; Zhou, W.; Huang, K.; Tang, Z.; Zhao, F.-J. Nitrate stimulates anaerobic microbial arsenite oxidation in paddy soils. *Environ. Sci. Technol.* **2017**, *51*, 4377–4386. [CrossRef] [PubMed]
23. Liu, L.; Shen, R.-L.; Zhao, Z.-Q.; Ding, L.-J.; Cui, H.-L.; Li, G.; Yang, Y.-P.; Duan, G.-L.; Zhu, Y.-G. How different nitrogen fertilizers affect arsenic mobility in paddy soil after straw incorporation? *J. Hazard. Mater.* **2022**, *436*, 129135. [CrossRef] [PubMed]
24. Islam, M.S.; Mostafa, M. Influence of chemical fertilizers on arsenic mobilization in the alluvial Bengal delta plain: A critical review. *AQUA—Water Infrastruct. Ecosyst. Soc.* **2021**, *70*, 948–970. [CrossRef]
25. IUSS Working Group WRB. *World Reference Base for Soil Resources 2014, Update 2015 International Soil Classification System for Naming Soils and Creating Legends for Soil Maps*; World Soil Resources Reports No. 106; FAO: Rome, Italy, 2015.
26. Pason, T.; Maita, Y.; Lalli, C. *A Manual of Chemical and Biological Methods for Seawater Analysis*; Pergamon Press: Oxford, UK, 1984.
27. Stookey, L.L. Ferrozine—A new spectrophotometric reagent for iron. *Anal. Chem.* **1970**, *42*, 779–781. [CrossRef]
28. Yin, D.; Wang, X.; Chen, C.; Peng, B.; Tan, C.; Li, H. Varying effect of biochar on Cd, Pb and As mobility in a multi-metal contaminated paddy soil. *Chemosphere* **2016**, *152*, 196–206. [CrossRef]
29. Dong, H.; Guan, X.; Lo, I.M. Fate of As (V)-treated nano zero-valent iron: Determination of arsenic desorption potential under varying environmental conditions by phosphate extraction. *Water Res.* **2012**, *46*, 4071–4080. [CrossRef] [PubMed]
30. García-Salgado, S.; García-Casillas, D.; Quijano-Nieto, M.A.; Bonilla-Simón, M.M. Arsenic and heavy metal uptake and accumulation in native plant species from soils polluted by mining activities. *Water Air Soil Pollut.* **2012**, *223*, 559–572. [CrossRef]
31. Yang, L.; Donahoe, R.J.; Redwine, J.C. In situ chemical fixation of arsenic-contaminated soils: An experimental study. *Sci. Total Environ.* **2007**, *387*, 28–41. [CrossRef] [PubMed]
32. Zhang, Q.; Zhu, L.; Wang, J.; Xie, H.; Wang, J.; Wang, F.; Sun, F. Effects of fomesafen on soil enzyme activity, microbial population, and bacterial community composition. *Environ. Monit. Assess.* **2014**, *186*, 2801–2812. [CrossRef]
33. Khadem, A.; Raiesi, F. Influence of biochar on potential enzyme activities in two calcareous soils of contrasting texture. *Geoderma* **2017**, *308*, 149–158. [CrossRef]
34. Stpniewska, Z.; Wolińska, A.; Ziomek, J. Response of soil catalase activity to chromium contamination. *J. Environ. Sci.* **2009**, *21*, 1142–1147. [CrossRef] [PubMed]
35. Andrews, S.S.; Karlen, D.; Mitchell, J. A comparison of soil quality indexing methods for vegetable production systems in Northern California. *Agric. Ecosyst. Environ.* **2002**, *90*, 25–45. [CrossRef]
36. Govaerts, B.; Sayre, K.D.; Deckers, J. A minimum data set for soil quality assessment of wheat and maize cropping in the highlands of Mexico. *Soil Tillage Res.* **2006**, *87*, 163–174. [CrossRef]
37. Bryla, D.R.; Shireman, A.D.; Machado, R.M. Effects of method and level of nitrogen fertilizer application on soil pH, electrical conductivity, and availability of ammonium and nitrate in blueberry. *VI Int. Symp. Miner. Nutr. Fruit Crops 868* **2008**, *868*, 95–102. [CrossRef]
38. Nikolajsen, M.T.; Pacholski, A.S.; Sommer, S.G. Urea ammonium nitrate solution treated with inhibitor technology: Effects on ammonia emission reduction, wheat yield, and inorganic N in soil. *Agronomy* **2020**, *10*, 161. [CrossRef]
39. Fendorf, S.; Herbel, M.; Tufano, K.; Kocar, B. Biogeochemical processes controlling the cycling of arsenic in soils and sediments. In *Biophysico-Chemical Processes of Heavy Metals and Metalloids in Soil Environments*; John Wiley & Sons, Inc.: Hoboken, NJ, USA, 2008; pp. 313–338.
40. Liu, Y.; Zhang, M.; Li, Y.; Zhang, Y.; Huang, X.; Yang, Y.; Zhu, H.; Xiong, H.; Jiang, T. Influence of Nitrogen Fertilizer Application on Soil Acidification Characteristics of Tea Plantations in Karst Areas of Southwest China. *Agriculture* **2023**, *13*, 849. [CrossRef]
41. Pan, S.; Wang, Y.; Qiu, Y.; Chen, D.; Zhang, L.; Ye, C.; Guo, H.; Zhu, W.; Chen, A.; Xu, G. Nitrogen-induced acidification, not N-nutrient, dominates suppressive N effects on arbuscular mycorrhizal fungi. *Glob. Change Biol.* **2020**, *26*, 6568–6580. [CrossRef]
42. Li, Y.; Muhammad, R.; Saba, B.; Xia, H.; Wang, X.; Wang, J.; Jiang, C. Can the co-application of biochar and different inorganic nitrogen fertilizers repress N2O emissions in acidic soil? *Environ. Res.* **2023**, *231*, 116217. [CrossRef] [PubMed]
43. Xie, H.; Jiang, R.; Zhang, F.; McGrath, S.; Zhao, F. Effect of nitrogen form on the rhizosphere dynamics and uptake of cadmium and zinc by the hyperaccumulator Thlaspi caerulescens. *Plant Soil* **2009**, *318*, 205–215. [CrossRef]
44. Jiang, Z.; Liu, Y.; Yang, J.; Brookes, P.C.; Gunina, A. Rhizosphere priming regulates soil organic carbon and nitrogen mineralization: The significance of abiotic mechanisms. *Geoderma* **2021**, *385*, 114877. [CrossRef]
45. Craft, C. Biogeochemistry of wetlands: Science and applications. *Soil Sci. Soc. Am. J.* **2009**, *73*, 692. [CrossRef]
46. Zhang, H.; Selim, H. Reaction and transport of arsenic in soils: Equilibrium and kinetic modeling. *Adv. Agron.* **2008**, *98*, 45–115.
47. Wang, J.; Shaheen, S.M.; Jing, M.; Anderson, C.W.; Swertz, A.-C.; Wang, S.-L.; Feng, X.; Rinklebe, J. Mobilization, methylation, and demethylation of mercury in a paddy soil under systematic redox changes. *Environ. Sci. Technol.* **2021**, *55*, 10133–10141. [CrossRef] [PubMed]
48. Han, X.-q.; Xiao, X.-y.; Guo, Z.-h.; Xie, Y.-h.; Zhu, H.-w.; Peng, C.; Liang, Y.-q. Release of cadmium in contaminated paddy soil amended with NPK fertilizer and lime under water management. *Ecotoxicol. Environ. Saf.* **2018**, *159*, 38–45. [CrossRef] [PubMed]
49. Reddy, K.R.; DeLaune, R.D.; Inglett, P.W. *Biogeochemistry of Wetlands: Science and Applications*; CRC Press: Boca Raton, FL, USA, 2022; 732p.
50. Becking, L.B.; Kaplan, I.R.; Moore, D. Limits of the natural environment in terms of pH and oxidation-reduction potentials. *J. Geol.* **1960**, *68*, 243–284. [CrossRef]

51. Kappler, A.; Bryce, C.; Mansor, M.; Lueder, U.; Byrne, J.M.; Swanner, E.D. An evolving view on biogeochemical cycling of iron. *Nat. Rev. Microbiol.* **2021**, *19*, 360–374. [CrossRef]
52. Ahmadi, K.; Omidi, H.; Soltani, E. Efficacy of KNO3 and priming duration for improving germination, seedling growth, and antioxidant enzymes of Keluss (Kelussia odoratissma Mozaff). *Acta Ecol. Sin.* **2023**, *43*, 1120–1128. [CrossRef]
53. Quan, G.-L.; Xie, K.-Y.; Tong, Z.-Y.; Li, X.-L.; Wan, L.-Q.; Bi, S.-Y.; Wan, X.-F. The effect of compound bio-fertilizers on soil physical and chemical properties and soil enzyme activity in Leymus chinensis steppe. *Acta Prataculturae Sin.* **2016**, *25*, 27.
54. Sun, J.; Chillrud, S.N.; Mailloux, B.J.; Stute, M.; Singh, R.; Dong, H.; Lepre, C.J.; Bostick, B.C. Enhanced and stabilized arsenic retention in microcosms through the microbial oxidation of ferrous iron by nitrate. *Chemosphere* **2016**, *144*, 1106–1115. [CrossRef]
55. Jahangir, M.; Nitu, T.; Uddin, S.; Siddaka, A.; Sarker, P.; Khan, S.; Jahiruddin, M.; Müller, C. Carbon and nitrogen accumulation in soils under conservation agriculture practices decreases with nitrogen application rates. *Appl. Soil Ecol.* **2021**, *168*, 104178. [CrossRef]
56. Alam, M.; Tokunaga, S.; Stagnitti, F. Removal of arsenic from contaminated soils using different salt extractants. *J. Environ. Sci. Health Part A* **2007**, *42*, 447–451.
57. Ascar, L.; Ahumada, I.; Richter, P. Influence of redox potential (Eh) on the availability of arsenic species in soils and soils amended with biosolid. *Chemosphere* **2008**, *72*, 1548–1552. [CrossRef]
58. Honma, T.; Ohba, H.; Kaneko-Kadokura, A.; Makino, T.; Nakamura, K.; Katou, H. Optimal soil Eh, pH, and water management for simultaneously minimizing arsenic and cadmium concentrations in rice grains. *Environ. Sci. Technol.* **2016**, *50*, 4178–4185. [CrossRef] [PubMed]
59. Goldberg, S. Competitive adsorption of arsenate and arsenite on oxides and clay minerals. *Soil Sci. Soc. Am. J.* **2002**, *66*, 413–421. [CrossRef]
60. Senn, D.B.; Hemond, H.F. Nitrate controls on iron and arsenic in an urban lake. *Science* **2002**, *296*, 2373–2376. [CrossRef]
61. Xiu, W.; Lloyd, J.; Guo, H.; Dai, W.; Nixon, S.; Bassil, N.M.; Ren, C.; Zhang, C.; Ke, T.; Polya, D. Linking microbial community composition to hydrogeochemistry in the western Hetao Basin: Potential importance of ammonium as an electron donor during arsenic mobilization. *Environ. Int.* **2020**, *136*, 105489. [CrossRef] [PubMed]
62. Yang, W.H.; Weber, K.A.; Silver, W.L. Nitrogen loss from soil through anaerobic ammonium oxidation coupled to iron reduction. *Nat. Geosci.* **2012**, *5*, 538–541. [CrossRef]
63. Weng, T.-N.; Liu, C.-W.; Kao, Y.-H.; Hsiao, S.S.-Y. Isotopic evidence of nitrogen sources and nitrogen transformation in arsenic-contaminated groundwater. *Sci. Total Environ.* **2017**, *578*, 167–185. [CrossRef]

sustainability

MDPI

Systematic Review

Do Soil Microbes Drive the Trade-Off Between C Sequestration and Non-CO_2 GHG Emissions in EU Agricultural Soils? A Systematic Review

Arianna Latini [1,*], Luciana Di Gregorio [1], Elena Valkama [2], Manuela Costanzo [1], Peter Maenhout [3], Marjetka Suhadolc [4], Francesco Vitali [5], Stefano Mocali [5], Alessandra Lagomarsino [5] and Annamaria Bevivino [1,*]

1. Department for Sustainability, Italian National Agency for New Technologies, Energy and Sustainable Economic Development (ENEA), Via Anguillarese 301, 00123 Rome, Italy; luciana.digregorio@enea.it (L.D.G.); manuela.costanzo@enea.it (M.C.)
2. Bioeconomy and Environment Unit, Natural Resources Institute Finland (Luke), Itäinen Pitkäkatu 4 A, 20520 Turku, Finland; elena.valkama@luke.fi
3. Plant Sciences Unit, Flanders Research Institute for Agriculture, Fisheries and Food (ILVO), Burg. Van Gansberghelaan 109, 9820 Merelbeke, Belgium; peter.maenhout@ilvo.vlaanderen.be
4. Biotechnical Faculty (ULBF), University of Ljubljana, SI-1000 Ljubljana, Slovenia; marjetka.suhadolc@bf.uni-lj.si
5. Research Centre for Agriculture and Environment, Council for Agricultural Research and Economics (CREA-AA), Via di Lanciola 12/A, 50125 Cascine del Riccio, Italy; francesco.vitali@crea.gov.it (F.V.); stefano.mocali@crea.gov.it (S.M.); alessandra.lagomarsino@crea.gov.it (A.L.)

* Correspondence: arianna.latini@enea.it (A.L.); annamaria.bevivino@enea.it (A.B.)

Abstract

The role of soil microbial communities in soil organic matter (OM) decomposition, transformation, and the global nitrogen (N) and carbon (C) cycles has been widely investigated. However, a comprehensive understanding of how specific agricultural practices and OM inputs shape microbial-driven processes across different European pedoclimatic conditions is still lacking, particularly regarding their effectiveness in mitigating greenhouse gas (GHG) emissions. This systematic review synthesizes current knowledge on the biotic mechanisms underlying soil C sequestration and GHG reduction, emphasizing key microbial processes influenced by land management practices. A rigorous selection was applied, resulting in 16 eligible articles that addressed the targeted outcomes: soil microorganism biodiversity, including microbiome composition and other common Biodiversity Indexes, C sequestration and non-CO_2 GHG emissions (namely N_2O and CH_4 emissions), and N leaching. The review highlights that, despite some variations across studies, the application of OM enhances soil microbial biomass (MB) and activity, boosts soil organic carbon (SOC), and potentially reduces emissions. Notably, plant richness and diversity emerged as critical factors in reducing N_2O emissions and promoting carbon storage. However, the lack of methodological standardization across studies hinders meaningful comparison of outcomes—a key challenge identified in this review. The analysis reveals that studies examining the simultaneous effects of agricultural management practices and OM inputs on soil microorganisms, non-CO_2 GHG emissions, and SOC are scarce. Standardized studies across Europe's diverse pedoclimatic regions would be valuable for assessing the benefits of OM inputs in agricultural soils. This would enable the identification of region-specific solutions that enhance soil health, prevent degradation, and support sustainable and productive farming systems.

https://doi.org/10.3390/su18010319

Keywords: soil microorganisms; soil microbiome; soil microbial diversity; C sequestration; GHG emissions; N_2O emissions; CH_4 emissions; N leaching; OM input

1. Introduction

Soil microorganisms play a central role in the decomposition and transformation of organic matter (OM) through diverse metabolic pathways; thus, they are crucial in responding to climate changes in agriculture, through nutrient cycles, sequestration of soil carbon (C) and greenhouse gas (GHG) emissions [1–3]. Agricultural fields represent a significant source of GHG emissions, mainly methane (CH_4) and nitrous oxide (N_2O); however, at the same time, they also hold an untapped potential for mitigation of these emissions, particularly through C sequestration, depending on the land use and management practices [4].

Soil C sequestration is the process by which CO_2 is removed from the atmosphere and stored in the soil C pool. This process is primarily mediated by plants through photosynthesis and rhizodeposition, with C stored in the form of soil organic C (SOC) [5]. The co-benefits of soil C sequestration include advancing food and nutritional security, improving water renewability and quality, enhancing biodiversity, and strengthening elemental recycling. Soils in different agroecosystems, however, are often strongly depleted of their SOC stock and degraded due to intensive land use, soil management practices and farming systems. Restoring soil quality requires increasing SOC stock through sustainable management practices (e.g., conservation agriculture), which create a positive C budget [6]. Soil microbial communities play a pivotal role in C cycling within soil systems, returning C that enters the soil from above-ground plant production to the atmosphere as CO_2 and contributing to the stabilization of organic C, thereby influencing soil C storage and turnover [7,8]. Notably, bacterial network interactions are positively associated with the rate of C fixation; for example, *Proteobacteria* spp. promote the activity of other microbes that contribute to CO_2 fixation [9].

On the other hand, microbial activity, particularly nitrification and denitrification, is mainly responsible for the production of soil nitrous oxide (N_2O), a gas with a global warming potential 298 times greater than that of CO_2 [10]. N_2O contributes to ozone depletion and is released by microbial processes. Various studies have linked N_2O emissions to microbial functional diversity [11]. The application of mineral fertilizers, particularly ammonium-based fertilisers, has significantly increased agricultural yields and economic returns to farmers worldwide. However, agricultural intensification and the associated rise in nitrogen (N) applications have also heightened the risk of systemic losses through N_2O emissions and N leaching [12,13]. Non-fertilized soils exhibit up to 78% less denitrification than ammonium nitrate fertilised soils [14]. During nitrification, the ammonium added as fertilizer, fixed from the atmosphere by legumes, or mineralized from soil OM, crop residues, or other inputs, is oxidized to nitrite and eventually to nitrate in a series of reactions that can also produce N_2O. Likewise, when denitrifiers use nitrate as an electron acceptor under low-oxygen conditions, N_2O is an intermediate product that can readily escape to the atmosphere [4]. When crops compete with microbes for available N, N_2O fluxes tend to be lower.

Nitrogen leaching is one of the most common forms of aquatic system pollution, leading to high nitrate concentrations in drinking water, which can cause serious health issues [15]. Additionally, N leaching results in the loss of a nutrient resource for plants, leading to lower yields and significant economic losses for farmers. Controlled utilization

https://doi.org/10.3390/su18010319

of N and irrigation inputs can minimize the losses and emissions, thereby improving water quality [16].

Terrestrial ecosystems also influence emissions and removal of the atmospheric methane (CH_4) in soils. More than one-third of global CH_4 emissions occur through the microbial breakdown of organic compounds in soil under anaerobic conditions. Both CH_4 production and oxidation occur simultaneously, since methanogenic (CH_4-producing) and methanotrophic (CH_4-consuming) microorganisms coexist in soils. Consequently, emissions of biogenic CH_4 from soils would be significantly higher if not for the methanotrophic microorganisms that oxidize CH_4 before it escapes the soil, reducing the amount of CH_4 released into the atmosphere [17]. In the global CH_4 budget, agricultural soils are typically considered minor net emitters of CH_4 [18], though there is ongoing debate regarding whether they act as a net source or sink for methane, in contrast to wetlands (regarded as the dominant natural source) and upland soils (regarded as the primary biological sink) [19].

Recently, effective and climate-friendly measures and technologies for C sequestration and GHG emission reduction have been proposed for farmland soils. These include the application of organic matter (OM) inputs, such as biochar, returning straw to the field, and applying organic fertilizer and biological soil improvers. Among sustainable agricultural strategies and management systems, practices such as crop rotation, cover crops, no tillage (NT) or reduced tillage, beyond biodiversity embracing, livestock and crop integration, and agroforestry, are those more commonly adopted. Soil microbes play a crucial role in the effectiveness of these mitigation measures and technologies, but studies on the role of microorganisms in GHG emissions from farmland are still insufficient [20], and the high microbial diversity in soils still remains underexplored.

The scientific literature has provided evidence regarding the specific relationships between OM inputs or certain agricultural management practices with soil microorganisms and C sequestration [21–23], or with soil microorganisms and non-CO_2 GHG emissions, particularly N_2O emissions, CH_4 emissions, and N leaching [24–26]. A recent review examining the impact of soil management practices on the trade-offs between soil carbon sequestration, N_2O and CH_4 emissions, and nitrogen leaching—while focusing on abiotic factors—highlighted a persistent gap in understanding these complex interactions [27].

Therefore, the effects of soil microorganisms on C and N dynamics need to be addressed to allow correct evaluation of the success of sustainable management practices in supplying their ecosystem services [28,29]. In the frame of the H2020 European Joint Programme on Soil (EJP-Soil), the ΣOMMIT project (Sustainable Management of soil Organic Matter to Mitigate Trade-offs between C sequestration and nitrous oxide, methane and nitrate losses) had the objective to evaluate trade-offs and synergies between soil C sequestration, nitrous oxide, methane and nitrate losses as affected by soil management options aimed at increasing soil C storage. This systematic review, as a specific task of the ΣOMMIT project, provides a literature survey focused on the relationships between soil microorganisms and microbiomes and the trade-offs between C sequestration and non-CO_2-GHG emissions, in farming systems applying OM inputs (e.g., organic wastes, biochar, digestate) or other management systems in European soils. New insights will help assess the potential effects of OM inputs and other practices in arable farming on the abundance, diversity, and activity of biotic drivers, ultimately contributing to climate change mitigation.

https://doi.org/10.3390/su18010319

2. Materials and Methods

2.1. Procedure for the Systematic Review

The PRISMA protocol ("Preferred Reporting Items for Systematic reviews and Meta-Analyses") [30] was followed [PRISMA 2020 checklist in Supplementary Material S1]. The main requirements of the PRISMA protocol, detailed below, are: (i) the accurate and precise definition of the research objective of the study; (ii) the selection of information sources; (iii) the definition of the eligibility criteria and the exclusion criteria.

(i) Definition of the research objective of the study

A clear research question regarding the objective of this review has been defined, including the key terms to perform the search. The research question was based on PICO [31], applied to a search addressing soil microorganisms in relation to the trade-offs between C sequestration, N_2O, and other non-CO_2 GHG emissions, as reported in Table 1.

Table 1. PICO acronym, general description and explication of the PICO components, key terms and sentences used in the research question for the search in the current review.

<table>
<tr><th>PICO Acronym</th><th colspan="2">General Description of the Component</th><th>PICO Component in the Search</th><th>Key Term/s Used in the Search</th><th>Sentence (Key Terms + Boolean Operators) Applied in the Search</th></tr>
<tr><td>P: Population</td><td colspan="2">The related term should represent the subject of the search</td><td>EU agricultural soils</td><td>soil *, agr *, farm * europe *, EU</td><td>[soil AND (agr * OR farm *) AND (Europe * OR EU)]</td></tr>
<tr><td>I: Intervention</td><td colspan="2">The related term should represent the applied treatment (as an independent variable of an experiment)</td><td>Addition of soil OM inputs</td><td>"crop residue *", "cover crop *", manure, mulch, slurry, compost, biochar, digestate, sludge, "organic matter", OM, amendment</td><td>"crop residue *" OR "cover crop *" OR manure OR mulch OR slurry OR compost OR biochar OR digestate OR sludge OR "organic matter" OR OM OR amendment *</td></tr>
<tr><td>C: Comparator</td><td colspan="2">The related term would allow the comparison of the interventions (treatments).</td><td>The comparator is omitted; when present it may be mineral fertilization or no fertilization</td><td>N.A.</td><td>N.A.</td></tr>
<tr><td rowspan="3">O: Outcome</td><td rowspan="3">The related term should represent what it is planned to be measured or improved (as a dependent variable of an experiment)</td><td>Outcome 1</td><td>Effects on soil microorganisms</td><td>microb *, microorganism *, arche *, bacteria *, fung *</td><td>microb * OR microorganism * OR arche * OR bacteri * OR fung *</td></tr>
<tr><td>Outcome 2</td><td>Effects on C sequestration.</td><td>Cseq *, "C seq *", "C stor *", "C cycl *", "carbon seq *", "carbon stor *", "carbon cycl *"</td><td>"C seq *" OR "C stor *" OR "C cycl *" OR "carbon seq *" OR "carbon stor *" OR "carbon cycl *"</td></tr>
<tr><td>Outcome 3</td><td>Effects on non-CO_2 GHG emissions (namely N_2O emissions or CH_4 emissions or N leaching)</td><td>"N_2O emission *", "nitrous oxide emission *"</td><td>"N_2O emission *" OR "nitrous oxide emission *"</td></tr>
</table>

"N.A." stands for "Not Available". "*" replaces 0 or more characters allowing to find any number or to indicate a character that may or not may be present.

The research question was formulated as follows: "What are the effects of OM inputs on soil microorganisms in relation to C sequestration and N_2O emissions in EU agricultural

https://doi.org/10.3390/su18010319

soil?". To avoid making the bibliographic search overly complex, the number of keywords has been minimized. In this regard, it is assumed that any full-text reporting on CH_4 emissions and/or N leaching would generally mention N_2O emissions. However, only N_2O emissions were explicitly reported (see Table 1), while no keywords were used to search for CH_4 emissions or N leaching.

(ii) Selection of information sources

The database selected for the search was Scopus (https://www.scopus.com, accessed on 10 November 2025), which was considered the most suitable for this study due to its broad journal coverage, advanced keyword search capabilities, and citation analysis tools focused on recent literature. This choice allowed for a comprehensive overview of all published research [32]. The search was conducted in March 2023.

(iii) Eligibility criteria

The eligibility screening of the numerous articles retrieved from the scientific literature involved the development, testing and application of the selection criteria listed in Table 2, following the guidelines in [33]. Preliminary searches across various databases revealed that the number of required keywords to address the research question was excessively high and difficult to manage. At the same time, including all selected keywords significantly limited the number of retrieved articles. Therefore, to avoid overlooking valuable information, in very few cases studies that had not been carried out in field were also included. Only a few studies reporting microcosm experiments were retained, as it is well-documented that, in the short-term, the microbial community modifications can be detected with greater precision, whereas specific biological effects in field experiments may be masked or influenced by pedoclimatic variations [34–36].

Table 2. Eligibility criteria for the inclusion and the exclusion of manuscripts.

Eligibility Criteria	Inclusion	Exclusion
Language	English	Other than English
Experiment location	Europe	Non-EU countries
Experiment type	Field trial, microcosm	Wet lab, greenhouse experiment, microcosm, modelling, and simulations
Land use	Agricultural, grassland	Pastures, rice field, forests and semi-natural areas, wetlands
OM input	Crop residues, cover crop (green manure/mulch), livestock manure, slurry, digestate, compost, biochar, sludge *	Different OM input or no OM input
Time period	From 2005	Manuscripts before 2005
Document type	Original article	Review, meta-analysis, book chapter, conference paper, book, and others

* Sometimes, even though a specific OM input was mentioned or discussed in the manuscript, the manuscript focus was on another specific agricultural management practice (i.e., conservation tillage and crop rotation). In these cases, the manuscript was finally included in this review because considered strictly pertinent.

Regarding the literature search, the following considerations are noteworthy. The OM inputs considered in this study, reported in Table 2, were like in Valkama et al. [37]. To search for microbial diversity, in addition to the general terms related to microorganisms (microb* OR microorganism* OR arche* OR bacteria* OR fung*), the following biological parameters and Biodiversity Indexes (BIs) related to microbial abundance and diversity

https://doi.org/10.3390/su18010319

were included: "Microbial biomass", "Microbial composition", "Shannon", "Simpson", "Fisher", "OTU", "ASV", "Richness", "Evenness", "Relative Abundance", "Phylogenetic diversity" and "Diversity". The search focused on studies localized in Europe. Selected studies ranged from 2005 to August 2022, the starting period was chosen due to the introduction of sequencing-by-synthesis technology by 454 Life Science in 2005 [38]. After evaluating the eligibility criteria through a preliminary analysis of titles and abstracts, the remaining articles were examined carefully, with respect to terms related to C sequestration, N_2O emissions and possibly other non-CO_2 GHG emissions, such as CH_4 emissions and N leaching. The advanced search in Scopus (https://www.scopus.com/search/form.uri?display=basic#basic, accessed on 10 November 2025), within "Documents" > "All fields", was carried out through the query sentences in Table 1, all of them linked by the Boolean operator "AND".

2.2. Abstract Bibliometric Analysis

The abstracts of the articles selected for this systematic review were extracted and examined using the Bibliometrix package within the R Studio (version 4.5.1) environment [39]. Bibliometrix uses the "co-occurrence method" to cluster keywords based on their frequency in articles, revealing thematic relationships [40]. Network analysis was performed using Walktrap algorithm to further clarify connections between key concepts in the abstracts, and to identify predominant topics and their interconnections. A word cloud was generated to visually represent the frequency of terms occurring in the abstracts. Furthermore, thematic mapping was utilized to categorize and visualize clusters of related terms and concepts within the abstracts. This technique helped identify overarching themes and patterns in the literature, aiding in the contextualization and interpretation of the findings. Raw data, including the BibTeX file, are provided in Supplementary Materials S2 and S3.

3. Results

3.1. Systematic Review and Descriptive Analysis

Following the schematic procedure reported in Figure 1, 16 manuscripts from the scientific literature were finally analysed [Supplementary Materials S4].

In detail, starting from 5273 total initial items, the following were removed serially: 173 for being in a non-English language; 510 because the "Source type" was not "Journal"; 49 for not being "Article" or "Review"; 42 for being published before 2005; 3760 for being from non-EU countries; and 120 for being from unrelated "Subject areas". After completing the screening and assessing eligibility criteria, 619 full-text articles were obtained (Table 2; Figure 1). The number of articles related to the study's objective showed an increasing trend from 2005, reaching 88 in 2022 [Figure 2]. More than a quarter of these articles originated from Germany [Figure 3A], 43.3% were in the "Agricultural and Biological Science" area, and 35.3% were in "Environmental Science" [Figure 3B]; 85% were articles and 15% were reviews [Figure 3C].

To refine the search for relevant manuscripts, a set of Biodiversity Indexes (BIs) was chosen for the microbial data collection template. As shown in Figure 1, of the initial 619 items that complied with the eligibility criteria, only 152 of them including BIs were extracted from the pool. In screening for the presence of terms related to BIs, 8 records resulted for "Shannon", 42 records for "Fisher", 47 records for "Richness", 73 records for "Simp-son", 10 records for "Evenness", 1 record for "Phylogenetic diversity" and 8 records for "Relative abundance". It has to be noted that, in some cases, the BI term was only present in the reference section of a manuscript. For example, many records contained the term "Simpson", referring to highly cited studies authored by "Simpson G".

After a preliminary analysis of the Title and the Abstract of these 152 articles, a general scarcity of research encompassing microbial composition and diversity, and at the same time both C sequestration and N_2O emissions, was observed. Only manuscripts with explicit evidence of a clear treatment—such as the application of one or more OM inputs or specific agricultural management practices—were selected, following a rigorous screening based on the eligibility criteria outlined in Table 2.

Finally, 15 research articles were selected from Scopus. The article selection process followed rigid criteria to identify studies that simultaneously reported information and results on soil microorganisms, C sequestration, and non-CO_2 GHG emissions. These 15 articles were considered representative of the state-of-the-art knowledge regarding the integrated effects of OM fertilization and other agricultural practices on soil microorganisms and the trade-offs between C seq and GHG non-CO_2 emissions, although not all of them provided insights into all three topics. For this reason, among the final selected articles, those studies including agricultural management systems (e.g., tillage) rather than OM input application were also kept for the systematic review.

Additionally, another manuscript [41], Johansen et al. 2013, was selected and included in this study, proceeding from the References of Bachmann et al. (2014) [42] [Figure 1; Supplementary Material S4]. For the 16 manuscripts, the main information on the geographical climatic location, the year(s) and duration of experiments, the crops, the OM inputs and/or the agricultural management practice, and the outcomes in terms of soil microbial data and BIs, C sequestration, and non-CO_2 emissions is summarized in Table 3. No studies related to Archaea were found, so the current review only encompasses Bacteria and Fungi.

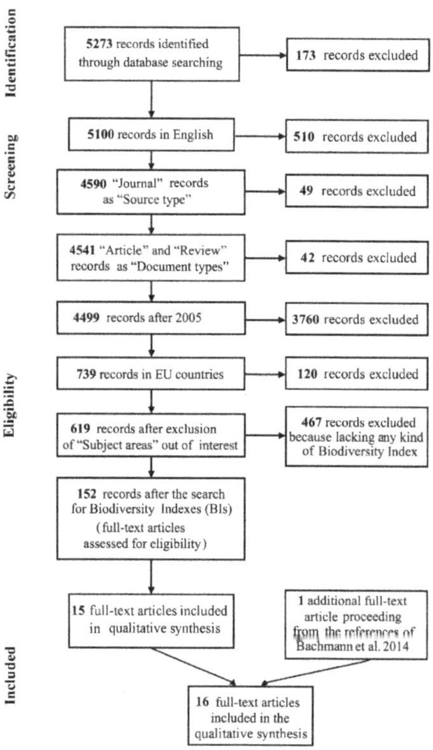

Figure 1. Flow of information through the different phases of a systematic review according to the PRISMA protocol [30] for the search carried out in the Scopus database. 1 additional full-text article proceeding from the references of Bachmann et al. 2014 [42].

https://doi.org/10.3390/su18010319

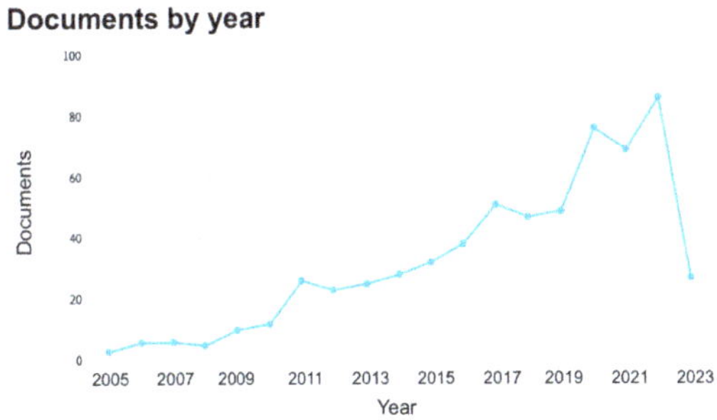

Figure 2. Number of documents (scientific journal articles) by year, from 2005 to March 2023, retrieved from the key word search in Scopus.

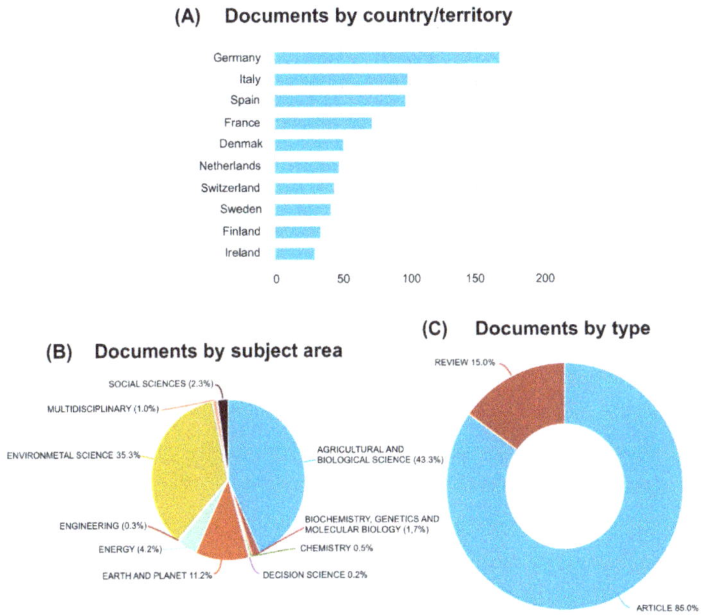

Figure 3. (**A**) Number of documents by affiliated country/territory; (**B**) number of documents by subject area; (**C**) number of documents by type.

Among the 16 studies included in the final selection, the majority examined the trade-offs between carbon sequestration and non-CO_2 greenhouse gas emissions (notably N_2O) and concurrently presented data related to soil microbial dynamics. However, during an in-depth analysis of the full texts, it was found that not all articles fully adhered to the inclusion criteria. In some cases, the articles focused specifically on the relationships between soil C sequestration and microorganisms or between non-CO_2 GHG emissions and microorganisms, with one of the three topics only discussed rather than directly evaluated at the experimental level. For example, Ribas et al. [43] explored the potential effect of plant diversity on yield and GHG exchanges in forage mixtures using slurry for fertilization. The authors measured CO_2, N_2O, and CH_4 emissions but did not assess soil microorganism variations, even though soil microorganism effects were often mentioned and discussed throughout the paper. This article was included in the current review because, more interestingly, it proposed plant diversity, particularly N-rich crops, as a tool to regulate gas balances.

https://doi.org/10.3390/su18010319

Table 3. Main information related to the studies reported in the 16 finally selected Original Research manuscripts from Scopus focused on in this systematic review.

Manuscript N°	Environmental Zone in Europe	Country, City, Location	Years of Experiment and Duration	Cultivated Crops	OM Input/Agricultural Management Practice	Microbial Data and Biodiversity Indexes (BIs)	C Sequestration and Related Data	N_2O Emissions and Related Data	CH_4 Emissions or N Leaching Data
1–Anastopoulous et al. 2019 [44]	Mediterranean	Greece, Pathos (Cyprus)	2017, 5 and 34 days after application of treatments	N.A.	Organic agricultural wastes from orange, mandarin, and banana peels vs. ammonium nitrate	Soil microbial communities were analysed through: - qPCR of denitrifiers (nir*K*, nir*S*, nos*Z I* and nos*Z II*) - 16S rRNA seq. - BI: Shannon Index	- Initial soil organic carbon - CO_2 emissions	- Initial soil N - Initial soil nitrate N - N_2O emissions	N.A.
2—Bachmann et al. 2014 [42]	Continental	Germany, Mecklenburg-West Pomerania	2008, 3 years	Maize	Digestate	Activity of enzymes connected to soil P cycle: - dehydrogenase - alkaline phosphatase - acid phosphatase	- Total C content - TOC - CO_2 efflux	- Total N - NH_4-N - N as calcium ammonium nitrate - N-uptake (in kg ha^{-1})	N.A.
3—Badagliacca et al. 2021 [29]	Mediterranean	Italy, Sicily, Pietranera Farmlocated in the North of Agrigento	2013–14, 23 years	Durum wheat (*Triticum durum*), faba bean (*Vicia faba* L.)	- Conservation system: NT vs. CT - Crop sequence: wheat monoculture vs. wheat-faba bean rotation	- Microbial Biomass C (MBC) - Microbial Biomass N (MBN) - Basal Respiration (BR) - Abundance of main microbial groups by phospholipid fatty acids (PLFAs) - Microbial quotient	- TOC - Labile Organic Carbon (LOC)	- Total N	N.A.

https://doi.org/10.3390/su18010319

Table 3. *Cont.*

Manuscript N°	Environmental Zone in Europe	Country, City, Location	Years of Experiment and Duration	Cultivated Crops	OM Input/Agricultural Management Practice	Microbial Data and Biodiversity Indexes (BIs)	C Sequestration and Related Data	N_2O Emissions and Related Data	CH_4 Emissions or N Leaching Data
4—Badagliacca et al. 2022 [45]	Mediterranean	Italy, Calabria	2019–20, 1 year	Durum wheat (*Triticum durum*)	Conservation system: NT vs. CT	- PFLA analysis - Activity of enzymes connected to C, N, P and S cycling. For C cycling: α-and β-glucosidase, α- and β-galactosidase, α- and β- mannosidase, β-1,4-glucanase, β-1,4-xylanase, α-arabinase, β-D-glucuronidase. For N cycling: N-acetyl-b-D-glucosaminidase, leucine amino-peptidase, trypsin-like protease	- Soil permanganate oxidizable C (POxC) as labile C	- Nitrate-N (NO_{3-}-N) and ammonium-N (NH_{4+}-N) - total soluble N (TSN) - extractable organic N	N.A.
5—Dicke et al. 2015 [46]	Continental	Germany, Berge	2012–13, (October 2021 to September 2013)	Winter wheat (*Triticum aestivum* L.)	Biochar (pyrolisis char and hydrochars) and digestate	Soil microbial communities were analysed through: - qPCR of *nosZ* denitrifiers	- C content in soil and in OM inputs (as %DM)	- N_2O emission (cumulative annual emission as of N_2O-N in kg ha^{-1} a^{-1}) - Soil inorganic N (NH_{4+}-N and NO_3—N in mg kg^{-1} soil) - In the OM inputs: N (%DM), NH_{4+}-N and NO_3-N in mg kg^{-1}	N.A.

https://doi.org/10.3390/su18010319

Table 3. *Cont.*

Manuscript N°	Environmental Zone in Europe	Country, City, Location	Years of Experiment and Duration	Cultivated Crops	OM Input/Agricultural Management Practice	Microbial Data and Biodiversity Indexes (BIs)	C Sequestration and Related Data	N_2O Emissions and Related Data	CH_4 Emissions or N Leaching Data
6—Drost et al. 2020 [47]	Microcosm experiment	The Netherlands, Wageningen	2017, short-term time-course, 50 days	Wheat was grown and harvested prior to soil collection	Cover crops [monocultures of oat (*Avena strigosa*), vetch (*Vicia sativa*), radish (*Raphanus sativus*), a mixture of 3 species and a mixture of 15 species	- Fungal biomass by ergosterol - Bacterial biomass by 16S rRNA gene qPCR - Fungal biomass C - Bacterial biomass C - Total (fungal + bacterial) microbial biomass C - Microbial functional diversity by Biolog ECO plates	- C content of the added plant material - CO_2 fluxes	- N_2O fluxes - Cumulative fluxes - Plant available N concentration	N.A.
7—Johansen et al. 2013 [41]	Microcosm experiment	Denmark, Taastrup	2011, short-term time-course, 9 days	Barley (*Hordeum vulgare*), perennial ryegrass (*Lolium perenne*), white clover (*Trifolium repens*), bread wheat (*Triticum aestivum*)	- OM inputs: (a) raw cattle slurry; (b) anaerobically digested cattle slurry/maize; (c) anaerobically digested cattle slurry/grass-clover; (d) fresh glass-clover (green manure) Crop rotation: - spring barley (Hordeum vulgare) undersown with grass (Lolium perenne) - clover (Trifolium repens), grass-clover, grass-clover and winter wheat (Triticum aestivum)	- PLFA analysis for microbial biomass and community composition - Catabolic response profiling (functional diversity)	- Available organic C - CO_2 emissions	- Mineral N - N_2O emissions	N.A.

https://doi.org/10.3390/su18010319

Table 3. *Cont.*

Manuscript N°	Environmental Zone in Europe	Country, City, Location	Years of Experiment and Duration	Cultivated Crops	OM Input/Agricultural Management Practice	Microbial Data and Biodiversity Indexes (BIs)	C Sequestration and Related Data	N_2O Emissions and Related Data	CH_4 Emissions or N Leaching Data
8—Lori et al. 2020 [48]	Terrestrial Model Ecosystems (TMEs)	For arable cropping system in Switzerland, Therwil; for mountain grasslands in France, Vercors; for agroforestry systems in Portugal, Montemor-o-Novo	Autumn 2015, 1 year		- Ecological intensive vs. conventional intensive, in three agricultural systems: - arable cropping (grassland in rotation) - mountain grassland - agroforestry - Different soil types, N fertilizer application (slurry, synthetic, cow manure), tillage system (CT, reduced tillage and NT), and vegetation cover (% of grasses, legumes and others)	- Enzyme activity involved in degradation of N containing molecules: leucine aminopeptidase activity (LAP) for protein degradation potential and β-1,4-N-acetylglucosaminidase (NAG) for chitin and peptidoglycan degradation. - Abundance of proteolytic microbial communities by qPCR on alkaline (apr) and neutral (npr) metallopeptidase genes. - Community and diversity composition of apr encoding communities. BI: richness, evenness and Shannon Index.	- Soil C content	- Applied N fertilization - Soil N content - Soil dissolved mineral and organic N	NO_3^- content in leachates

Table 3. *Cont.*

Manuscript N°	Environmental Zone in Europe	Country, City, Location	Years of Experiment and Duration	Cultivated Crops	OM Input/Agricultural Management Practice	Microbial Data and Biodiversity Indexes (BIs)	C Sequestration and Related Data	N_2O Emissions and Related Data	CH_4 Emissions or N Leaching Data
9—Lubbers et al. 2020 [49]	Microcosm experiment	Netherland, Droevendaal Agricultural	120 days	Absent	Hay residue	- Microbial Biomass N (MBN)	- Dissolved organic C - CO_2 flux measurement	- N_2O flux - Key physico-chemical factors influencing N_2O emissions, including total dissolved N, ammonia (NH^+), nitrate and nitrite (NO^- and NO^-), dissolved organic N	N.A.
10—Niklaus et al 2016 [50]	Continental	Germany, Jena	2007–2008, long-term (2 years) field experiment	Experimental grassland communities established in 2002	Plant species richness Synthetic NPK fertilization	- Nitrifyng (NEA) and denitrifying (DEA) enzyme activities - Bacterial nitrifiers by qPCR of nirK (copper nitrite reductase) and nirS (cd1 nitrite reductase) genes - Bacterial denitrifiers by qPCR of nitrite oxido-reductase gene (nxrA) from Nitrobacter-like nitrite-oxidizing bacteria (NOB). - Ammonia-oxidizing bacteria (AOB) qPCR	*Authors suggest that plant diversity could influence C storage by affecting soil microbial communities and their ability to process OM, which indirectly impacts carbon sequestration.*	- Soil-atmosphere N_2O fluxes - Soil inorganic N concentrations (NH_{4+} and NO_{3-})	Soil-atmosphere CH_4 fluxes

https://doi.org/10.3390/su18010319

Table 3. *Cont.*

Manuscript N°	Environmental Zone in Europe	Country, City, Location	Years of Experiment and Duration	Cultivated Crops	OM Input/Agricultural Management Practice	Microbial Data and Biodiversity Indexes (BIs)	C Sequestration and Related Data	N_2O Emissions and Related Data	CH_4 Emissions or N Leaching Data
11—Panico et al. 2020 [51]	Mediterranean	Italy, Ponticelli, near Naples	2018, 6 months	Sorghum (*Sorghum bicolor* L.), sunflower (*Helianthus annuus* L.)	Conventional agricultural practice	- Microbial biomass expressed ad microbial C - Microbial respiration (CO_2 evolution from soil samples) - Total fungal biomass expressed as fungal C - Soil microbial quotient (qCO_2) - Soil microbial activity and functional diversity - BIs: Catabolic Evenness and Shannon Index	- Soil C content - Soil organic C content	- Soil inorganic N concentrations (NH_{4+} and NO_{3-}) - N_2O emissions	N.A.
12—Ribas et al. 2015 [43]	Mediterranean	Spain, Catalan Central Depression, Castellnou d'Ossó	Spring 2008, 4 months	Tall fescue (*Festuca arundinacea*), alfalfa (*Medicago sativa*), chicory (*Cichorium intybus*)	- Plant species diversity. - Three N fertilization levels with filtered pig slurry	*Indirect assessment of microbial activity through the effects of plant diversity on soil conditions that influence gas emissions*	- CO_2 gas exchange rate	- Soil inorganic N content - N_2O gas exchange rate - NH_3 gas exchange rate	CH_4 gas exchange rate

https://doi.org/10.3390/su18010319

Table 3. *Cont.*

Manuscript N°	Environmental Zone in Europe	Country, City, Location	Years of Experiment and Duration	Cultivated Crops	OM Input/Agricultural Management Practice	Microbial Data and Biodiversity Indexes (BIs)	C Sequestration and Related Data	N_2O Emissions and Related Data	CH_4 Emissions or N Leaching Data
13—Rosinger and Bonkowski2021 [52]	Continental	Garzweiler mines, ca. 10 kmsouth of Mönchengladbach	2019, 7 years	Alfalfa (*Medicago sativa*)	- Reclaimed soils from agricultural post-mining chronosequence. - Wheat and barley crop rotation - Mineral fertilization (NPK and calcium ammonium nitrate) - Compost addition	- Microbial Biomass C (MBC) - Microbial Biomass N (MBN) - Basal Respiration (BR)	- Dissolved organic carbon - Soil organic carbon	- Total dissolved nitrogen	N.A.
14—Rosinger et al. 2022 [53]	Continental	Garzweiler mines, ca. 10 km south of Mönchengladbach	2018, 33 years	Alfalfa (*Medicago sativa*)	- Reclaimed soils from agricultural post-mining chronosequence - Wheat and barley crop rotation - Mineral fertilization (NPK and calcium ammonium nitrate) - Compost addition	- Microbial Biomass C (MBC) - Microbial Biomass N (MBN) - Basal Respiration (BR)	- Dissolved organic carbon - Soil organic carbon	- Total dissolved nitrogen	N A.
15—Rummel et al. 2020 [54]	Continental	Germany, South of Göttingen	2016, 22 days	Maize (*Zea mays*)	- Two N fertilizer levels (N1, N2) - Three litter addition treatments: control, root, root+shoot)	- Bacterial community structure by 16S rRNA gene sequencing - BIs: Shannon, Simpson, PD Index	- Organic carbon as water-extractable organic carbon (WEOC)	- N_2O fluxes - Soil mineral N - Net N mineralization	N.A.

Table 3. *Cont.*

Manuscript N°	Environmental Zone in Europe	Country, City, Location	Years of Experiment and Duration	Cultivated Crops	OM Input/Agricultural Management Practice	Microbial Data and Biodiversity Indexes (BIs)	C Sequestration and Related Data	N_2O Emissions and Related Data	CH_4 Emissions or N Leaching Data
16—Zistl-Schlingmann et al. 2020 [55]	Continental	Germany, the Graswang and Fendt sites	2017, 6 years	Montane grassland	- ^{15}N cattle slurry application - Extensive vs. intensive management - Control climate vs. reduced precipitation (+2 °C)	- Microbial Biomass N (MBN)	- Soil organic C	- Soil inorganic N concentrations (NH_{4+} and NO_{3-}) - Dissolved organic N - Total N	N leaching rate

"N.A." stands for "Not Available".

https://doi.org/10.3390/su18010319

In the surveyed literature, soil microbial status was represented by various parameters, often by microbial biomass (e.g., microbial biomass carbon and microbial biomass nitrogen, MBC and MBN) and/or by basal respiration (BR). Soil microbial communities and their abundances were frequently investigated by 16S rRNA or by phospholipid fatty acids (PLFAs) to characterize the main microbial groups. When microbial biodiversity was specifically addressed, the most used BIs were the Shannon and Simpson indexes, richness, and evenness. Some studies also investigated bacterial nitrifiers and denitrifiers, as well as N-cycling related genes via qPCR. C sequestration was mainly evaluated using SOC (sometimes equally defined as TOC for total organic carbon) and total carbon and more indirectly through CO_2 fluxes. N_2O emissions were almost always measured as fluxes, although in some cases, indirect assessment of total N, inorganic N, dissolved and organic N, was conducted, as these factors influence N_2O emissions [56] [Table 3]. One study focused on N leaching by measuring NO_3- content in leachates [48], and another measured N leaching rate [55]. Finally, two articles analysed CH_4 fluxes [43,50]. In Table 3, the last four columns on the right contain the main parameters measured in the 16 articles related to microorganisms, C sequestration, N_2O emissions, CH_4 emissions, and N leaching.

3.2. Network Analysis and Word Cloud

A semantic network map based on word co-occurrence in the 16 finally selected articles was created to highlight how the different research areas interrelate and overlap [Figure 4A]. The nodes clustered into three main groups, suggesting distinct aspects within the relationships between soil microorganisms and greenhouse gas emissions.

The co-occurrence network analysis revealed three main clusters of keywords [Figure 4B]. A strong association between "microbial biomass" and "organic carbon" (Cluster 1) was found, suggesting that organic matter (OM) inputs enhance microbial activity, promoting carbon sequestration and potentially reducing greenhouse gas (GHG) emissions. Similarly, "organic matter" (Cluster 2) showed high betweenness centrality, emphasizing its central role in soil quality and functional diversity. Environmental topics such as "microbial community" and "climate change" were grouped in another cluster (Cluster 3). Moreover, nodes like "microbial community" had high PageRank, indicating their significant influence and centrality in studying microbial dynamics under environmental changes [Figure 4B; Supplementary Material S2].

The thematic map analysis showed the complex interactions between the soil microbiome, agricultural practices, and greenhouse gas emissions [Figure 5]. At its core, "soil microbial", including "microbial community" and "microbial biomass", emerged as the motor theme. This highlighted the important role of microbial diversity and processes in regulating soil health and carbon dynamics, with high centrality values indicating strong connections to other research areas [Supplementary Material S3]. Nearby, the "organic matter" cluster (including "soil quality" and "greenhouse gas") was closely linked but distinct, suggesting a complementary focus on how organic inputs influence microbial activity and emissions. Niche themes, such as "climate change", "intensive management", and "organic nitrogen", appeared well-developed but less connected to the central topics, possibly reflecting specialized studies on environmental impacts or specific management practices. On the periphery, emerging themes such as "nitrous oxide" and "transformation processes" suggested growing interest in targeted areas like greenhouse gas mitigation and biochemical pathways in soil.

3.3. Results Classified by Agricultural and Fertilization Practices

The inclusion of other studies beyond the initial focus on OM input, based on the literature keyword search described in the Materials and Methods (Section 2), provided additional insights into soil microorganisms and trade-offs in European agricultural soils. Among the 16 studies, different OM inputs and some agricultural management practices were encountered [Table 3]. Hence, results are further categorized according to the experimented agricultural and fertilization practices, including the application of agricultural wastes (e.g., fruit peels, hay residues), digestate and biochar, slurry (including cattle and anaerobic digested slurry), tillage system, crop rotation, cover crops, and litter [Table 4]. Additionally, in three articles, the contribution of plant diversity in regulating nutrient cycles was assessed [43,50,51], highlighting a complex relationship with soil carbon storage and GHG emissions.

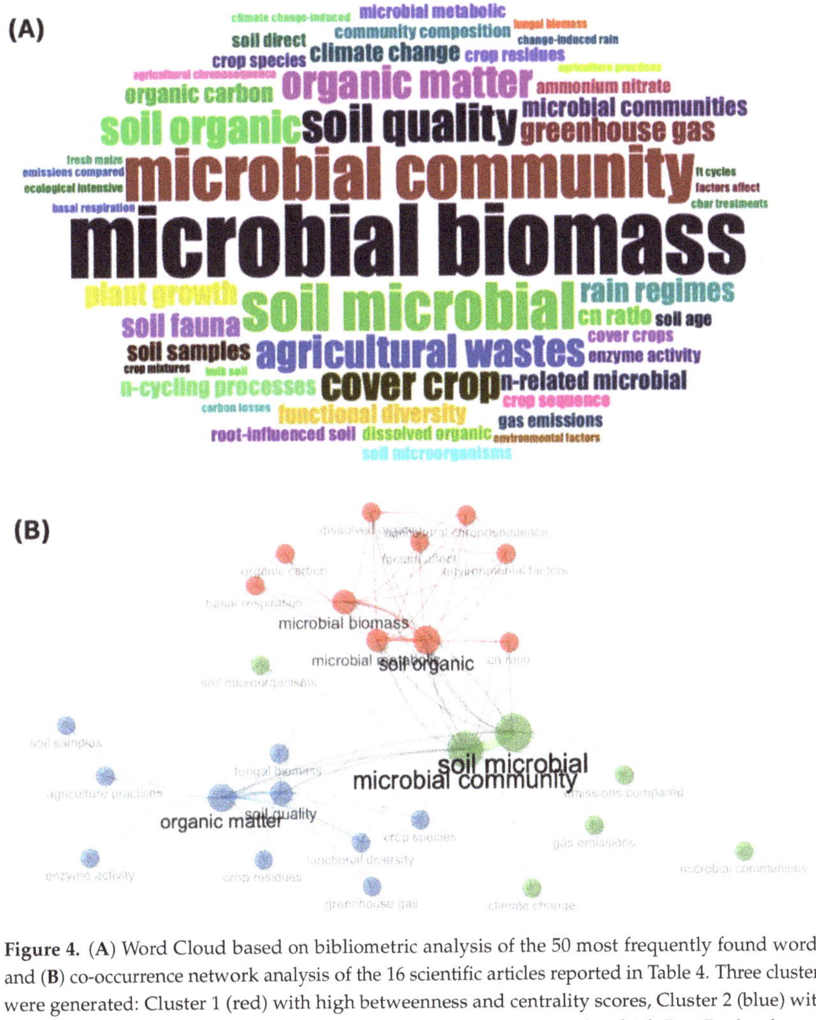

Figure 4. (**A**) Word Cloud based on bibliometric analysis of the 50 most frequently found words, and (**B**) co-occurrence network analysis of the 16 scientific articles reported in Table 4. Three clusters were generated: Cluster 1 (red) with high betweenness and centrality scores, Cluster 2 (blue) with closeness centrality, and Cluster 3 (green) with influential nodes based on high PageRank values.

3.3.1. Agricultural Organic Wastes

Agricultural organic wastes (AOWs) are outputs from agricultural activities, primarily consisting of crop residues and livestock waste. In particular, crop residues should be viewed as a valuable resource. They serve as a nutrient source for beneficial fungi, bacteria, and insects, help reduce evaporation from the soil surface, and contribute to maintaining

https://doi.org/10.3390/su18010319

Table 4. *Cont.*

	Article N° (Based in Table 3), First Author and Year of Publication	Type of Agricultural Management Practice or OM Input	Effects on C Sequestration	Effects on N_2O Emissions	Effects on Microbiome
Plant diversity	10-Niklaus et al. 2016 [50]	Plant species richness + NPK fertilization	N.A.	↓ N_2O emissions by plant species richness (unless fertilization is applied)	Plant species richness ↑ nitrifying enzyme activity; ↔ microbial gene abundance
	11-Panico et al. 2020 [51]	Plant species richness, soil tillage and mineral fertilization	↑ C_{tot}, soil respiration and qCO_2 in root-influenced soils more than in bulk soils	↑ N_2O emissions with plant growth stage	↓ microbial diversity in root-influenced soils
	12-Ribas et al. 2015 [43]	Plant species diversity vs. monocolture	↑ CO_2 emissions	↓ N_2O emissions	↑ microbial amount and functionality

Even if they are not part of the soil microbiome, the observation on biodiversity increase in soil earthworms has been reported since they are known to profoundly shape the soil microbiome. "↑" indicates an increase; "↓" indicates a decrease; "↔" indicates little or no effect; "⟹" denotes therefore.

3.3.2. Biochar, Digestate and Slurry

Biochar, digestate, and slurry are organic materials derived from agricultural and waste processing. Biochar is produced by pyrolyzing organic matter, digestate is the by-product of anaerobic digestion, and slurry is a water-based mixture of organic waste, often used as fertilizers or soil amendments. Dicke et al. [46] determined that biochar-treated soil samples produced significantly lower N_2O emissions with respect to digestate-treated soil samples and controls. However, despite different tested conditions, the abundance of the *nosZ* gene was not affected, in contrast with other studies. Interestingly, the authors detected that the *nosZ* gene copy number was significantly higher in October 2012 than in June 2013, showing a major influence of temperature and the crop growing season.

Digestate is generally characterized by valuable nutrient (N, P, K) content and low OM [58]. As expected, the chemical composition and the nutrient availability in soils treated with digestate are subjected to changes due to the anaerobic digestion process. In an on-farm field trial lasting 3 years, Bachmann et al. [42] found that CO_2 efflux from the soil surface and N uptake by maize were not significantly affected by the application of anaerobic digestion slurry residues. However, in digestate-amended soil, they observed a 50% reduction in the activity of soil microorganisms (dehydrogenase, acid- and alkaline-phosphatase) and a decrease in C turnover.

Johansen et al. [41] compared various amendments, including raw cattle slurry, anaerobically digested cattle slurry/maize, anaerobically digested cattle slurry/grass-clover, and fresh grass-clover, all applied to soil undergoing organic rotation. They observed only slight changes in microbial community composition with slurry-based amendments, while N_2O emissions were lower compared to those from fresh grass-clover. Additionally, they found that anaerobically digested cattle slurry resulted in 30–40% more NO_3- than raw cattle slurry.

Studying the effects of plant diversity on yield and CO_2 and N_2O emissions (also discussed in the Plant diversity subsection), Ribas et al. [40] tested three different fertilization levels using filtered pig slurry. They found that increasing the slurry application reduced soil ammonium content, but no significant changes in emissions were observed, suggesting that this outcome was likely due to their specific fertilization technique.

In addition, in a ^{15}N-tracing study conducted in C- and N-rich montane grassland soil, Zistl-Schlingmann et al. [55] investigated the N balance following the addition of cattle slurry under two different management conditions, namely extensive and inten-

https://doi.org/10.3390/su18010319

sive, while also comparing climate-controlled conditions vs. +2 °C heating and reduced precipitation conditions. Both management practices led to high plant biomass, with N uptake by plants primarily derived from soil organic matter mineralization due to high microbial immobilization. Differently, fertilizer N was only lightly exploited by plants, determining significant gaseous N losses, particularly under climate change conditions. The study further revealed that more intensive management practices caused higher N mining, depleting organic N stocks in soil and resulting in lower biomass production and SOC stocks. Ultimately, the intensive application of cattle slurry as fertilizer was linked to higher N_2O emissions and to SOC weakening.

3.3.3. Conservation System and Crop Rotation

No-tillage (NT) is widely reported as the most effective conservation system for improving soil organic matter due to no soil disturbance. This characteristic of no-tillage is extremely beneficial because surface residues and soil organic matter are left undisturbed, slowing decomposition and maximizing soil organic matter gains. Moreover, NT improves moisture retention by reducing evaporation and enhancing water infiltration, making more water available for crop production [59]. In semi-arid Mediterranean regions, where soils are particularly prone to OM depletion, Badagliacca et al. [29] found that long-term NT increased SOC, providing a greater availability of organic substrates, and consequently enhancing microbial biomass and microbial quotient, thus ameliorating soil quality. Interestingly, the increase in microbial biomass carbon (MBC) in NT vs. conventional tillage (CT) was mainly attributed to bacterial communities rather than fungi; a result that contrasts with several other studies that reported a greater fungal presence in NT systems. Notwithstanding, in a previous study conducted in the same field location in Sicily (Italy) by the same Italian research team [60], an almost 50% increase in total N_2O emissions, together with enhanced activity of denitrifying enzymes, was observed during the faba bean growing season (crop rotation) in NT compared to CT. In a later study conducted in semi-arid Mediterranean regions, Badagliacca et al. [45] investigated the effects of NT on wheat performance, revealing an initial reduction in grain yield, along with changes in soil nutrient levels, microbial communities, and enzyme dynamics, which were likely due to the reduced availability of crop residues and slower decomposition. Overall, their findings suggested that NT application could lead to a decreased release of labile N and C forms into the soil, due to the lower crop residue mineralization, which in turn may affect wheat yield and soil microbial community.

In the postmining agricultural sequence analysed in [52,53], where fields typically underwent wheat and barley crop rotation, high-carbon soils were found to be more vulnerable to microbial losses compared to low-carbon soil. In other words, SOC content was associated with less microbial variation, primarily affecting MBC and basal respiration.

3.3.4. Cover Crops and Litter

In farming management, cover crops are plants grown primarily to protect and improve soil health, reduce erosion, and enhance soil fertility during the off-season between main crop cycles. Among the analysed manuscripts, Drost, et al. [47] explored the ability of different cover crops, both monocultures and mixtures proceeding from diverse species, to stimulate soil microbial functional diversity by providing a variety of OM inputs and to mitigate GHG emissions. They observed a similar increase in fungal biomass across all cover crop treatments tested. The metabolic potential of the microbial community, as assessed by Biolog ECO plates, was higher in cover crop treatments formed by multiple species, supporting the hypothesis that plant species richness promotes greater microbial diversity. Concerning the GHG emissions, they found an immediate reduction in both

https://doi.org/10.3390/su18010319

CO_2 and N_2O emissions from the first day of soil collection after the cover crop addition, regardless of the cover crop species applied.

In the above-mentioned [41], the soil application of grass-clover provided a higher amount of readily degradable organic C with respect to raw or digested cattle slurry, leading to an increase in microbial biomass. However, this also resulted in an undesirable increase of up to 10 times in CO_2 and N_2O emissions compared to all other examined treatments.

Several studies have proved that plant residues can increase N_2O emissions upon incorporation into soils [61]. The term "litter" refers to the layer of decomposing plant and animal material that accumulates on the soil surface. The chemical composition of soil litter plays a critical role in C decomposition and availability for biological processes of nitrogen transformation. In [54], a comparison of GHG emissions from fresh maize root and shoot litter revealed that shoot litter significantly increased both CO_2 and N_2O emissions compared to root litter. Additionally, they observed a significant correlation between total CO_2 and N_2O emissions and the soil bacterial community composition, with reduced bacterial diversity in the presence of higher N levels and higher available C. Remarkably, the changes observed in bacterial community structure reflected the degradability of the analysed maize litter types. A lower diversity due to higher C and N availability favoured the presence of fast-growing C-cycling and N-reducing bacteria, such as *Actinobacteria*, *Chloroflexi*, *Firmicutes*, and *Proteobacteria*.

In a comprehensive analysis on selected Terrestrial Model Ecosystems (TMEs) used in agricultural systems, the abundance, diversity and activity of N-related microbes, as well as abiotic N-related soil indicators (such as NH_4^+, NO_3^{-f}, NO_2 and dissolved organic N) were not significantly affected by the management strategy. However, the strategy did influence the N-related microbial community composition in connection with the N-cycling processes of forage-N uptake and NO_3^- leaching [48]. Furthermore, soil OM positively influenced N leaching by decreasing NO_3^- leaching, enhancing microbial communities and at the same time increasing forage-N uptake.

3.3.5. Plant Diversity

At least three of the analysed manuscripts dealt with plant diversity as a driver of changes in soil microbial structure and composition, subsequently affecting soil C content and or non-CO_2 GHG emissions. Valuable insights into the relationships between the soil microbiome and N_2O and CH_4 fluxes were provided in Niklaus et al. [50]. Although the authors focused on how plant diversity—specifically grass species richness—affects these emissions, they also carefully examined and linked the abundance of bacterial nitrifiers and denitrifiers to these fluxes. Soil N_2O emissions decreased with species richness in the absence of mineral fertilization but increased when fertilizer was added, which turned out to be directly proportional to the fraction of legumes in plant communities. Concerning methane, plant diversity was inversely related to soil CH_4 uptake, regardless of fertilization. Moreover, plant diversity exhibited significant effects on soil microbial processes and the abundance of microbial groups involved in these processes.

The effects of plant diversity on yield and GHG exchanges in a forage mixture, measuring N_2O, CH_4, NH_3 and CO_2 fluxes in comparison with yield and soil inorganic N, were explored in [43]. In addition to yield, they observed a significant modulation of soil NO_3^- and NH_4^+, with plant diversity reducing N_2O emissions. The authors highlighted that nitric and ammonium nitrates could impact soil microbial communities both directly and indirectly, although they made these associations without directly measuring microbial populations. Consistent with the findings in [50], legumes were responsible for the highest emission rates for all GHGs.

https://doi.org/10.3390/su18010319

Lastly, Panico et al. [51] investigated the effects of two different plant species (sorghum and sunflower) on soil and its microbial communities, measuring the induced changes in soil physical–chemical characteristics and microbiota composition and functionality. Soil mineral fertilization was found to influence N-cycling processes based on soil N content and N_2O emissions. In this context, bacterial communities were more affected than fungal communities.

The average effects on C sequestration, N_2O emissions and soil microorganisms resulting from the different types of agricultural management practices or OM inputs discussed in the 16 analysed articles are tentatively summarized in Table 4. Agricultural organic wastes led to an increase in soil CO_2 emissions, thereby reducing the soil's carbon sequestration potential, but also had a positive effect on N_2O emissions, which were associated with a significant increase in soil microbial diversity and/or microbial biomass. The effects and trade-offs driven by biochar, digestate and slurry resulted in much more complex and varied outcomes across the analysed studies. Additionally, tillage system, crop rotation, cover crops and litter had differentiated effects on C sequestration and N_2O emissions, although most of the analysed studies indicated a shared gain in soil microorganisms. Finally, with a few exceptions, plant diversity was generally associated with increased soil CO_2 and N_2O emissions, with varied response at the soil microbial level [Table 4].

4. Discussion

Since the strong anthropogenic impact on biogeochemical cycles at the global level has been ascertained, current research focuses on developing and implementing strategies to manage nutrient cycling and, consequently, balance GHG emissions. In agricultural soils, C and N cycles are intricately linked, with efforts to enhance soil C sequestration having a direct impact on the balance of GHG gases, particularly N_2O, as well as CH_4 emissions and N leaching. The aim of this study was to identify the effects of the application of OM inputs and other management practices on soil microbes in relation to C sequestration and non-CO_2 GHG emissions in EU agricultural soil, through an analysis of scientific literature. Based on the preliminary article search from the main data sources, it became apparent that integrated data on soil microorganisms and trade-offs were somewhat limited. Moreover, the studies varied considerably, with only a small amount of data available from several independent studies, making it challenging to conduct a robust statistical analysis.

Among the parameters used to assess soil microbial status and composition, microbial biomass (MB) and activity are often considered as potential indicators of soil quality and functionality as an ecosystem service provider. Indeed, both MB and microbial activity are strictly related to the soil's living component and are highly responsive to management practices that alter the soil microbiota [62,63]. Indeed, Badagliacca et al. [29] demonstrated that soil MB and the main microbial groups showed varied or no significant responses to the different tillage systems, highlighting the diverse outcomes observed in different studies. Hence, the response of soil microorganisms is likely to be site-specific, depending on the context in which tillage systems are adopted, based on climate, soil type, fertility, and other management practices.

Soil microorganisms play a crucial role in regulating N_2O emissions in agricultural systems. An updated description of the key bacterial taxonomic groups involved in various biochemical reactions of the N cycle is provided in [64], highlighting the potential of N_2O reductase (*nosZ*) as a target for N_2O mitigation, enabling soils to act as a N_2O sink. A general positive effect on N_2O emission reduction has been ascertained in most of the analysed articles, with a few exceptions, such as [41], which reported a huge increase in N_2O emissions due to enhanced microbial activity following the application of grass-clover,

https://doi.org/10.3390/su18010319

in contrast with the well-known ability of clover to fix atmospheric N through symbiosis with *Rhizobium* bacteria in its root nodules [65]. These unexpected differences likely arise from the complexity of such studies, where a high number of both natural and artificial variables can influence the outcomes.

Based on the analysed manuscripts, only limited information could be extracted regarding the specific role of soil microbiome in modulating the trade-off between C se-questration and non-CO_2 GHG emissions. Nevertheless, evidence from the more recent literature suggests that soil microbial communities promoting C sequestration while limit-ing GHG emissions are characterized by (i) a high abundance of CO_2-fixing and C-accumulating taxa; (ii) a predominance of methane-oxidizing bacteria prevailing over methanogenic archaea; (iii) denitrifier communities enriched in *nosZ*, enabling the complete reduction of N_2O and N_2 [66]; (iv) high phylogenetic and functional diversity across C and N metabolic pathways; and (v) functional gene profiles favouring oxidation and assimilation over production of CH_4 and N_2O (e.g., higher *pmoA* relative to *nirS/nirK* gene abundances) [67–69].

It is worth noting that the literature search for the current systematic review was conducted in March 2023, during the execution of the EU ΣOMMIT project, and since then, several new studies and advancements have been made. Among the various OM inputs, integrating agricultural organic waste (AOW) into soil management practices can significantly influence C sequestration, N_2O emissions, and soil microbial dynamics. A result of the articles analysed in this review is that organic amendments used to enhance soil OM stock and promote C sequestration can potentially offset GHG emissions [70]. Although N_2O emissions were lower in the analysed papers [44,49], the impact of AOW on N_2O emissions is variable, depending on factors such as C/N ratios and soil conditions. This is in line with some studies reporting increased emissions due to enhanced microbial activity [71]. At the same time, OM inputs can stimulate microbial diversity, including N_2O-reducing bacteria that help mitigate emissions [64].

Biochar promotes C storage by stabilizing C in soils and may reduce N_2O emissions by enhancing soil structure [72]. On the other hand, digestate and slurry contribute to C sequestration but can increase N_2O emissions, especially under high N conditions, by altering microbial communities [37,73]. In a similar way, the results reported on the analysed articles show that biochar tends to mitigate N_2O emissions, while slurry can enhance microbial activity but also elevate emissions.

More recently, outside the timeframe of this systematic review, numerous publications have reported on the effects of biochar on N_2O emissions in relation to changes in microbial communities. This growing body of evidence, extending beyond the period covered by the reviewed papers, prompted the authors to undertake a deeper analysis of these newer studies with the support of Elicit AI (https://elicit.com/, accessed on 10 November 2025), an advanced research assistant for streamlining the analysis of scientific literature. A main output from the analysis of ten manuscripts [37,74–82] is that biochar application to EU agricultural soils increases soil organic carbon by 30–150% and reduces N_2O emissions by 13–54% through a combination of enhanced microbial functional gene expression favoring complete denitrification, increased microbial biomass, and physical entrapment of N_2O in biochar pores, with effectiveness dependent on application rates above 20 t/ha and pyrolysis temperatures exceeding 500 °C. The global meta-analysis [78] revealed that biochar increased *amoA* gene abundance by 39.4% and *nosZ* genes by 41.7% while reducing *nirS* gene abundance by 17.8%. This pattern suggests enhanced nitrification capacity alongside increased potential for complete reduction of N_2O to N_2. Furthermore, biochar significantly alters bacterial community composition across European sites: for example,

https://doi.org/10.3390/su18010319

Italian sites demonstrated enrichment of *Protobacteria* and *Gemmatimonadetes*. The complete report of this analysis is annexed to this manuscript as Supplementary Material S5.

Among the management practices, conservation agriculture may have beneficial effects on both soil C sequestration and N_2O emissions, even though NT is typically linked to lower crop yields. Interestingly, in Badagliacca et al. [45] it was found that NT reduces the release of labile C (i.e., C readily available for microbial activity), which in turn slows microbial decomposition rates. This process encourages the accumulation of OM in the soil, potentially enhancing long-term C sequestration. Additionally, the reduced availability of easily decomposable C may limit soil respiration, further supporting carbon storage over time. The same study also found that NT was associated with a decrease in labile N, which may decrease microbial activity, including the processes that contribute to N_2O production (such as nitrification and denitrification). Since labile N is a key substrate for these processes, its reduction could result in lower N_2O emissions. However, if N availability becomes too limited, it could hinder crop growth [83,84], potentially leading to a trade-off between reduced emissions and overall productivity.

In a study summarizing 57 publications, it was found that in plots with cover crops, over the long term, there was a 9% increase in organic matter, a 41% increase in microbial biomass in the soil, and about 50% less N leaching [85]. According to [86], cover crops, compared to monocropping enhance soil biodiversity and nutrient cycling, prevent runoff and N leaching, and improve soil physical properties and C sequestration over the long term. In a meta-analysis, it was concluded that although the effects of cover crop species and residue management on microbial properties vary across different soil and climatic conditions, cover crops overall can enhance biological soil health by enhancing microbial community abundance compared to soils without cover crops [87]. In another meta-analysis, it was found that cover crops generally enhanced soil microbial abundance, activity, and, to a lesser extent, diversity [88].

Plants release a variety of compounds through their roots into the soil, known as root exudates. These compounds, which include sugars, organic acids, amino acids, and secondary metabolites, serve as food sources for soil microbes. The type and amount of exudates vary depending on the plant species present in the ecosystem. Greater plant diversity typically results in a wider range of root exudates, thereby supporting a more diverse microbial community and beneficial microbial functions such as decomposition, nutrient cycling, and organic matter stabilization [89]. Concerning N_2O emissions, plant diversity influences N cycling by modifying the availability of labile nitrogen in the soil, which is crucial for understanding how plant diversity can both increase and decrease N_2O emissions, depending on the balance of microbial processes in the soil [90].

Importantly, the findings of this review are consistent with a recent meta-analysis on SOC [91], assessing that, especially in the complex agricultural setup, only a limited number of high-quality meta-analyses were available in scientific literature, leading to uncertain conclusions and potentially unreliable recommendations for scientists, policy-makers, and farmers. Notwithstanding, even though the current review provides evidence that our knowledge is still limited, insightful considerations may be drawn related to the effects driven by the application of OM inputs to agricultural soils or by implementing management practices like conservation agriculture, crop rotation, and others on soil microorganisms, C sequestration and N_2O emissions.

5. Limitations

While this study provides useful insights into the trade-offs between soil microbes and the interplay of carbon sequestration and non-CO_2 greenhouse gas emissions in EU agricultural soils, some main limitations are recognized. First, the sample size of included

https://doi.org/10.3390/su18010319

studies is relatively small, consisting of only 16 scientific peer-reviewed manuscripts, selected following the PRISMA procedure. This limited sample size reflects several technical factors mentioned throughout the manuscript, particularly the complexity and number of required keywords to capture three outcome terms. We selected Scopus as the sole database because it allows flexible keyword combinations, whereas other databases present more restrictions and a more difficult keyword handling; however, most Society journals are not indexed in Scopus. Future studies could benefit from refining or reframing complex, multi-output research questions into testable hypotheses to increase the significance of the results.

Second, the selected studies are predominantly from a few EU countries, such as Germany and Italy, resulting in limited geographical and pedoclimatic representation across European agricultural soils. Therefore, the results should be interpreted within the context of the available evidence. There is a clear need for future studies in underrepresented European regions to improve spatial coverage and validate observed trends across diverse pedoclimatic conditions. Extending this research to a global scale would further enhance the generalizability of the findings and provide a more mechanistic understanding of the relationships among soil microbial communities, carbon sequestration, and non-CO_2 GHG emissions.

6. Conclusions

The data from the 16 scientific articles reviewed in this study are valuable; however, the lack of robust long-term studies conducted across diverse EU regions and pedoclimatic conditions remains a significant gap. Further research is needed to better understand the effects of different management practices—such as soil conservation and OM-based fertilization—on soil microbial diversity, carbon sequestration, and N_2O and other GHG emissions.

Given the complexity of these studies, with numerous parameters involved and different outputs to analyse, the low level of standardization in relation to the field experiment, methodology used and data measured should be addressed to allow comparisons among different climate/regional locations. The EU Mission 'A Soil Deal for Europe' (Mission Soil), under the EU Horizon Europe Program, has the goal to create 100 Living Labs (LLs) and Lighthouses by 2030 to promote sustainable land and soil management in urban and rural areas (https://mission-soil-platform.ec.europa.eu/about/mission-soil, accessed on 10 November 2025). Through these LLs, Mission Soil can play a pivotal role in demonstrating the potential of EU-funded soil microbiome research to enhance biodiversity and support in situ measurements of GHG fluxes and carbon sequestration. At this aim, the use of soil indicators—chemical, physical and biological—and the establishment of general experimental principles and protocols, starting from the field design up to the soil sample collection, soil microbiome analysis (16S rRNA sequencing) and GHG emissions measurement, could help to make the comparisons and assessments of variations in soil microbiota and the trade-offs between carbon sequestration and non-CO_2 GHG emissions less complex.

Most of the selected studies addressed one or more management practices, including tillage systems, crop rotations, and the addition of litter, digestate, organic waste, or biochar. The application of OM inputs was generally found to have positive effects on microbial biomass (MB), influencing the functional diversity dynamics of soil microorganisms. At the same time, a general increase in soil organic carbon (SOC) content and potential reductions in GHG emissions were observed. Overall, despite a few contrasting results, the main findings advocate for soil biodiversity conservation to promote carbon sequestration and mitigate N_2O emissions and encourage sustainable agricultural practices. Another relevant factor frequently considered in these studies, in addition to the feedstock origin of the

https://doi.org/10.3390/su18010319

OM, is the cultivated plant species. Plant richness and diversity appear to reduce soil N_2O emissions and drive C sequestration by shaping the activity and composition of soil microorganisms, which play a pivotal role in N cycling and C storage.

Supplementary Materials: The following supporting information can be downloaded at: https://www.mdpi.com/article/10.3390/su18010319/s1, Supplementary Material S1: PRISMA 2020 Checklist; Supplementary Material S2: Co-occurrence analysis of keywords with centrality metrics (Betweenness, Closeness, and PageRank) to identify clusters and relationships within microbial and soil science concepts; Supplementary Material S2: Thematic map of co-occurrence analysis with centrality metrics (Betweenness, Closeness, and PageRank) to identify basic, emerging, and niche themes; Supplementary Material S3: List of the 16 finally selected original articles from Scopus picked out in this systematic review; Supplementary Material S4: Report generated by Elicit AI on the effects of biochar inputs on soil microorganisms in relation to C sequestration and N_2O emissions in EU agricultural soil.

Author Contributions: All authors discussed the results and contributed to the final manuscript. A.L. (Arianna Latini): conceived and developed the theoretical framework, collected the data, performed the analysis, wrote the original draft, reviewed, and edited the paper. L.D.G.: collected the data, performed the analysis, reviewed and edited the paper. M.C.: collected the data, performed the analysis and reviewed the paper. E.V.: reviewed and edited the paper, contributed to the implementation of the research. A.L. (Alessandra Lagomarsino): performed data curation, reviewed and edited the paper, funding acquisitions. M.S., P.M., F.V. and S.M.: contributed to the interpretation of the results and reviewed the paper. A.B.: conceptualization, data validation, reviewed and edited the paper, supervision, funding acquisition. All authors have read and agreed to the published version of the manuscript.

Funding: This research was funded by the European Union's Horizon 2020 research and innovation program under grant agreement No. 862695 (European Joint Programme SOIL) "Towards climate-smart sustainable management of agricultural soils". The research was developed in the framework of the EJP SOIL internal project ΣOMMIT—SUstainable Management of soil Organic Matter to MItigate Trade-offs between C sequestration and nitrous oxide, methane and nitrate losses" (https://projects.au.dk/ejpsoil/soil-research/ommit, last accessed on 15 December 2025). This manuscript reflects only the authors' views and opinions; neither the European Union nor the European Commission can be considered responsible for them. The authors also gratefully acknowledge funding from the Italian project "SoilHUB2—Creazione di una rete italiana di competenze per la gestione sostenibile del suolo." (D.M. MASAF—DSR 4—Prot. N° 0672924 del 20 December 2024); the European Union ECO-READY project "Achieving Ecological Resilient Dynamism for the European food system through consumer-driven policies" (GA No. 101084201) under the HORIZON-CL6-2022 Research and Innovation Programme; the European Union DELISOIL project "Delivering safe, sustainable, tailored & societally accepted soil improvers from circular food production for boosting soil health" (GA No. 101112855) under the Horizon Europe Programme; and the "Strengthening the MIRRI Italian Research Infrastructure for Sustainable Bioscience and Bioeconomy" SUS-MIRRI.IT project funded by the European Union—NextGeneration EU, PNRR—Mission 4 "Education and Research" Component 2: from research to business, Investment 3.1: Fund for the realization of an integrated system of research and innovation infrastructures—IR0000005 (D.M. Prot. n.120 del 21/06/2022).

Institutional Review Board Statement: Not applicable.

Informed Consent Statement: Not applicable.

Data Availability Statement: All information required to access the manuscripts analysed in this review has been provided. The data supporting the findings are reported in the Supplementary Material S3 and the References section.

Acknowledgments: The authors wish to thank the entire team of the EJP SOIL internal project ΣOMMIT (https://projects.au.dk/ejpsoil/soil-research/ommit, last accessed on 15 December 2025) for insightful discussions and valuable suggestions.

https://doi.org/10.3390/su18010319

Conflicts of Interest: The authors declare no conflicts of interest.

Abbreviations

The following abbreviations are used in this manuscript:

amoA	Ammonia Monooxygenase gene
AOB	Ammonia-Oxidizing Bacteria
AOW	Agricultural Organic Waste
BI	Biodiversity Index
BR	Basal Respiration
C	Carbon
CH4	Methane
CO2	Carbon dioxide
CT	Conventional Tillage
DM	Dry Matter
GHG	Greenhouse Gas
LOC	Labile Organic Carbon
MBC	Microbial Biomass Carbon
MBN	Microbial Biomass Nitrogen
N	Nitrogen
N_2O	Nitrous Oxide
nirK and nirS	Nitrite Reductase genes (commonly used as functional marker of denitrifying bacteria)
nosZ I and nosZ II	Nitrous Oxide Reductase genes of clade I and II, respectively
NT	No Tillage
OM	Organic Matter
PLFAs	Phospholipid Fatty Acids
pmoA	Particulate Methane Monooxygenase gene
PRISMA	Preferred Reporting Items for Systematic reviews and Meta-Analyses
qCO_2	Soil microbial quotient
SOC	Soil Organic carbon
TME	Terrestrial Model Ecosystem
TOC	Total Organic Carbon

References

1. Kannojia, P.; Sharma, P.K.; Sharma, K. Chapter 3—Climate Change and Soil Dynamics: Effects on Soil Microbes and Fertility of Soil. In *Climate Change and Agricultural Ecosystems*; Choudhary, K.K., Kumar, A., Singh, A.K., Eds.; Woodhead Publishing: Sawston, UK, 2019; pp. 43–64. [CrossRef]
2. Suman, J.; Rakshit, A.; Ogireddy, S.D.; Singh, S.; Gupta, C.; Chandrakala, J. Microbiome as a key player in sustainable agriculture and human health. *Front. Soil Sci.* **2022**, *2*, 821589. [CrossRef]
3. Wang, Q.; Chai, Q.; Dou, X.; Zhao, C.; Yin, W.; Li, H.; Wei, J. Soil microorganisms in agricultural fields and agronomic regulation pathways. *Agronomy* **2024**, *14*, 669. [CrossRef]
4. Paustian, K.; Lehmann, J.; Ogle, S.; Reay, D.; Robertson, G.P.; Smith, P. Climate-smart soils. *Nature* **2016**, *532*, 49–57. [CrossRef]
5. Ontl, T.A.; Schulte, L.A. Soil Carbon Storage. *Nat. Educ. Knowl.* **2012**, *3*, 35. Available online: https://www.nature.com/scitable/knowledge/library/soil-carbon-storage-84223790/ (accessed on 15 December 2025).
6. Lal, R.; Negassa, W.; Lorenz, K. Carbon sequestration in soil. *Curr. Opin. Environ. Sustain.* **2015**, *15*, 79–86. [CrossRef]
7. Bhattacharyya, S.S.; Ros, G.H.; Furtak, K.; Iqbal, H.M.; Parra-Saldívar, R. Soil carbon sequestration—An interplay between soil microbial community and soil organic matter dynamics. *Sci. Total Environ.* **2022**, *815*, 152928. [CrossRef]
8. Wu, H.; Cui, H.; Fu, C.; Li, R.; Qi, F.; Liu, Z.; Yang, G.; Xiao, K.; Qiao, M. Unveiling the crucial role of soil microorganisms in carbon cycling: A review. *Sci. Total Environ.* **2023**, *909*, 168627. [CrossRef]
9. Li, Z.; Tong, D.; Nie, X.; Xiao, H.; Jiao, P.; Jiang, J.; Li, Q.; Liao, W. New insight into soil carbon fixation rate: The intensive co-occurrence network of autotrophic bacteria increases the carbon fixation rate in depositional sites. *Agric. Ecosyst. Environ.* **2021**, *320*, 107579. [CrossRef]

https://doi.org/10.3390/su18010319

10. Intergovernmental Panel on Climate Change (IPCC). *Climate Change 2007: The Physical Science Basis. Contribution of Working Group I to the Fourth Assessment Report of the IPCC*; Cambridge University Press: Cambridge, UK, 2007. Available online: https://archive.ipcc.ch/publications_and_data/ar4/wg1/en/ch2s2-10-2.html (accessed on 15 December 2025).
11. Bahram, M.; Espenberg, M.; Pärn, J.; Lehtovirta-Morley, L.; Anslan, S.; Kasak, K.; Kõljalg, U.; Liira, J.; Maddison, M.; Moora, M.; et al. Structure and function of the soil microbiome underlying N2O emissions from global wetlands. *Nat. Commun.* **2022**, *13*, 1430. [CrossRef]
12. Maraseni, T.N.; Qu, J. An international comparison of agricultural nitrous oxide emissions. *J. Clean. Prod.* **2016**, *135*, 1256–1266. [CrossRef]
13. Lebender, U.; Senbayram, M.; Lammel, J.; Kuhlmann, H. Effect of mineral nitrogen fertilizer forms on N_2O emissions from arable soils in winter wheat production. *J. Plant Nutr. Soil Sci.* **2014**, *177*, 722–732. [CrossRef]
14. Wang, J.; Chadwick, D.R.; Cheng, Y.; Yan, X. Global analysis of agricultural soil denitrification in response to fertilizer nitrogen. *Sci. Total Environ.* **2018**, *616–617*, 908–917. [CrossRef]
15. De Laporte, A.; Banger, K.; Weersink, A.; Wagner-Riddle, C.; Grant, B.; Smith, W. Economic and environmental nitrate leaching consequences of 4R nitrogen management practices including use of inhibitors for corn production in Ontario. *J. Environ. Manag.* **2021**, *300*, 113739. [CrossRef]
16. Kodur, S.; Shrestha, U.B.; Maraseni, T.N.; Deo, R.C. Environmental and economic impacts and trade-offs from simultaneous management of soil constraints, nitrogen and water. *J. Clean. Prod.* **2019**, *222*, 960–970. [CrossRef]
17. Reeburgh, W.S. 4.03—Global methane biogeochemistry. In *Treatise on Geochemistry*; Elsevier: Amsterdam, The Netherlands, 2007; Volume 4, pp. 65–89. [CrossRef]
18. Cowan, N.; Maire, J.; Krol, D.; Cloy, J.M.; Hargreaves, P.; Murphy, R.; Carswell, A.; Jones, S.K.; Hinton, N.; Anderson, M.; et al. Agricultural soils: A sink or source of methane across the British Isles? *Eur. J. Soil Sci.* **2020**, *72*, 1842–1862. [CrossRef]
19. Guo, J.; Feng, H.; Peng, C.; Chen, H.; Xu, X.; Ma, X.; Li, L.; Kneeshaw, D.; Ruan, H.; Yang, H.; et al. Global climate change increases terrestrial soil CH_4 emissions. *Glob. Biogeochem. Cycles* **2023**, *37*, e2021GB00. [CrossRef]
20. Wang, H.; Zhao, R.; Zhao, D.; Liu, S.; Fu, J.; Zhang, Y.; Dai, N.; Song, D.; Ding, H. Microbial-mediated emissions of greenhouse gas from farmland soils: A Review. *Processes* **2022**, *10*, 2361. [CrossRef]
21. Six, J.; Frey, S.D.; Thiet, R.K.; Batten, K.M. Bacterial and fungal contributions to carbon sequestration in agroecosystems. *Soil Sci. Soc. Am. J.* **2006**, *70*, 555–569. [CrossRef]
22. Gougoulias, C.; Clark, J.M.; Shaw, L.J. The role of soil microbes in the global carbon cycle: Tracking the below-ground microbial processing of plant-derived carbon for manipulating carbon dynamics in agricultural systems. *J. Sci. Food Agric.* **2014**, *94*, 2362–2371. [CrossRef]
23. Basile-Doelsch, I.; Balesdent, J.; Pellerin, S. Reviews and syntheses: The mechanisms underlying carbon storage in soil. *Biogeosciences* **2020**, *17*, 5223–5242. [CrossRef]
24. Lazcano, C.; Zhu-Barker, X.; Decock, C. Effects of organic fertilizers on the soil microorganisms responsible for N_2O emissions: A review. *Microorganisms* **2021**, *9*, 983. [CrossRef] [PubMed]
25. Abalos, D.; Recous, S.; Butterbach-Bahl, K.; de Notaris, C.; Rittl, T.F.; Topp, C.F.E.; Petersen, S.O.; Hansen, S.; Bleken, M.A.; Rees, R.M.; et al. A review and meta-analysis of mitigation measures for nitrous oxide emissions from crop residues. *Sci. Total Environ.* **2022**, *828*, 154388. [CrossRef]
26. Forey, O.; Sauze, J.; Piel, C.; Gritti, E.S.; Devidal, S.; Faez, A.; Ravel, O.; Nahmani, J.; Rouch, L.; Blouin, M.; et al. Earthworms do not increase greenhouse gas emissions (CO_2 and N_2O) in an ecotron experiment simulating a three-crop rotation system. *Sci. Rep.* **2023**, *13*, 21920. [CrossRef]
27. Maenhout, P.; Di Bene, C.; Cayuela, M.L.; Govednik, A.; Keuper, F.; Mavsar, S.; Mihelic, R.; O'Toole, A.; Schwarzmann, A.; Suhadolc, M.; et al. Trade-offs and synergies of soil carbon sequestration: Addressing knowledge gaps related to soil management strategies. *Eur. J. Soil Sci.* **2024**, *75*, e13515. [CrossRef]
28. Chabert, A.; Sarthou, J.-P. Conservation agriculture as a promising trade-off between conventional and organic agriculture in bundling ecosystem services. *Agric. Ecosyst. Environ.* **2020**, *292*, 106815. [CrossRef]
29. Badagliacca, G.; Laudicina, V.A.; Amato, G.; Badalucco, L.; Frenda, A.S.; Giambalvo, D.; Ingraffia, R.; Plaia, A.; Ruisi, P. Long-term effects of contrasting tillage systems on soil C and N pools and on main microbial groups differ by crop sequence. *Soil Tillage Res.* **2021**, *211*, 104995. [CrossRef]
30. Moher, D.; Shamseer, L.; Clarke, M.; Ghersi, D.; Liberati, A.; Petticrew, M.; Shekelle, P.; Stewart, L.A.; PRISMA-P Group. Preferred reporting items for systematic review and meta-analysis protocols (PRISMA-P) 2015 statement. *Syst. Rev.* **2015**, *4*, 1. [CrossRef] [PubMed]
31. Schardt, C.; Adams, M.B.; Owens, T.; Keitz, S.; Fontelo, P. Utilization of the PICO framework to improve searching PubMed for clinical questions. *BMC Med. Inform. Decis. Mak.* **2007**, *7*, 16. [CrossRef]

https://doi.org/10.3390/su18010319

32. Falagas, M.E.; Pitsouni, E.I.; Malietzis, G.A.; Pappas, G.; Kouranos, V.D.; Arencibia-Jorge, R.; Karageorgopoulos, D.E.; Reagan-Shaw, S.; Nihal, M.; Ahmad, N.; et al. Comparison of PubMed, Scopus, Web of Science, and Google Scholar: Strengths and weaknesses. *FASEB J.* **2008**, *22*, 338–342. [CrossRef]
33. Frampton, G.K.; Livoreil, B.; Petrokofsky, G. Eligibility screening in evidence synthesis of environmental management topics. *Environ. Evid.* **2017**, *6*, 27. [CrossRef]
34. Ranjard, L.; Echairi, A.; Nowak, V.; Lejon, D.P.H.; Nouaã¯m, R.; Chaussod, R. Field and microcosm experiments to evaluate the effects of agricultural Cu treatment on the density and genetic structure of microbial communities in two different soils. *FEMS Microbiol. Ecol.* **2006**, *58*, 303–315. [CrossRef]
35. Treonis, A.M.; Wall, D.H.; Virginia, R.A. Field and microcosm studies of decomposition and soil biota in a cold desert soil. *Ecosystems* **2002**, *5*, 159–170. [CrossRef]
36. Eller, G.; Krüger, M.; Frenzel, P. Comparing field and microcosm experiments: A case study on methano- and methylo-trophic bacteria in paddy soil. *FEMS Microbiol. Ecol.* **2005**, *51*, 279–291. [CrossRef]
37. Valkama, E.; Tzemi, D.; Esparza-Robles, U.R.; Syp, A.; O'TOole, A.; Maenhout, P. Effectiveness of soil management strategies for mitigation of N_2O emissions in European arable land: A meta-analysis. *Eur. J. Soil Sci.* **2024**, *75*, e13488. [CrossRef]
38. Mishra, A.; Singh, L.; Singh, D. Unboxing the black box—One step forward to understand the soil microbiome: A systematic review. *Microb. Ecol.* **2023**, *85*, 669–683. [CrossRef]
39. Aria, M.; Cuccurullo, C. bibliometrix: An R-tool for comprehensive science mapping analysis. *J. Informetr.* **2017**, *11*, 959–975. [CrossRef]
40. Klarin, A. How to conduct a bibliometric content analysis: Guidelines and contributions of content co-occurrence or co-word literature reviews. *Int. J. Consum. Stud.* **2024**, *48*, e13031. [CrossRef]
41. Johansen, A.; Carter, M.S.; Jensen, E.S.; Hauggard-Nielsen, H.; Ambus, P. Effects of digestate from anaerobically digested cattle slurry and plant materials on soil microbial community and emission of CO_2 and N_2O. *Appl. Soil Ecol.* **2013**, *63*, 36–44. [CrossRef]
42. Bachmann, S.; Gropp, M.; Eichler-Löbermann, B. Phosphorus availability and soil microbial activity in a 3 year field experiment amended with digested dairy slurry. *Biomass-Bioenergy* **2014**, *70*, 429–439. [CrossRef]
43. Ribas, A.; Llurba, R.; Gouriveau, F.; Altimir, N.; Connolly, J.; Sebastià, M.T. Plant identity and evenness affect yield and trace gas exchanges in forage mixtures. *Plant Soil* **2015**, *391*, 93–108. [CrossRef]
44. Anastopoulos, I.; Omirou, M.; Stephanou, C.; Oulas, A.; Vasiliades, M.A.; Efstathiou, A.M.; Ioannides, I.M. Valorization of agricultural wastes could improve soil fertility and mitigate soil direct N_2O emissions. *J. Environ. Manag.* **2019**, *250*, 109389. [CrossRef]
45. Badagliacca, G.; Presti, E.L.; Ferrarini, A.; Fornasier, F.; Laudicina, V.A.; Monti, M.; Preiti, G. Early effects of no-till use on Durum wheat (*Triticum durum Desf.*): Productivity and soil functioning vary between two contrasting mediterranean soils. *Agronomy* **2022**, *12*, 3136. [CrossRef]
46. Dicke, C.; Andert, J.; Ammon, C.; Kern, J.; Meyer-Aurich, A.; Kaupenjohann, M. Effects of different biochars and digestate on N_2O fluxes under field conditions. *Sci. Total Environ.* **2015**, *524–525*, 310–318. [CrossRef]
47. Drost, S.M.; Rutgers, M.; Wouterse, M.; de Boer, W.; Bodelier, P.L. Decomposition of mixtures of cover crop residues increases microbial functional diversity. *Geoderma* **2020**, *361*, 114060. [CrossRef]
48. Lori, M.; Piton, G.; Symanczik, S.; Legay, N.; Brussaard, L.; Jaenicke, S.; Nascimento, E.; Reis, F.; Sousa, J.P.; Mäder, P.; et al. Compared to conventional, ecological intensive management promotes beneficial proteolytic soil microbial communities for agro-ecosystem functioning under climate change-induced rain regimes. *Sci. Rep.* **2020**, *10*, 7296. [CrossRef]
49. Lubbers, I.M.; Berg, M.P.; De Deyn, G.B.; van der Putten, W.H.; van Groenigen, J.W. Soil fauna diversity increases CO_2 but suppresses N_2O emissions from soil. *Glob. Change Biol.* **2020**, *26*, 1886–1898. [CrossRef]
50. Niklaus, P.A.; Le Roux, X.; Poly, F.; Buchmann, N.; Scherer-Lorenzen, M.; Weigelt, A.; Barnard, R.L. Plant species diversity affects soil–atmosphere fluxes of methane and nitrous oxide. *Oecologia* **2016**, *181*, 919–930. [CrossRef] [PubMed]
51. Panico, S.C.; Esposito, F.; Memoli, V.; Vitale, L.; Polimeno, F.; Magliulo, V.; Maisto, G.; De Marco, A. Variations of agricultural soil quality during the growth stages of sorghum and sunflower. *Appl. Soil Ecol.* **2020**, *152*, 103569. [CrossRef]
52. Rosinger, C.; Bonkowski, M. Soil age and soil organic carbon content shape biochemical responses to multiple freeze–thaw events in soils along a postmining agricultural chronosequence. *Biogeochemistry* **2021**, *155*, 113–125. [CrossRef]
53. Rosinger, C.; Clayton, J.; Baron, K.; Bonkowski, M. Soil freezing-thawing induces immediate shifts in microbial and resource stoichiometry in Luvisol soils along a postmining agricultural chronosequence in Western Germany. *Geoderma* **2022**, *408*, 115596. [CrossRef]
54. Rummel, P.S.; Pfeiffer, B.; Pausch, J.; Well, R.; Schneider, D.; Dittert, K. Maize root and shoot litter quality controls short-term CO_2 and N_2O emissions and bacterial community structure of arable soil. *Biogeosciences* **2020**, *17*, 1181–1198. [CrossRef]
55. Zistl-Schlingmann, M.; Kengdo, S.K.; Kiese, R.; Dannenmann, M. Management intensity controls nitrogen-use-efficiency and flows in grasslands—A ^{15}N tracing experiment. *Agronomy* **2020**, *10*, 606. [CrossRef]

https://doi.org/10.3390/su18010319

56. Smith, K.; Lex, B.; Braatz, B. N_2O: Direct Emissions from Agricultural Soils. Background Paper to IPCC Good Practice Guidance and Uncertainty Management in National Greenhouse Gas Inventories, 1999, 361–380. Available online: https://www.ipcc-nggip.iges.or.jp/public/gp/bgp/4_5_N2O_Agricultural_Soils.pdf (accessed on 15 December 2025).
57. Li, W.; Zhang, Y.; Mao, W.; Wang, C.; Yin, S. Functional potential differences between *Firmicutes* and *Proteobacteria* in response to manure amendment in a reclaimed soil. *Can. J. Microbiol.* **2020**, *66*, 689–697. [CrossRef]
58. Möller, K.; Müller, T. Effects of anaerobic digestion on digestate nutrient availability and crop growth: A review. *Eng. Life Sci.* **2012**, *12*, 242–257. [CrossRef]
59. Waring, E.R.; Pederson, C.; Lagzdins, A.; Clifford, C.; Helmers, M.J. Water and soil quality respond to no-tillage and cover crops differently through 10 years of implementation. *Agric. Ecosyst. Environ.* **2024**, *360*, 108791. [CrossRef]
60. Badagliacca, G.; Benítez, E.; Amato, G.; Badalucco, L.; Giambalvo, D.; Laudicina, V.A.; Ruisi, P. Long-term no-tillage application increases soil organic carbon, nitrous oxide emissions and faba bean (*Vicia faba* L.) yields under rain-fed Mediterranean conditions. *Sci. Total Environ.* **2018**, *639*, 350–359. [CrossRef]
61. Chen, H.; Li, X.; Hu, F.; Shi, W. Soil nitrous oxide emissions following crop residue addition: A meta-analysis. *Glob. Change Biol.* **2013**, *19*, 2956–2964. [CrossRef] [PubMed]
62. Panettieri, M.; de Sosa, L.L.; Domínguez, M.T.; Madejón, E. Long-term impacts of conservation tillage on Mediterranean agricultural soils: Shifts in microbial communities despite limited effects on chemical properties. *Agric. Ecosyst. Environ.* **2020**, *304*, 107144. [CrossRef]
63. Bandyopadhyay, S.; Maiti, S.K. Steering restoration of coal mining degraded ecosystem to achieve sustainable development goal-13 (climate action): United Nations decade of ecosystem restoration (2021–2030). *Environ. Sci. Pollut. Res.* **2022**, *29*, 88383–88409. [CrossRef]
64. Hiis, E.G.; Vick, S.H.W.; Molstad, L.; Røsdal, K.; Jonassen, K.R.; Winiwarter, W.; Bakken, L.R. Unlocking bacterial potential to reduce farmland N2O emissions. *Nature* **2024**, *630*, 421–428. [CrossRef]
65. Harris, C.; Ratnieks, F.L.W. Clover in agriculture: Combined benefits for bees, environment, and farmer. *J. Insect Conserv.* **2022**, *26*, 339–357. [CrossRef]
66. Xiao, W.; Zhang, Q.; Zhao, S.; Chen, D.; Zhao, Z.; Gao, N.; Huang, M.; Ye, X. Combined metabolomic and microbial community analyses reveal that biochar and organic manure alter soil C-N metabolism and greenhouse gas emissions. *Environ. Int.* **2024**, *192*, 109028. [CrossRef] [PubMed]
67. Jiang, Z.; Yang, S.; Pang, Q.; Abdalla, M.; Karbin, S.; Qi, S.; Hu, J.; Qiu, H.; Song, X.; Smith, P. Metagenomic insights into the influence of soil microbiome on greenhouse gas emissions from paddy fields under varying irrigation and fertilisation regimes. *J. Environ. Manag.* **2025**, *393*, 127129. [CrossRef] [PubMed]
68. Parasar, B.J.; Agarwala, N. Unravelling the role of biochar-microbe-soil tripartite interaction in regulating soil carbon and nitrogen budget: A panacea to soil sustainability. *Biochar* **2025**, *7*, 37. [CrossRef]
69. Shakoor, A.; Pendall, E.; Macdonald, C.A. Microbial mechanisms of interactive climate-driven changes in soil N2O and CH4 fluxes: A global meta-analysis. *J. Environ. Manag.* **2025**, *376*, 124380. [CrossRef]
70. Wong, C.T.F.; Falcone, M.; Rich, G.; Stubler, C.; Malama, B.; Lazcano, C.; Decock, C. Short-term effects of increasing compost application rates on soil C and greenhouse gas (N_2O and CO_2) emissions in a California central coast vineyard. *Front. Environ. Sci.* **2023**, *11*, 1123510. [CrossRef]
71. Hassan, M.U.; Aamer, M.; Mahmood, A.; Awan, M.I.; Barbanti, L.; Seleiman, M.F.; Bakhsh, G.; Alkharabsheh, H.M.; Babur, E.; Shao, J.; et al. Management strategies to mitigate N_2O emissions in agriculture. *Life* **2022**, *12*, 439. [CrossRef]
72. Lehmann, J.; Joseph, S. *Biochar for Environmental Management: Science, Technology and Implementation*, 2nd ed.; Lehmann, J., Joseph, S., Eds.; Routledge: London, UK, 2015; p. 976. [CrossRef]
73. Häfner, F.; Ruser, R.; Claß-Mahler, I.; Möller, K. Field application of organic fertilizers triggers N_2O emissions from the soil N pool as indicated by 15N-labeled digestates. *Front. Sustain. Food Syst.* **2021**, *4*, 614349. [CrossRef]
74. Taghizadeh-Toosi, A.; Clough, T.; Condron, L.; Sherlock, R.; Anderson, C.; Craigie, R. Biochar incorporation into pasture soil suppresses in situ nitrous oxide emissions from ruminant urine patches. *J. Environ.* Qual. **2011**, *40*, 468–476. [CrossRef]
75. Scotti, C.; Bertora, C.; Valagussa, M.; Borrelli, L.; Cabassi, G.; Tosca, A. Agroenvironmental performances of biochar application in the mineral and organic fertilization strategies of a maize–ryegrass forage system. *Agriculture* **2022**, *12*, 925. [CrossRef]
76. Prommer, J.; Wanek, W.; Hofhansl, F.; Trojan, D.; Offre, P.; Urich, T.; Schleper, C.; Sassmann, S.; Kitzler, B.; Soja, G.; et al. Biochar decelerates soil organic nitrogen cycling but stimulates soil nitrification in a temperate arable field trial. *PLoS ONE* **2014**, *9*, e86388. [CrossRef]
77. Jenkins, J.R.; Viger, M.; Arnold, E.C.; Harris, Z.M.; Ventura, M.; Miglietta, F.; Girardin, C.; Edwards, R.J.; Rumpel, C.; Fornasier, F.; et al. Biochar alters the soil microbiome and soil function: Results of next-generation amplicon sequencing across Europe. *Glob. Change Biol. Bioenergy* **2016**, *9*, 591–612. [CrossRef]

https://doi.org/10.3390/su18010319

78. Bai, J.; de Almeida Moreira, B.R.; Bai, Y.; Nadar, C.G.; Feng, Y.; Yadav, S. Assessing biochar's impact on greenhouse gas emissions, microbial biomass, and enzyme activities in agricultural soils through meta-analysis and machine learning. *Sci. Total Environ.* **2025**, *963*, 178541. [CrossRef] [PubMed]
79. Harter, J.; Guzman-Bustamante, I.; Kuehfuss, S.; Ruser, R.; Well, R.; Spott, O.; Kappler, A.; Behrens, S. Gas entrapment and microbial N_2O reduction reduce N_2O emissions from a biochar-amended sandy clay loam soil. *Sci. Rep.* **2016**, *6*, 39574. [CrossRef]
80. Quin, P.; Joseph, S.; Husson, O.; Donne, S.; Mitchell, D.; Munroe, P.; Phelan, D.; Cowie, A.; Van Zwieten, L. Lowering N_2O emissions from soils using eucalypt biochar: The importance of redox reactions. *Sci. Rep.* **2015**, *5*, 16773. [CrossRef]
81. Kotuš, T.; Šimanský, V.; Drgoňová, K.; Illéš, M.; Wójcik-Gront, E.; Balashov, E.; Buchkina, N.; Aydın, E.; Horák, J. Combination of biochar with N–fertilizer affects properties of soil and N_2O emissions in maize crop. *Agronomy* **2022**, *12*, 1314. [CrossRef]
82. Sun, Z.; Sänger, A.; Rebensburg, P.; Lentzsch, P.; Wirth, S.; Kaupenjohann, M.; Meyer-Aurich, A. Contrasting effects of biochar on N_2O emission and N uptake at different N fertilizer levels on a temperate sandy loam. *Sci. Total Environ.* **2017**, *578*, 557–565. [CrossRef]
83. Marschner, H. *Marschner's Mineral Nutrition of Higher Plants*, 3rd ed.; Marschner, P., Ed.; Academic Press: Cambridge, MA, USA, 2012.
84. Xu, G.; Fan, X.; Miller, A.J. Plant nitrogen assimilation and use efficiency. *Annu. Rev. Plant Biol.* **2012**, *63*, 153–182. [CrossRef] [PubMed]
85. Shackelford, G.E.; Kelsey, R.; Dicks, L.V. Effects of cover crops on multiple ecosystem services: Ten meta-analyses of data from arable farmland in California and the Mediterranean. *Land Use Policy* **2019**, *88*, 104204. [CrossRef]
86. Yousefi, M.; Dray, A.; Ghazoul, J. Assessing the effectiveness of cover crops on ecosystem services: A review of the benefits, challenges, and trade-offs. *Int. J. Agric. Sustain.* **2024**, *22*, 2335106. [CrossRef]
87. Muhammad, I.; Wang, J.; Sainju, U.M.; Zhang, S.; Zhao, F.; Khan, A. Cover cropping enhances soil microbial biomass and affects microbial community structure: A meta-analysis. *Geoderma* **2021**, *381*, 114696. [CrossRef]
88. Kim, N.; Zabaloy, M.C.; Guan, K.; Villamil, M.B. Do cover crops benefit soil microbiome? A meta-analysis of current research. *Soil Biol. Biochem.* **2020**, *142*, 107701. [CrossRef]
89. Lange, M.; Eisenhauer, N.; Sierra, C.A.; Bessler, H.; Engels, C.; Griffiths, R.I.; Mellado-Vázquez, P.G.; Malik, A.A.; Roy, J.; Scheu, S.; et al. Plant diversity increases soil microbial activity and soil carbon storage. *Nat. Commun.* **2015**, *6*, 6707. [CrossRef] [PubMed]
90. Stevens, C.J.; Dise, N.B.; Mountford, J.O.; Gowing, D.J. Impact of nitrogen deposition on the species richness of grasslands. *Science* **2004**, *303*, 1876–1879. [CrossRef]
91. Fohrafellner, J.; Zechmeister-Boltenstern, S.; Murugan, R.; Valkama, E. Quality assessment of meta-analyses on soil organic carbon. *Soil* **2023**, *9*, 117–140. [CrossRef]

https://doi.org/10.3390/su18010319

MDPI AG
Grosspeteranlage 5
4052 Basel
Switzerland
Tel.: +41 61 683 77 34

Sustainability Editorial Office
E-mail: sustainability@mdpi.com
www.mdpi.com/journal/sustainability

www.ingramcontent.com/pod-product-compliance
Lightning Source LLC
Chambersburg PA
CBHW061458210926
51850CB00008B/1755
* 9 7 8 3 7 2 5 8 7 7 8 5 0 *